GEOMORPHOLOGY IN THE ANTHROPOCENE

The Anthropocene – referring to the epoch that began when human activities started to have a significant impact on Earth's systems – is a major new concept in the Earth sciences. This book examines the effects on geomorphology during this latest period of Earth's history. Drawing on examples from many different environments around the world, this comprehensive volume demonstrates that the human impact on landforms and land-forming processes has and continues to be profound, driven by a number of forces, including: the use of fire; the extinction of fauna; the development of agriculture, urbanization, and globalization; and new methods of harnessing energy. This human impact has accelerated since the Industrial Revolution and – in particular – since the Second World War. The book explores the ways in which future climate change in response to anthropogenic causes may further magnify the human effect on geomorphology, with respect to future hazards such as floods and landslides, the state of the cryosphere, and global sea levels. The book concludes with a consideration of the ways in which landforms are now being managed and protected. Covering all major aspects of geomorphology, this book is ideal for undergraduate and graduate students taking courses in geomorphology, environmental science, and physical geography, and for all researchers of geomorphology.

ANDREW S. GOUDIE is Emeritus Professor of Geography at the University of Oxford. He specializes in the study of desert processes and climate change, and has worked in the Middle East, India and Pakistan, East Africa, Southern Africa, Australia, and the United States. From 2005 to 2009, Professor Goudie was President of the International Association of Geomorphologists and he has also been President of the Geographical Association, President of Section E of the British Association, and Chairman of the British Geomorphological Research Group. He was the recipient of the Farouk El-Baz Prize for Desert Research from the Geological Society of America in 2007, and the Founder's Medal of the Royal Geographical Society in 1991.

HEATHER A. VILES is Professor of Biogeomorphology and Heritage Conservation at the University of Oxford. Her research focuses on understanding weathering and rock breakdown in coastal, arid, and urban environments, and applying that knowledge to conserving heritage sites. She also works extensively on the links between geomorphology and ecology. Professor Viles has been Chairman of the British Society for Geomorphologists, Vice President (fieldwork) of the Royal Geographical Society, and is currently on the executive committee of the International Association of Geomorphologists. She received the Ralph Alger Bagnold Medal from the European Geosciences Union in 2015 for establishing the field of biogeomorphology.

"Among contemporary physical geographers, there are none who are the equal of Andrew Goudie and Heather Viles in their ability to synthesize vast areas of the literature and to bring out new meanings from the avalanche of data that is published each week... This is the first book that explores, in depth, the relation between the Anthropocene epoch and landscape science (geomorphology)... [It] can be recommended to any serious student of the global implications of human modification of Earth's surface... [as well as to the] geoscience and environmental science communities, from geographers to geologists and geophysicists"

- Olav Slaymaker, University of British Columbia

"What an interesting topic! What a good book! It presents the geomorphological evidence for the concept of the Anthropocene... With great clarity the authors give a wonderful review of the issues and a very clear account of the problems involved in selection of the start point and character of the possible new stratigraphical unit. Breathtaking in scope, it also gives a fine account of geomorphological processes and landforms linked to human achievements."

- Denys Brunsden, King's College, London

"In this comprehensive examination of human impacts on diverse landscapes, Goudie and Viles provide numerous examples and details of how human activities have altered and continue to alter Earth's surface. This book provides a valuable reference and thorough overview for students and professionals."

- Ellen Wohl, Colorado State University

"In today's climate of media-induced alarm about what mankind is doing to our planet, this book stands out as a calm and considered appraisal of human impacts on Earth resources and natural systems. Few are better placed than these authors to interpret the scientific data on human and natural forces driving those rapid changes currently challenging sustainability of Earth systems."

- Michael Crozier, Victoria University of Wellington

GEOMORPHOLOGY IN THE ANTHROPOCENE

ANDREW S. GOUDIE
University of Oxford

HEATHER A. VILES
University of Oxford

CAMBRIDGE
UNIVERSITY PRESS

University Printing House, Cambridge CB2 8BS, United Kingdom

Cambridge University Press is part of the University of Cambridge.

It furthers the University's mission by disseminating knowledge in the pursuit of
education, learning and research at the highest international levels of excellence.

www.cambridge.org
Information on this title: www.cambridge.org/9781107139961

© Andrew S. Goudie and Heather A. Viles 2016

First published 2016

Printed in the United States of America by Sheridan Books, Inc.

A catalogue record for this publication is available from the British Library

Library of Congress Cataloging-in-Publication Data
Names: Goudie, Andrew, author. | Viles, Heather A., author.
Title: Geomorphology in the Anthropocene / Andrew S. Goudie, University of Oxford ;
and Heather A. Viles, University of Oxford.
Description: Cambridge, United Kingdom ; New York : Cambridge University Press, 2016.
| Includes bibliographical references and index.
Identifiers: LCCN 2016015452 | ISBN 9781107139961 (hardback)
Subjects: LCSH: Geomorphology. | BISAC: SCIENCE / Earth Sciences / Geography.
Classification: LCC GB401.5 .G68 2016 | DDC 551.41–dc23 LC record available at
https://lccn.loc.gov/2016015452

ISBN 978-1-107-13996-1 Hardback

Contents

Acknowledgments

This book is the result of many influences. As a Cambridge undergraduate, Andrew was exposed to the study of the Quaternary and environmental change as a result of being taught by people like Dick Grove, Richard Hey, Richard West, and Nick Shackleton. Since then, he has gained a great deal from links with Alayne Perrott, Adrian Parker, and David Anderson. He has also been grateful to archaeologists who have given him a long time perspective on human affairs, including Raymond and Bridget Allchin and David Price Williams. His Oxford colleagues Michael Williams and Frank Emery taught him much about historical geography, adding to what he had gleaned from the lectures of Clifford Darby at Cambridge. Environmental management is a field to which he was introduced by John Doornkamp, David Jones, Ron Cooke, and Denys Brunsden, while John Davey encouraged him to write about the human impact. He has also benefited over the years from collaboration or discussions with Stan Trimble, Cleide Rodriguez, Piotr Migoń, and Olav Slaymaker. We are both grateful to NASA, Google Earth, and Google Scholar for making so many resources available, and to Elsevier for permission to reproduce Figures 6.4, 7.4, 7.6, 7.7, 7.10, 7.11, and 9.9; John Wiley and Sons for Figures 5.5 and 9.6; Springer for Figures 4.4, 6.11, and 6.13; Nature Publishing Group for Figure 3.14; and CSIRO for Figure 6.7. We thank John Wiley and Sons for permission to use Table 7.2. We thank Ailsa Allen for drawing some of the figures.

1

Introduction to the Anthropocene and Anthropogeomorphology

1.1 The Anthropocene

This book examines how humans have modified landforms and the processes which formed them during the Anthropocene. It takes a long time perspective and draws on examples from many different environments and countries. It demonstrates how extensive and significant human impacts on geomorphology have been, and how these impacts are likely to increase in future. The Anthropocene is itself a contested concept, both in terms of whether or not it exists and when it began. We argue that geomorphological evidence for the Anthropocene has been underplayed, but may be crucial in assessing the reality and scope of the Anthropocene. This chapter introduces the major debates over the Anthropocene, the field of Anthropogeomorphology, and the framework used in the rest of the book.

There are four key areas of debate surrounding the Anthropocene. First, there is debate surrounding what the Anthropocene actually is – what the concept means. Second, there is debate over whether the Anthropocene is a real entity and something that can be identified in the geological record. Third, if it is real, there is much debate over when the Anthropocene started and whether there can be a clear "golden spike" which marks its beginning. Finally, there is a rich and complex debate over what the Anthropocene means for humans and our relationship with the Earth.

The word "Anthropocene," which has Greek roots, is a new term for an older concept and a great deal of argument concerns how it can be differentiated, if at all, in terms of a boundary with the Holocene. Those who propose that the Anthropocene should become formally established as part of the geological timescale do so on the grounds that human activities now dominate the Earth System, and have led to a marked shift in its state. Those who oppose such a move note the difficulty of establishing a "golden spike" marking the beginning of the Anthropocene, and

1

doubt whether the concept is necessary or desirable. It is agreed, however, that the human impact on the environment and the Earth System has been increasing hugely in the last few centuries and that humans are now a very potent geomorphological force as part of this.

It is also apparent that in coming centuries a combination of population increases, land cover changes, climatic changes, and new technologies will increase this force still further. The burgeoning interest in the topic is reflected in the recent appearance of three new journals – *Anthropocene* (Elsevier), *The Anthropocene Review* (Sage), and *Elementa: The Science of the Anthropocene* (BioOne) and is fully discussed by Castree (2014a, b, and c), and Castree (2015).

The term "Anthropogene" was much used by Russian and some other European scholars in the twentieth century, more or less as a synonym for the Quaternary (see discussion in Gerasimov, 1979), but the word "Anthropocene" is largely a product of the twenty-first century. However, the recognition that humans have had a major suite of impacts on the natural environment has a much longer history (see Goudie, 2013b). Glacken (1967) pointed that out in a scholarly monograph, and others have recently reviewed the history of terminology and concepts surrounding the Anthropocene, such as its use in 1922 by the Russian geologist Aleksei Pavlov (Lewis and Maslin, 2015). An important stimulus to such ideas arose in the seventeenth and eighteenth centuries as Europeans became aware of the ravages inflicted in the tropics by their overseas expansion (Grove and Damodoran, 2006). In the nineteenth century George Perkins Marsh (1864; Figure 1.1) wrote his remarkable *Man and Nature*, the first full-length study in the English language of how humans were transforming the Earth's surface by deforestation and other processes (Lowenthal, 2016). Subsequently, many historical geographers became concerned with such activities as the use of fire, the clearing of woodland, and the drainage of wetlands (see Whitaker, 1940 for a review of early work), and in 1956 many of these issues were considered in a symposium on *Man's Role in Changing the Face of the Earth* edited by William L. Thomas, and in a masterful review by Turner et al. (1990). Ter-Stephanian (1988) sought to float the term "Technogene" for the accumulated significant effects of humans on the Earth System, but this seems to have been largely forgotten.

At the beginning of the twenty-first century, Steffen et al. (2004) reviewed ways in which biogeochemical systems interact at a global scale and the term "Earth System Science" started to be employed widely. It was against this background that the term Anthropocene was introduced by Crutzen (2002) as a name for a new epoch in Earth's history – an epoch when human activities have "become so profound and pervasive that they rival, or exceed the great forces of Nature in influencing the functioning of the Earth System" (Steffen, 2010, p. 443). In the last 300 years, Steffen et al. (2007) suggest, we have moved from the Holocene into the

Figure 1.1 George Perkins Marsh (1801–1882), from the Library of Congress Prints and Photographs Division, Washington, DC (http://loc.gov/pictures/resource/cwpbh.02223/; accessed November 17, 2015).

Anthropocene. They identify three stages in the Anthropocene. Stage 1, which lasted from c. 1800 to 1945, they call "The Industrial Era." Stage 2, extending from 1945 to c. 2015, they call "The Great Acceleration," and Stage 3, which may perhaps now be starting, is a stage when people have become aware of the extent of the human impact and may thus start stewardship of the Earth System (see Chapter 11).

However, many scientists argue that the Anthropocene has a much longer history than this scheme suggests, with early humans causing major environmental changes through such processes as the use of fire and the hunting of wild animals (e.g. Ruddiman, 2003). Indeed, one of the great debates surrounding the Anthropocene is when it started and whether it should be regarded as a formal stratigraphic unit with the same rank as the Holocene (Zalasiewicz et al., 2011a; Rull, 2013; Bostock et al., 2015; Edgeworth et al., 2015; Zalasiewicz et al., 2015). Walker et al. (2015), for example, consider the possibility that the Anthropocene might be designated a unit of lesser rank, that is, of stage, age, or even substage/sub-age status, and hence could become a subdivision of the Holocene rather than an epoch in its own right. On the other hand, some even think that the Anthropocene should replace the Holocene, which would become downgraded and reclassified as the final stage of the Pleistocene (Lewis and Maslin, 2015). Conversely, there are those who think the Anthropocene started with the Industrial Revolution and that 1800 AD is a logical start date for the new epoch (Steffen et al, 2011; Zalasiewicz et al., 2011b). At the other extreme, there are those, including archaeologists (Balter, 2010, 2013), who believe that substantial human impacts go back considerably further (see examples in Chapter 11) and have drawn attention to the deep history of widespread human impacts (Ellis et al. 2013a, b; Braje and Erlandson, 2014; Braje et al., 2014; Albert, 2015; Braje, 2015; Piperno et al., 2015). This case was made powerfully by Ruddiman et al. (2015, p. 38) who argued,

Does it really make sense to define the start of a human-dominated era millennia after most forests in arable regions had been cut for agriculture, most rice paddies had been irrigated, and CO_2 and CH_4 concentrations had been rising because of agricultural and industrial emissions? And does it make sense to choose a time almost a century after most of Earth's prairie and steppe grasslands had been plowed and planted? Together, forest cutting and grassland conversion are by far the two largest spatial transformations of Earth's surface in human history. From this viewpoint, the "stratigraphically optimal" choice of 1945 as the start of the Anthropocene does not qualify as "environmentally optimal."

Foley et al. (2013) have proposed the term "Palaeoanthropocene" for the period between the first signs of human impact and the start of the Industrial Revolution, whereas Glikson (2013) suggested a sub-division of the Anthropocene into three phases. He regarded the discovery of ignition of fire as a turning point in biological evolution and termed it the Early Anthropocene. The onset of the Neolithic he referred to as the Middle Anthropocene, while the onset of the industrial age since about 1750 AD he referred to as the Late Anthropocene. Smith and Zeder (2013) argued that the Anthropocene started around 10,000 years ago at the Holocene/Pleistocene boundary, with the initial domestication of plants and animals and the development of agricultural economies. Ruddiman (2013, 2014), on the other hand, argued that early deforestation and agriculture caused large greenhouse gas emissions slightly later, but nevertheless quite early in the Holocene. In China, Zhuang and Kidder (2014) have identified the importance and extent of gully erosion on slopes and sedimentation in valleys that developed in the Neolithic, when, they argue, Ancient China saw its own version of the Great Acceleration. Certini and Scalenghe (2011) preferred to put the lower boundary at around 2,000 years ago when major civilizations flourished, but Gale and Hoare (2012) have argued that the worldwide diachroneity of human impact makes it impossible to establish a single chronological datum for the start of the Anthropocene. It is certainly dangerous to think that in all places the human impact has shown a continually increasing trajectory, for there are many examples of ravages in one era being followed by phases of restoration, recovery, and stability in another. Trimble (2013) demonstrates this in the context of land use and land degradation history in the American Midwest.

Lewis and Maslin (2015) review the evidence for a "golden spike" which might provide an incontrovertible, globally relevant mark in the sedimentary record of the start of the Anthropocene. They find two candidates: a dip in atmospheric CO_2 levels around 1610 as recorded in high-resolution Antarctic ice cores, and a spike in ^{14}C concentrations in 1964 as recorded within tree-rings of a dated pine tree in Poland. According to Lewis and Maslin (2015), the evidence is most convincing for the 1610 date, which they prefer but do not go so far as to recommend. Rose (2015) argued that a stratigraphic marker for the start of the Anthropocene was provided by spheroidal carbonaceous fly ash particles (SCPs) – by-products of industrial fossil-fuel combustion. He found that data from over 75 lake sediment records showed a global, synchronous, and dramatic increase in particle accumulation starting in c. 1950 and driven by the increased demand for electricity and the introduction of fuel-oil combustion, in addition to coal, as a means to produce it. He argued that SCPs are morphologically distinct and solely anthropogenic in origin, providing an unambiguous marker. However, the validity of a search for these sorts of golden spike has been rejected by Hamilton (2015) and the controversy rumbles on.

Geomorphological change is an important component of the Anthropocene, though its effects will have varied greatly in space and time, and it is often neglected in accounts of human impacts on the Earth System (Brown et al., 2013). For example, in Central Europe the initial deforestation of a slope in the Neolithic may have been the most important geomorphological event since the end of the Pleistocene, while in Dubai it is the alteration of the coastline in just the last few decades (see Chapter 3). In this context, Fuller et al. (2015, pp. 266–7) provide an interesting analysis of Anthropocene changes in the rivers of New Zealand. They found that the nature and timing of human impact in New Zealand's river catchments are highly variable between regions and catchments, and this makes any attempt at formally defining the Anthropocene problematic at best because there is no ubiquitous, synchronous marker in New Zealand river catchments that marks the start of the Anthropocene:

In catchments draining the Southern Alps, natural processes are far more significant in determining erosion, sedimentation, and river activity. The clearest evidence for Polynesian impact is found in Northland's catchments in the form of increased floodplain sedimentation. Here, the start of the Anthropocene could be considered to equate with Māori occupation c. 1280 AD, with further augmentation associated with European settlement in the 1800s and 1900s. Farther east, in the East Coast Region of the

North Island, the start of the Anthropocene could be taken as c. 1920 when European clearance of indigenous vegetation in the Waipaoa and Waiapu catchments exposed a highly erodible terrain to a range of erosion processes, which resulted in erosion rates exceeding by an order of magnitude those estimated at the end of the Last Glacial Maximum. Each catchment and region must be recognised as unique in its response to human disturbance. New Zealand challenges the notion that the Anthropocene can be defined simply by a critical regime change in which human impact becomes the dominant controlling factor in the environment and overwhelms the forces of nature. New Zealand's highly active tectonic and climatic regime largely mitigates against Mankind becoming the dominant factor controlling river activity and alluvial sedimentation in most of its naturally dynamic catchments. The exception is Northland and the East Coast Region, where a regime change has been identified by these systems having been overwhelmed by sediment generated as a result of human impact resulting in rapid valley-floor sedimentation.

Whenever key anthropogenic changes may have taken place, however, their overall impact on the Earth System at a global scale is now immense. Hooke et al. (2012) provide estimates of the land area modified by human action in 2007 and suggest that more than 50 percent of Earth's ice-free land area has now been directly modified by human action. As Phillips (1997) perceptively pointed out, however, the significance of human actions depends on how big they are in comparison with natural changes, how they relate to the relaxation times of systems, how long in duration they are, and how frequent they are.

Smil (2015, p. 28) has cautioned against both exaggerating the power of humans and of rushing into accepting the creation of the Anthropocene as a new division of geological time. As he asked:

But is our control of the planet's fate really so complete? There is plenty of counter evidence. Fundamental variables that make life on Earth possible – the thermonuclear reactions that power the sun, suffusing the planet with radiation; the planet's shape, rotation, tilt, the eccentricity of its orbital path (the "pacemaker" of the ice ages), and the circulation of its atmosphere – are all beyond any human interference. Nor can we ever hope to control the enormous terraforming processes, the Earth's plate tectonics driven by internal heat and resulting in slow but constant creation of new ocean floor, forming, reshaping, and elevating landmasses whose distributions and altitudes are key determinants of climate variability and habitability. Similarly, we are mere bystanders watching volcanic eruptions, earthquakes, and tsunamis, the three most violent consequences of plate tectonics . . . let us wait before we determine that our mark on the planet is anything more than a modest microlayer in the geologic record.

Humans will continue to modify the Earth System in coming decades. New technologies will be developed and applied in areas like agriculture and mining, and increasing population levels will lead to further changes in land cover and in the exploitation of natural resources. The effects of land cover changes, as Slaymaker et al. (2009) point out, may be at least as important as the changes that will be caused by future climatic change. However, as the various reports of the

Intergovernmental Panel on Climate Change (IPCC) have shown (e.g. IPCC, 2013), global warming will greatly modify biomes, lead to massive changes in the cryosphere, and cause sea levels to rise. One major current concern is that certain key boundaries, thresholds, or tipping points may be crossed (Rockstrom et al., 2009).

There are signs that we are progressing into the era of environmental stewardship (see Chapter 11). Wohl (2013) argues that there are many ways in which geomorphologists can contribute to the management of what is now being called "the critical zone." This is Earth's near surface layer from the tops of the trees down to the deepest groundwater, where most human interactions with the Earth's surface take place and the locus of most geomorphological activity. Some examples of such management are discussed in Chapter 11. Harden et al. (2013) indicate some of the issues and focal points that the geomorphological community need to address for understanding human–landscape interactions in the Anthropocene, and these include the study of boundaries, thresholds, and feedbacks. Chin et al. (2014) have also argued that there needs to be a greater concern with feedbacks between society and the geomorphological environment, while Hamilton et al. (2015) explore the implication of the Anthropocene concept for the social sciences and the humanities. The whole way in which human geographers consider human/nature relationships may need to be reevaluated (Lorimer, 2012).

1.2 Anthropogeomorphology: Its History

Anthropogeomorphology, a term invented by Golomb and Eder (1964), is the study of the human role in creating landforms and modifying the operation of geomorphological processes. It thus focuses on many key aspects of geomorphological processes within the Anthropocene. Most of the classic textbooks of geomorphology, including those from the last few decades, however, ignore it totally.

Some Milestones in Anthropogeomorphology

Marsh, G. P. (1864) *Man and Nature.* New York: Scribner. The pioneer work on the human impact on the environment.

Woeikof, A. (1901). De l'influence de l'homme sur la terre. *Annales de Geographie*, 10, 97–114; 193–215. An influential Russian work on the multitude of ways in which humans modify the surface of the Earth.

Shaler, N. S. (1912). *Man and the Earth.* New York: Duffield. One of a number of works by an author who was keenly aware of soil erosion and other human impacts on the landscape of the United States.

Gilbert, G. K. (1917). Hydraulic-mining debris in the Sierra Nevada. *USGS Professional Paper* 105. A detailed study by one of America's greatest geomorphologists on the consequences of gold mining inland from San Francisco.

Bennett, H. H. (1938). *Soil Conservation*. New York: McGraw Hill. A massive survey by the head of the Soil Conservation Service, which contains a great deal of quantitative data on the effects of land use change on erosion rates.

Jacks, G. V. and Whyte, O. (1939). *The Rape of the Earth: A World Survey of Soil Erosion*. London: Faber and Faber. A popular global survey and polemic on the global menace of soil erosion.

Happ, S. C., Rittenhouse, G., and Dobson, G. C. (1940). Some principles of accelerated stream and valley sedimentation. *US Department of Agriculture Technical Bulletin*, 695. An impressive example that studies the links between erosion on slopes and alluviation in valleys.

Thomas, W. F. (ed.) (1956). *Man's Role in Changing the Face of the Earth*. Chicago: University of Chicago Press. A multi-author report of a ground-breaking symposium on the human impact.

Brown, E. H. (1970). Man makes the Earth. *Geographical Journal* 136, 74–85. A thoughtful and largely neglected study of anthropogeomorphology.

Trimble, S. W. (1974). *Man-Induced Soil Erosion on the Southern Piedmont*. Ankeny, IA: Soil Conservation Society of America. A masterful historical survey of the effects of land use change on erosion.

Nir, D. (1983). *Man, a Geomorphological Agent: An Introduction to Anthropic Geomorphology*. Jerusalem: Keter. A thorough review of knowledge by an Israeli geographer.

Slaymaker, O., Spencer, T. and Embleton Hamann, C. (eds.) (2009). *Geomorphology and Global Environmental Change*. Cambridge: Cambridge University Press. An edited work that places anthropogeomorphology in the context of global change.

Szabó, J., David, L., and Lóczy, D. (eds.) (2010). *Anthropogenic Geomorphology*. Dordrecht: Springer. A largely Hungarian review that is especially strong on constructed and excavated landforms.

Nevertheless, anthropogeomorphology has a long history of study (Goudie, 2013b). Research on torrents in the European Alps, undertaken in the late eighteenth and early nineteenth centuries, deepened immeasurably the realization of human capacity to change the environment (Surell, 1841). Similarly, de Saussure (1796) showed that Alpine lakes had suffered a lowering of water levels in recent times because of deforestation. In Venezuela, von Humboldt, with his partner Bonpland (see Humboldt and Bonpland, 1815), concluded that the level of Lake Valencia in

Figure 1.2 A newly-formed ravine that developed at Milledgeville, Georgia, following deforestation (from Lyell, 1875, p. 338).

1800 was lower than it had been in previous times, and that deforestation, the clearing of plains, irrigation, and the cultivation of indigo, were among the causes of the gradual drying up of the basin (Boussingault, 1845; Cushman, 2011).

Lyell, in later editions of the *Principles of Geology* (Lyell, 1875, p. 338), noted the effects of recent deforestation in Georgia and Alabama in the United States. This had produced numerous ravines with considerable rapidity (Figure 1.2). The extent of human influence on the environment was explored in detail by Marsh (1864), who dealt at length with human influence on the woods, the waters, and the sands, and discussed such issues as accelerated erosion, flooding, and coastal dune movement.

A major phase of anthropogeomorphological research, based on the study of the threat posed by soil erosion, took place in the 1930s and 1940s. Notable here was the advocacy of people like Bennett (1938) and Lowdermilk (1934, 1935), who were associated with the early days of the Soil Conservation Service in the United States but who toured the world advocating the importance of soil conservation. Their work was a stimulus to Dale and Carter's (1955) *Topsoil and Civilisation*, which was a global discussion of soil erosion over the last 6,000 years. Also, at the same time, there was a concern about soil erosion and the means of soil conservation in the British colonies (Tempany et al., 1944), and in particular in Africa (Anderson, 1984; Beinart, 1984).

Wind erosion was seen as a big threat, and of particular note in this respect was the work conducted by W. S. Chepil and his collaborators at the Wind Erosion Research Center at Kansas State University, established in 1947 (e.g. Chepil and Woodruff, 1963). They were concerned with establishing the fundamentals of soil movement by wind, the properties of soils which influenced their susceptibility to wind erosion, the sedimentary characteristics of dust storms, and the effects of various land cover treatments (mulches, field size, maintenance of crop residues, type of ploughing). They also developed technology for advancing aeolian research, including dust samplers and portable wind tunnels.

1.3 Direct and Indirect Anthropogeomorphological Influences

Some geomorphological features are produced by direct anthropogenic processes. These tend to be relatively obvious in form and are frequently created deliberately. They include landforms produced by construction (e.g. spoil tips, embankments, sea walls), excavation (e.g. mines), hydrological interference (e.g. reservoirs and canals), and farming (e.g. terraces); see Table 1.1.

By contrast, landforms produced by indirect and inadvertent anthropogenic processes are often more difficult to recognize because they involve the acceleration of natural processes rather than the operation of new ones. It is the indirect and inadvertent modification of process and form that is the most crucial aspect of anthropogeomorphology. Rates of weathering may be modified because of the acidification of precipitation caused by accelerated nitrate and sulfate emissions or because of accelerated salinization in areas of irrigation (Goudie and Viles, 1997; see Chapter 5). By modifying land cover, humans have accelerated erosion and sedimentation (Jones and Marcus, 2006; Wilkinson and McElroy, 2007); caused sedimentation on floodplains and estuaries, and in lakes and elsewhere (see Chapters 6 and 7); triggered landslides and debris flows (see Chapter 6); changed river channel forms (Chapter 7); and created ground subsidence and had an influence on the operation of earthquakes through the impoundment of reservoirs (Meade, 1991; see Chapter 4).

Table 1.1 *Deliberately created landforms.*

Feature	Cause
Artificial islands	Creation of new land
Artificial reefs	Fishing, trawl protection, surfing, habitat restoration
Banks along roads	Noise abatement
Broads	Peat exploitation
Canals	Transport, irrigation, drainage
City mounds (*tells*)	Human occupation
Craters	War, *qanat* construction
Cuttings and sunken lanes	Transport
Dikes, polders	River and coast management
Embankments	Transport, river, and coast management
Moats	Defense
Mounds and tumuli	Defense, memorials and burial
Pits and ponds	Mining, marling
Reservoirs	Water management, cooling basins
Ridge and furrow	Ploughing
Spoil heaps	Mining, waste disposal
Subsidence depressions	Mineral and water extraction
Terracing, lynchets	Water and erosion control, provision of flat land for agriculture

Indirect Anthropogenic Processes

Acceleration of Erosion and Sedimentation
Vegetation clearance
Engineering activities, especially road construction and urbanization
Increased fires
Hydrological regime modification

Subsidence
Drainage of organic soils and peat
Hydraulic (e.g. through groundwater and hydrocarbon pumping)
Melting of permafrost (thermokarst)
Underground mining (e.g. of coal and salt)

Slope Failure
Loading
Lubrication
Shaking
Undercutting

Continued

Earthquake Generation
Fracking (hydro-fracturing)
Loading by water impounded in reservoirs
Lubrication by water along fault planes

Weathering
Accelerated salinization, producing salt weathering
Acidification of precipitation by nitrate, sulphate, and CO2 emissions
Laterization (bowalization) by vegetation removal

Further, there are situations where humans may deliberately and directly alter landforms and processes but thereby initiate unintended or unanticipated events. For instance, there are records of attempts to reduce coastal erosion, often using hard engineering techniques, which, far from solving it, only exacerbated the problem. Examples include dune stabilization schemes in North Carolina (Dolan et al., 1973), the role of sea walls in causing beach scour (Bird, 1979), and downdrift coastal erosion as a result of updrift "protection" schemes (see Chapter 9).

1.4 Techniques in Anthropogeomorphology

A range of techniques has been employed in anthropogeomorphology. These include analyzing evidence from cores from lakes, bogs, and the like, which can give a long-term, often well-dated, high-resolution picture of environmental change which, when compared with data on potential drivers of change, can be used to identify the human roles (e.g. O'Hara et al., 1993; Jones et al., 2015). Palaeosols with a distinct human signature may also be utilized (e.g. Beach et al., 2015). Related to this is the presence of anthropogenic materials, such as artefacts, coal, fly ash, slag, human-generated isotopes, or pottery, in sedimentary sequences (e.g. Stinchcomb et al., 2013; Tang et al., 2015). Also important has been the collection of archival material (Trimble and Cooke, 1991; Trimble, 2008b), including old ground photographs of, for example, glaciers, gullies, soil erosion, debris flows, and arroyos (Cerney, 2010; Webb et al., 2010; Frankl et al., 2011), and the use of sequential maps and remotely sensed images (e.g. Bakoariniaina et al., 2006). The study of sequences of air photographs has particular value (Micheletti et al., 2015). Equally, long-term monitoring, when it exists, can indicate the timing and scale of landscape changes (Burt, 1994). Numerical modeling is an increasingly important approach to the understanding of systems behavior and the prediction of possible future changes (e.g. Nearing et al., 2005). For instance, literally hundreds of studies have employed the Soil and Water Assessment Tool (SWAT),

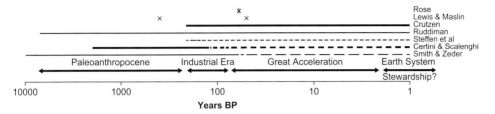

Figure 1.3 Timelines of the Anthropocene.

which is a river basin scale model developed to quantify the impact of land management practices in large, complex watersheds (Douglas-Mankin et al., 2010). In recent years, several technological developments have led to enhanced resolution and coverage of datasets on, for example, topography and its change over time, which allow an unprecedented insight into the changing face of the Earth and the human imprints upon it.

1.5 Our Approach

The relationships between the Anthropocene and geomorphology have not been explored in depth before, and this is one of the key aims of this book. There are two main ways in which geomorphology and the Anthropocene may be intertwined. First, human impacts on geomorphology (or Anthropogeomorphology) may be accelerated, complicated, and become much more serious within the context of the Anthropocene. Second, geomorphological processes provide an important vector by which key human impacts on global biogeochemical cycles operate – thus geomorphology may mediate and amplify many key Anthropocene processes.

In this book we take a long-term view of the Anthropocene because, although we recognize the increasing tempo of change since the Industrial Revolution, and especially since 1945, there are many examples of the potent impact of humans in previous millennia and in particular places. Figure 1.3 presents some different timelines of the Anthropocene drawn from reviewing key recent papers, and highlights the four broadly defined sub-sections that we focus on (Palaeoanthropocene from 7 ka BP until c. 1750; The Industrial Era from c. 1750 to c. 1945; the Great Acceleration from c. 1945 to c. 2000; and Earth System Stewardship from c. 2000 to the present). The exact timespans covered in the book vary according to the topics covered.

The organization of the book is as follows: In Chapter 2 we review the main driving forces of anthropogenic changes in their historical sequence from the use of fire to modern-day globalization. In Chapter 3 we discuss landforms that have been formed deliberately as a result of excavation and construction since at least the

Neolithic. In Chapter 4 we evaluate human contributions to subsidence and its geomorphological and Earth System consequences. Chapters 5 and 6 focus on human influences on weathering and hillslope processes and their consequences for geomorphology and the Earth System. In Chapters 7–10 we examine primarily non-deliberate anthropogenic changes in landforms and the rate of operation of geomorphological processes in the context of particular environments (fluvial, aeolian, coastal, and cryospheric). In each chapter we not only look back at the history of change over the different phases of the Anthropocene, but also look at the possible future consequences of changes in climate and in land use and land cover. Finally, in Chapter 11 we bring much of this material together, reviewing the impressive geomorphological changes wrought by humans during the Anthropocene, exploring their antiquity, discussing the future in the context of climate change and of other drivers, and drawing attention to the growing importance of environmental stewardship. We conclude with a consideration of some of the ways in which geomorphological changes impact upon the Earth System. Our aim is to demonstrate that the Anthropocene is a crucial issue for geomorphologists and also that geomorphologists can increasingly contribute to understanding the Anthropocene and mitigating some of the deleterious human impacts on the Earth system.

2

Drivers of Anthropogeomorphological Change

2.1 Humans Arrive

It is a truism that there could be no Anthropocene before humans arrived on the face of the Earth. The geomorphological impact of humans has changed greatly over time in response to their global distribution, numbers, and adoptions of new technologies. Figure 2.1 gives a timeline of some of the major changes that may have impacted upon geomorphology in the Quaternary era.

Fieldwork in Turkana, Kenya, has recently identified evidence of early hominin tool-making activity at Lomekwi 3, a 3.3-million-year-old archaeological site (Harmand et al., 2015). However, the first recognizable human, *Homo habilis,* evolved about 2.5–2.8 million years ago, more or less at the time that the Pleistocene ice ages were developing in mid-latitudes. The oldest remains of *Homo* have been found either in sediments from the rift valleys of East Africa (as in the Afar region of Ethiopia) (Villmoare et al., 2015) or in caves in South Africa. Since that time the human population has spread over virtually the entire land surface of Earth, reaching Asia by around 2 million years ago (Zhu et al., 2008) and Europe not much later (Moncel, 2010). In southern Europe there are stone tools with *Homo* that date back to 1.3–1.7 million years ago (Carbonell et al., 2008). In northwest Europe and Britain the earliest dates for human occupation are >0.78 million years (Parfitt et al., 2010).

Modern humans, *Homo sapiens*, appeared in Africa around 160,000 years ago (Stringer, 2003) and then spread out of Africa to other parts of the world. However, in some parts of the world they and their impacts arrived late. The Americas and Australia were probably virtually uninhabited until about 15,000 and 50,000 years ago, respectively. Some of these dates are controversial, and this was especially true of Australia, where a date of c. 50,000 years ago is now widely accepted (Balme, 2013). There is also considerable uncertainty about when humans arrived in the Americas (Goebel et al., 2008). Many authorities have argued that the first

Years before present	Driving force
	The Great Acceleration
100	
	Internal combustion engine
	Industrial revolution
	European colonization of Americas, Australia, etc.
1000	Peopling of New Zealand, Madagascar, Oceania, etc.
	The classical era
	Secondary products revolution
	Irrigation
	Metals and mining
	Settlements and urbanization
	Domestication, agriculture, land clearance
10,000	
	Pleistocene extinctions
	Peopling of Americas and Australia
100,000	
	Modern humans
1,000,000	Use of fire and stone tool manufacture
	Arrival of *Homo*

Figure 2.1 Timeline of human driving forces involved with geomorphological change.

colonizers of North America, equipped with so-called Clovis spears, arrived via the Bering land-bridge from Asia around 12,000 years ago. However, some earlier dates exist for the Yukon (Yesner, 2001) and South America (Lahaye et al., 2015), which may imply an earlier phase of colonization (Dillehay, 2003). The settlement of Oceania occurred relatively late, with colonization of the western archipelagos of Micronesia and eastern Melanesia taking place at c. 3500–2800 BP, of central and eastern Micronesia at 2200–2000 BP, and of eastern and southern Polynesia at 1100–700 BP (Anderson, 2009).

At the peak of the Last Glacial Maximum around 20,000 years ago, the human population of the Old World was possibly between about 2 and 8 million (Gautney and Holliday, 2015), but estimates of past populations are difficult to assess. Figure 2.2 gives an indication of world population totals for the last 12,000 years.

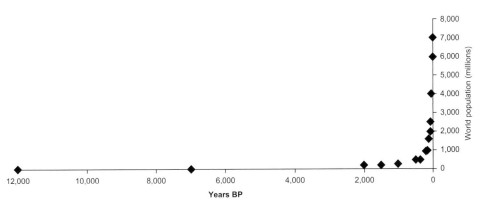

Figure 2.2 World population trends, 12,000 BP to the present day. (Sources: data in the Hyde History Database of the Global Environment, http://themasites.pbl.nl/tridion/en/themasites/hyde/index.html, accessed March 9, 2015, and Gautney and Holliday, 2015).

Note the miniscule levels at the start of this period, the mid-Holocene value of between 5 and 24 million, the value of c. 200 million at the end of the Roman period, the late medieval value of between c. 400 and 500 million, and the value of around 900–1000 million at the start of the Industrial Revolution. The total has now surged past 7000 million.

2.2 Fire

At an early stage in their history humans started using fire (Glikson, 2013; Bowman, 2014; Albert, 2015), a major agent by which they have influenced their environment. The date at which fire was first deliberately employed is controversial (Caldararo, 2002). It may have been used very early in South Africa, where Beaumont (2011) and Berna et al. (2012) found some traces of repeated burning events from Acheulean cave sediments dating back to more than a million years ago, and from East Africa, where Gowlett et al. (1981) found evidence for deliberate manipulation of fire from over 1.4 million years ago. After around 400,000 years ago evidence for the association between humans and fires becomes compelling.

Large parts of the Americas suffered fires at regular intervals prior to European settlement (Parshall and Coster, 2002). Abrams and Nowacki (2015) have reviewed the evidence for what they term the early Anthropocene burning hypothesis by Native Americans in the context of an area that is now the eastern United States and suggest that the impact of humans on vegetation occurred quite early in the Holocene but increased as the Holocene progressed. Fire was, and still is, central to the way of life of Australian aboriginal peoples (Hope, 1999) and fire scars are widespread in areas such as the Great Sandy Desert (Figure 2.3). In neighboring New Zealand, Polynesians carried out extensive firing of vegetation in pre-European settlement times after their arrival in c. 1280 AD (McWethy et al., 2009). McWethy et al. (2014) have documented the astonishing rate of deforestation at the hands of small transient Polynesian populations in New Zealand. The fires continued over a period of about a thousand years up to the period of European settlement (Mark and McSweeney, 1990). The changes in vegetation that resulted were substantial (McGlone and Wilmshurst, 1999), with forest cover being reduced from about 79 percent to 53 percent. It is apparent from many parts of the world that even small hunter-gatherer populations can cause great environmental changes through the use of fire (Lightfoot and Cuthrell, 2015).

Fire, whether natural or man-induced, is a major cause of slope instability, erosion, and debris flow generation (see Chapter 6) by removing or reducing protective vegetation, liberating hydrophobic substances, increasing peak stream flows, and leading to larger soil moisture contents and soil-water pore pressures (because of reduced interception of rainfall and decreased moisture loss by transpiration) (Wondzell and King, 2003).

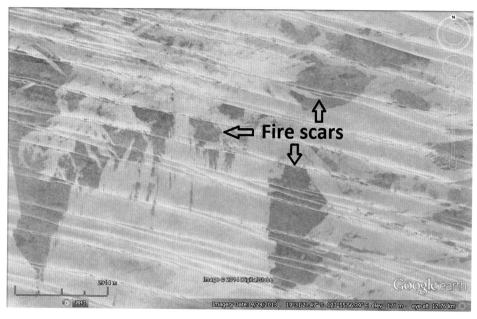

Figure 2.3 Fire scars, probably produced by aboriginal burning in the Great Sandy Desert of Australia. Scale bar c 2.9 km. © 2014 Digital Globe.

2.3 Tool Production

Early humans used stone tools produced by quarrying and splitting of rock outcrops and boulders. There are some locations where large "factory sites" occur. One of these is the Messak Settafet, a sandstone massif in the Sahara in Libya. Foley and Lahr (2015, p. 1) wrote:

It is littered with Pleistocene stone tools on an unprecedented scale and is, in effect, a man-made landscape. Surveys showed that parts of the Messak Settafet have as much as 75 lithics per square meter and that this fractured debris is a dominant element of the environment. The type of stone tools – Acheulean and Middle Stone Age – indicates that extensive stone tool manufacture occurred over the last half million years or more. The lithic-strewn pavement created by this ancient stone tool manufacture possibly represents the earliest human environmental impact at a landscape scale and is an example of anthropogenic change.

Huge amounts of waste material from tool manufacture, *débitage*, occur at other sites throughout Africa and the volumes of rock involved are colossal, particularly in areas with appropriate lithologies. Foley and Larh cite areas in the Zebra River Valley in Western Namibia which have an average lithic density of over 2 million per km^2, artifact-bearing Plio-Pleistocene sediments at Koobi Fora in Kenya which have a density of 40,000 lithics per km^2, and the Nubian Desert which yielded densities of between $1–12 \times 10^6$ per km^2. Such prolific factory sites are by no means restricted to Africa, and chert outcrops at the Rohri Hills in Sind, Pakistan, for instance, have extensive and dense spreads of material dating from the Paleolithic to the Harappan (Allchin, 1976).

2.4 Pleistocene Overkill

Another important early human impact was a spasm of animal extinction called the Pleistocene Overkill (Figure 2.4). Between 50,000 and 10,000 years ago, most large mammals became extinct everywhere except Africa and it is possible that human hunters were responsible, though the role of climate change cannot be dismissed (Boulanger et al., 2014; Faith, 2014). Megafaunal extinctions in Australia may have occurred around 41,000 years ago, shortly after human arrivals (Roberts et al., 2001). North and South America were stripped of large herbivores between 12,000 and 10,000 years ago. Extinctions continued into the Holocene on oceanic islands (Turvey, 2009), and in the Galapagos Islands virtually all extinctions took place after the first human contact in AD 1535 (Steadman et al., 1991). The possible geomorphological impacts of this have not received much attention, but the extinction of large herbivores such as elephants will have had an impact on vegetation cover, nutrient cycling, trampling activity, and track generation, all of which could have had geomorphological ramifications (Haynes, 2012). It may also

Figure 2.4 Reconstruction of an extinct giant herbivore from the La Brea Tar Pits, Los Angeles, California.

have had an impact on global climates because of the changes in vegetation cover that ensued (Doughty et al., 2010). More recently, the extirpation of animals, such as the American bison in the nineteenth century, probably had a major impact on buffalo wallows in the American plains (McMillan et al., 2011), while the extirpation of beavers had a major impact on river channels (Butler, 2006).

2.5 Agriculture and Domestication

Before the agricultural revolution some 10,000 years ago, the causes of which remain controversial (Barker, 2006), human groups lived by hunting and gathering. Population densities were low, possibly with a mean density of around 0.12 individuals per km^2, and the optimum territory for a band of hunter-gatherers in the Middle East would have been 300 and 2000 km^2 (Bar-Yosef, 1998). At that time the world population may have been around 2 to 8 million people (Gautney and Holliday, 2015, see also Figure 2.1). However, the agricultural revolution probably enabled an expansion of the total human population to about 200–300 million by the time of Christ, and to 500 million by AD 1650.

The first steps toward domestication of crops in the eastern Mediterranean lands and the Hilly Flanks of the Middle East may have taken place around 12,000 years ago (Özkan et al., 2011), while the domestication of millet in China has been extended back to 10,000 years ago (Lu et al., 2009). In Mesoamerica, maize, beans, and squashes may have been domesticated as early as 10,000–9,000 years ago (Zizumbo-Villarreal and Colunga-García Marín, 2010).

At much the same time, domestication of animals permitted the development of pastoral economies, and led to the gathering together of large numbers of animals, around, for example, wells and settlements. Domestic stock have geomorphological implications. They can damage soil structure through trampling and compaction, they produce heavily grazed lands that tend to have considerably lower infiltration capacities than those found in ungrazed lands, and they can change the nature and density of vegetation cover, thereby affecting erosion rates (see Chapter 6), including wind erosion (Aubult et al., 2015).

Most importantly of all, however, the spread of agriculture transformed land cover at a global scale. The deliberate removal of forest, whether by fire or cutting, is one of the most longstanding and significant ways in which humans have modified the environment (Bhagwat et al., 2014). Pollen analysis shows that large tracts of European forests, including those around the Mediterranean, were removed in Mesolithic and Neolithic times and at an accelerating rate thereafter (Innes et al., 2010). Forests were reduced to near-modern levels by 2,500 years BP in England and 2,200 years BP in France (Zennaro et al., 2015). With regard to the equatorial rain forests, Flenley (1979) and others have indicated that forest clearance for agriculture has been going on since at least 3000 BP in Africa, 7000 BP in South and Central America, and possibly since 9000 BP or earlier in India and New Guinea. Recent studies by archaeologists and paleoecologists have tended to show that prehistoric human activities were more extensive in the tropical forests than previously thought (Willis et al., 2004). It is also important to remember the scale of deforestation that took place in the Mediterranean lands in classical times and in Central Europe in medieval times (see Williams, 2003).

Land use and cover changes brought about by humans are fundamental drivers of geomorphological change, and in some regions there is evidence of markedly higher rates of erosion associated with Neolithic clearances (see Chapter 6), such as those in China (Rosen et al., 2015). It is also possible (Ruddiman, 2003, 2014) that mid-Holocene deforestation and land cover change, including rice irrigation in China, modified global climates by releasing carbon dioxide and methane into the atmosphere, and by altering albedo (He et al. 2014).

However, this process of land cover change has accelerated over the Anthropocene into the Industrial Era. In the last 300 years the areas of cropland and pasture have increased by around five-fold to six-fold (Goldewijk, 2001). Richards (1991: 164) calculated that since 1700 about 19 percent of the world's forests and woodlands have been removed. Over the same period the world's cropland area has increased more than four and a half times, and between 1950 and 1980 the rate of increase amounted to more than 100,000 km^2 per year. Detailed data on different areas are provided by Pongratz et al. (2008). They indicated that by AD

800, 2.8 million km^2 of natural vegetation had been transformed to agricultural land, which is about 3 percent of the area potentially covered by vegetation. By AD 1700, the agricultural area had extended to 7.7 million km^2. Over the next 300 years, the total agricultural area rose to 48.4 million km^2, especially as a result of the expansion of pasture. They concluded that between AD 800 and AD 1700, 4.9 million km^2 of natural vegetation had been brought under agricultural use, compared to 40.7 million km^2 in the following three centuries.

2.6 Irrigation and Water Management

One highly important development in agriculture was irrigation and the adoption of riverine agriculture. This came rather later than domestication. One of the earliest pieces of evidence of artificial irrigation is the mace-head of the Egyptian Scorpion King. It shows one of the last predynastic kings ceremonially cutting an irrigation ditch around 5,050 years ago (Butzer, 1976; see Figure 2.5), although it is possible that irrigation at Sumer in Iraq started even earlier. The construction of terraces, dams, and reservoirs were important because of their impacts on runoff, erosion, and sedimentation.

The global irrigated area in 1900 amounted to less than 50 million ha, but it increased hugely during the Great Acceleration, from 94 million ha in 1950–311 million ha in 2009 (www.world**watch**.org/global-**irrigated-area**-record-levels-expansion-slow accessed June 18, 2015; www.earth-policy.org/datacenter/xls/book_pb4_ch2_8.xls, accessed June 18, 2015).

Figure 2.5 An illustration from ancient Egypt of the construction of an irrigation channel. The original is in the Ashmolean Museum, Oxford.

2.7 Secondary Products Revolution

The Secondary Products Revolution (Greenfield, 2010) occurred in the Old World in the mid-Holocene. In the 1970s zoologists realized that the earliest domesticated economic animals – sheep and goats, cattle, and pigs – were butchered at a young age for two millennia or more after the initiation of domestication, and so were used almost solely for meat. Sherratt (1983) recognized that this conclusion carried an important implication: that secondary animal products – wool, dairy foods, traction power, and transport – were discovered later, and that this discovery should have had profound effects on human economies across the Old World. The ensuing expansion of herding, including that of horses, literally cleared the European landscape of its unbroken forests.

The plow (Figure 2.6) was particularly important in this process, being the first application of animal power to the mechanization of agriculture. The domestication of oxen in Mesopotamia and the Indus valley provided farmers with

Figure 2.6 One feature of the Secondary Products Revolution was the introduction of the animal-drawn plow. This greatly added to the power of humans to change the soil. The example here is from Sind, Pakistan.

much greater draft power than humans themselves could provide. Plows greatly changed the structure of soils and exposed them to erosion.

Evidence of wheeled vehicles appears from the late Neolithic near-simultaneously in Mesopotamia (Sumerian civilization), the Northern Caucasus (Maykop culture), and Central Europe. Animal-drawn carts both permitted more intensive farming and enabled the transportation of its products. Furthermore, the development of textiles from animal fibers afforded, for the first time, a commodity which could be produced for exchange in areas where arable farming was not the optimal form of land use. Moreover, the use of milk provided a means whereby large herds could use marginal or exhausted land, encouraging the development of the pastoral sector with transhumance or nomadism.

2.8 Urbanization

Although modest communal settlements may have occurred before domestication (Watkins, 2010), it was within a few thousand years of the adoption of cereal agriculture that people began to gather into ever larger settlements (cities) and into more institutionalized social formations (states). Around 6,000 years ago, cities developed in the basin of the Tigris and Euphrates, and more followed by c. 5,000 years ago in the coastal Mediterranean, the Nile valley, the Indus plain, and coastal Peru. In due course, cities had evolved which had considerable human populations. By 2015 there were over 500 urban areas with a population greater than 1 million, and 12 with a population over 20 million. The urban population in 2014 accounted for 54 percent of the total global population, up from 34 percent in 1960 (www.who.int/gho/urban_health/situation_trends/urban_population_growth_text/en/, accessed October 12, 2015).

Large cities and urban agglomerations have their own environmental effects (Douglas, 1983; Chin, 2006), sucking in resources and materials and exporting vast amounts of waste (see the discussion of tells and mounds in Chapter 3). Erosion rates, runoff, channel forms, and flooding are also affected by urbanization (see, e.g. the study of Sao Paulo by da Luz and Rodrigues, 2015). Particularly high rates of erosion are produced in the city construction phase, when there is a large amount of exposed ground and much disturbance produced by vehicle movements and excavations (see Chapter 6).

2.9 Mining and Metals

The mining of ores and the smelting of metals was one further development which increased human power (Roberts et al., 2009). Evidence for copper smelting occurs at Catal Hüyük in Turkey and in Jordan from the sixth millennium BC (Grattan et al., 2007) and in Serbia from the fifth millennium BC (Radivojević et al., 2010). The spread of metal working into other areas was rapid, particularly in the second half of the fifth millennium (Muhly, 1997), and by 2500 BC bronze products were in use from Britain in the West to northern China in the East. The smelting of iron ores may date back to the late third millennium BC. The question of when iron smelting occurred in Africa is still debated (Killick, 2015), but once adopted it required both large amounts of wood to fire the furnaces (Figure 2.7) and also enabled land clearance with iron tools. It may thus have led to a spasm of deforestation and associated erosion (see, e.g. Eriksson et al., 2000; Heckmann, 2014), though this is difficult to prove (Lane, 2009).

Smelting may produce intense air and soil pollution that can lead to the creation of industrial barrens, where little vegetation grows and erosion may be intense (Kozlov and Zvereva, 2007).

Taking overburden into account, the amount of material moved by the mining industry globally is probably at least 28 billion tons – about 1.7 times the estimated amount of sediment carried each year by the world's rivers (Young, 1992). Mining has led to the creation of large excavations, the erection of great piles of waste

Figure 2.7 A reconstructed Iron Age furnace from Zimbabwe, showing also the bellows and tuyeres.

(see Chapter 3), and the deposition of sediments into stream channels (see Chapter 7). Much waste is stored in tailing dams, but should these fail they can have major impacts on stream sediments and water chemistry (Kossoff et al., 2014). Nineteenth- and twentieth-century coal mining seems to have caused great changes in valley sedimentation and erosion in areas like the eastern United States, and Stinchcomb et al. (2013) speak of the Mammoth Coal Event as a major stratigraphic event in the Anthropocene.

The rate at which many major minerals are being mined is steadily increasing. This can be illustrated by the statistics for iron ore. In 1974 the world figure was c. 900 million tons. By 2012 this figure had risen to 3,000 million tons, an increase of more than three times. The comparable figures for coal are 3,000 million tons in 1974, and 7,924 million tons in 2012, an increase of 2.6 times (www.bgs.ac.uk/mineralsuk/statistics/home.html, accessed, November 13, 2014). As concerns about pollution and global warming grow, and alternative sources of energy become available, coal may become a less important source of energy. This is already the case in the United Kingdom, where current production has dropped to one-tenth of what it was at its peak in 1913.

2.10 Globalization

The perfecting of sea-going ships in the sixteenth and seventeenth centuries was the time when mainly self-contained regions of the world coalesced. Grove (1996) has shown how concerns were raised that tropical deforestation could lead to climatic and hydrological changes, promoting early attempts at conservation. A consequence of this globalization was that the natural faunal and floral realms broke down, and the introduction of invasive plants and animals transformed some environments, including oceanic islands. Biotic homogenization has taken place. Such invasions have had many geomorphological impacts (Tickner et al., 2001; Dukes and Mooney, 2004; Stromberg et al., 2007; Fei et al., 2014) on, for example, river channels (see Chapter 7), coastal dunes (see Chapter 8), salt marshes (see Chapter 9), and sediment mobilization (as in the case of the invasive signal crayfish, *Pacifastacus leniusculus*, in Europe) (Harvey et al., 2011).

2.11 Harnessing of Energy

The invention of the steam engine in the late eighteenth century, and of the internal combustion engine and high energy explosives in the late nineteenth century, massively increased human access to energy and lessened dependence on animals, wind, and water. Modern technologies have immense power output. A pioneer steam engine in AD 1800 might rate at 8–16 kilowatts. Modern railway diesels top

3.5 megawatts, a twenty-coach Eurostar train tops 12.2 megawatts, and a large aero engine 75 megawatts. The first John Deere tractors produced in the 1930s had a power of c. 7.5–15 kilowatts, whereas some of them now have a power of c. 372–450 kilowatts. There are now many types of equipment that enable earth materials to be moved in great quantities, the history of which is discussed by Haddock (1998) and Haycraft (2002). There has also been a great expansion in the number of machines being employed. For example, sales of tractors in India tripled from 122,000 in 1989–90, to 369,000 in 2009–10 (Sarkar, 2013). The number of tractors in the world as a whole increased from 0.3 million in 1920 to over 26 million in 2000 (McNeill, 2000, table 7.2).

The invention of the gasoline-powered chainsaw, especially after 1950, was another important development. This cutting-edge technology allowed humans to fell trees 100 or 1,000 times faster than with axes (McNeill, 2000, p. 307).

The burning of fossil fuels is, of course, the major driver of anthropogenic climatic changes and it is since the Industrial Revolution that the concentrations of greenhouse gases have risen so sharply and inexorably. The preindustrial level of carbon dioxide may have been only 260–70 parts per million by volume (ppmv). The 2014 level was around 396 ppmv. At the present rate it will reach 500 ppmv by the end of the twenty-first century. Ice core studies and recent direct observations suggest that until the beginning of the Industrial Revolution background levels of methane were relatively stable at around 600 parts per billion by volume (ppbv), although they may have been increased prior to that by rice farming and other agricultural activities (Ruddiman and Thomson, 2001; Ruddiman, 2014). They rose steadily between AD 1700 and 1900, and then increased still more rapidly, attaining levels that averaged 1,300 ppbv in the early 1950s and over 1,840 in 2014. This is three times over background levels.

The many effects of future anthropogenic climatic changes in geomorphology and the environment is a major component of studies of the Anthropocene, whether it be the reactivation of dunes on desert margins (see Chapter 8), the creation of thermokarst in melting permafrost regions (see Chapter 10), or the shrinkage of glaciers and ice caps (see Chapter 10).

Increasing amounts of energy are also being harvested from hydropower schemes, which require the construction of major dams (Zarfl et al., 2015), the consequences of which for river flow and sediment trapping are very major (see Chapter 7). At least 3,700 dams, each with a capacity of more than 1 megawatt, are currently either planned or under construction, primarily in countries with emerging economies. The Amazon and La Plata basins in Brazil will have the largest total number of new dams in South America, whereas the Ganges–Brahmaputra basin (mainly India and Nepal) and the Yangtze basin (China) will face the highest dam construction activity in Asia. The number of our planet's remaining free-flowing large rivers will decrease by about 21 percent.

Increasing energy demand and the search for clean sources of supply also has geomorphological implications. For instance, the increasing use of biofuels causes land cover changes which will modify local rates of erosion (Khanal et al., 2013). Moreover, if increased US maize production for ethanol were to reduce soybean production and exports, changes in world prices might provide an incentive to increase soybean production on land previously used as pasture in Brazil which in turn might push cattle farming into Amazonia (Marshall et al., 2011) with numerous hydrological and geomorphological consequences. New techniques to obtain gas from shale by hydraulic fracturing (fracking) are likely to be widely adopted in coming decades, but have implications for seismic activity (Rahm, 2011), and exploitation of geothermal energy can cause land subsidence (Allis et al., 2009; Chapter 4). Tidal barrage schemes have also been opposed because of their influence on sediment budgets and mud flat exposure (Kadiri et al., 2012). Finally, the process of construction of wind farms, particularly in areas of blanket peats, can create landsliding and erosion (Natural England, 2009).

2.12 The Great Acceleration

As will be clear from this discussion, anthropogenic influences on environments and landscapes have a long history. Fire, extinctions, domestication, pastoralism, agriculture, deforestation, introductions and invasions of flora and fauna, and water management date back a long way, and sometimes as far as the Paleolithic. However, since the start of the Industrial Revolution 300 years ago, a great acceleration in human impact has occurred, and Williams (2003) writes of "The Great Onslaught" that has occurred on forests since 1945. As McNeill (2000; 2003) has pointed out, the twentieth century was a time of extraordinary change. The human population increased from 1.5 to 6 billion, the world's economy increased fifteenfold, energy use increased thirteen- to fourteenfold, freshwater use increased ninefold, and the irrigated area by fivefold. In the hundred centuries from the dawn of agriculture to 1900, McNeill calculates that humanity only used about two-thirds as much energy (most of it from biomass) that it used in the twentieth century. Indeed, he argued that humankind used more energy in the twentieth century than in all preceding human history put together. The forest and woodland area shrank by about 20 percent accounting for perhaps half the net deforestation in world history. Kates et al. (1990), attempted to make quantitative comparisons of the human impact on ten "component indicators of the biosphere." For each component they defined total net change clearly induced by humans to be 0 percent for 10,000 years ago and 100 percent for 1985. They estimated dates by which each component had reached successive quartiles (i.e. 25, 50, and 75 percent) of its 1985 total change. They believed that about half of the components have changed more in the single generation since 1950 than in the whole of human history before that date. Hooke et al. (2012) provide data on the landscape of Earth and the changes in land cover and land use that have been achieved (Figure 2.8). While the data may be open to question in detail, the broad figures show that just over 51 percent of the Earth's land surface (excluding that occupied by ice sheets) has now been modified by human activities, with agriculture and forestry accounting for just over 44 percent. Steffen et al. (2015) show the trends of various socioeconomic and environmental indicators from 1750 onward, illustrating the dramatic changes during the Great Acceleration in many of these. We have appropriated a large amount of the world's biomass for our own use and Smil (2011) has estimated that through harvesting, deforestation, and conversion of grasslands and wetlands, humans have reduced the stock of global terrestrial plant mass by as much as 45 percent in the last 2000 years, with one-third of this being achieved in the twentieth century.

Some of the identified trends may continue into the future, though predictions are difficult. A major trend is that of human population. According to the highest

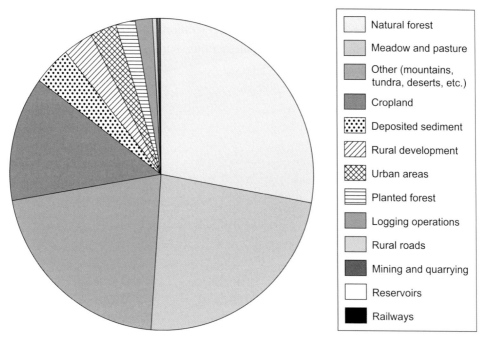

Figure 2.8 Land area modified by human action as of 2007 (compiled using data from Hooke et al, 2012, table 1).

UN estimate, this may rise to 16.6 billion by 2100; according to the lowest estimate, it may decline to 6.75 billion (http://esa.un.org/unpd/wpp/index.htm, accessed January 28, 2015). Growing and wealthier populations will demand a wide range of resources, including energy, minerals, and water, and agriculture will continue to be increasingly intensive in order to feed the growing population (Hall, 2015). All of this has geomorphological implications.

2.13 Conclusions

Homo has lived on Earth for several millions of years and has gradually spread outwards from Africa, reaching the Americas, Australasia, and parts of Oceania only in the late Pleistocene or Holocene. Early on, humans used fire and contributed to the demise of the megafauna. In the early and middle Holocene, domestication and the adoption of agriculture and pastoralism were major technological advances that led to enormous changes in land cover and land use, and thus in the state of the geomorphological environment. They were followed by four other major developments: the adoption of irrigation, the Secondary Products Revolution, urbanization, and the exploitation of minerals. Finally, in the late Holocene we have seen globalization and the harnessing of energy. In the last three centuries, since the Industrial Revolution started, and particularly since 1950 in what has been called the Great Acceleration, huge changes have been wrought to the land surface by human actions with great geomorphological consequences.

3

Construction and Excavation

3.1 Introduction: Humans as Earth Movers

Humans move large amounts of material which they either excavate or deposit to create artificial landforms. This was appreciated by Sherlock (1922), whose detailed quantitative study of the history of excavation and earth moving in Britain covers a period when earth-moving equipment was still ill-developed. He estimated that from about 1500 to 1922 mining, quarrying, the building of canals, railways, roads, docks, harbors, and buildings caused the excavation and movement of some 30.5 km^3 of material (Table 3.1). This equates to around 38–76; billion tons (depending on the density). On the basis of his calculations and detective work, Sherlock was able to state that "at the present time, in a densely peopled country like England, Man is many times more powerful, as an agent of denudation, than all the atmospheric denuding forces combined" (p. 333).

The most notable change since Sherlock wrote those words has taken place in the production of aggregates for concrete. Demand for these materials in the United Kingdom grew from 20 million tons per annum in 1900–202 million tons in 2001, a tenfold increase. Douglas and Lawson (2001) estimated that in Britain the total deliberate shift of earth-surface materials at the present day is between 688 and 972 million tons per year, depending on whether or not the replacement of overburden in opencast mining is taken into account. Some 38.5 km^3 have been removed as a result of mineral extraction in the UK in the 160 years since 1850. Price et al. (2011, p. 1056) estimated that in Great Britain over the last 200 years, "people have excavated, moved and built up the equivalent of at least six times the volume of Ben Nevis." They suggested that in industrial Manchester and Salford in northwest England, such activities have produced up to 10 m of artificial ground, while in the historic city of York in northeast England, deposits reflecting over 2,000 years of occupation are now up to 8 m thick. Jordan et al. (2016) have demonstrated that in another English town, Great Yarmouth, there is up to c. 5 m of artificial ground.

Table 3.1 *Humans as excavators and movers of earth materials*

Area	Cause	Timespan	Amount moved			Source
			Km³	Billion tons	Billion tons pa	
Great Britain	Mining and quarrying	c. 1500–1922	c. 27			Sherlock (1922)
		1850–2010	c. 38.5			Price (2011)
	Roads and infrastructure	c. 1500–1922		c. 4.5–8.5*		Sherlock (1922)
		1922–2010		c. 11		Price (2011)
		Present day			0.155**	Price (2011)
	All deliberate human excavation	c. 1500–1922	c. 30.5	c. 38–76*		Sherlock (1922)
	All deliberate human excavation	**Present day**			**0.972**	Douglas and Lawson (2001)
USA	Mining	Present day			3.8	Hooke (1994)
	Roads and infrastructure	Present day			3.8	Hooke (1994)
	All deliberate human excavation	**Present day**			**7.6**	Hooke (1994)
World	**All deliberate human excavation**	**Present day**			**30**	Hooke (1994)
	All deliberate human excavation	**Present day**			**57**	Douglas and Lawson (2001)
	River transport to oceans	Present day			14	Hooke (1994)
	River transport to oceans	Present day			8.3–51.1	Walling (2006)

Notes * Converted from cubic yards, assuming density of 1.25 to 2.5 g cm^{-3}.
** Based on data for 2000.

32

Hooke (1994) produced some useful data on the significance of deliberate human earth-moving actions in the United States and globally. Table 3.1 illustrates that, according to these estimates, over 7.5 times as much material is moved by human activity in the United States in comparison with the United Kingdom. Hooke (1994) calculated that deliberate human earth moving causes 30 billion tons to be moved per year on a global basis (Table 3.1) or around 200 tons of rock and soil moved per person on Earth (see also, Hooke, 1999 and 2000; and Hooke et al., 2012). However, Douglas and Lawson (2001) give a somewhat larger total figure – 57 billion tons yr^{-1}. For comparison, it has been estimated that the amount of sediment carried into the ocean by the world's rivers each year amounts to between 8.3 and 51.1 billion tons yr^{-1} (Hooke 1994; Douglas and Lawson, 2001; Walling, 2006). Thus, the amount of material moved by humans is rather greater than that moved by the world's rivers to the oceans – up to seven times as great, depending on which estimates of the two quantities are used (Table 3.1). As technology changes, this earth-moving ability increases still further (Haff, 2010), and new construction sites, such as major airports, require very considerable quantities of earth moving (Douglas and Lawson, 2003). Equally, as the size of ships increases, there is an increasing need for channel dredging and for resultant dredging spoil disposal. These earth-moving activities can lead to notable geomorphological changes (e.g. Pringle, 1996).

3.2 Landforms Produced by Construction and Dumping: Introduction

The process of constructing embankments, mounds, and the reclamation of land from the sea is longstanding. In the Middle East and other areas of long-continued human urban settlement the accumulated debris of life has gradually raised the level of the land surface and produced occupation mounds (*tells*) (Menze and Ur, 2012). In Britain, the 40 m high burial mound at Silbury Hill in Wiltshire, together with many long barrows, dates back to the Neolithic, while the pyramids of Central America, Egypt, and the Far East are even more spectacular early feats of landform creation. Likewise in the Americas, Native Indians, prior to the arrival of Europeans, created large numbers of mounds of different shapes and sizes for temples, burials, settlements, and effigies (Denevan, 1992: 377). Similarly, hydrological and erosion management has involved, over many centuries, the construction of massive banks, terraces, and walls – the ultimate result being the landscape of the present-day Netherlands. Shortage of land has led to extensive reclamation of land from the sea in locations like Singapore and Hong Kong. Transport developments have also necessitated the creation of large constructional forms, but probably the most important features are those resulting from the dumping of waste materials, especially those derived from mining. The ocean floors are also being affected because of the vast bulk of waste material that humankind is creating.

We now move on to consider some specific constructional landform types produced by humans.

3.3 Artificial Islands

One way of making new land is to create artificial islands. This has greatly developed recently because of the availability of powerful means to dredge, drain, and dump sediment. The world's largest artificial island is probably the Flevo-polder in the Netherlands, which has an area of 970 km^2 and was completed in 1968. Even more extraordinary is the construction of a series of islands on the shores of the Persian/Arabian Gulf (Figure 3.1) (https://en.wikipedia.org/wiki/List_of_artificial_islands, accessed November 17, 2015). The Pearl-Qatar lies off Doha and has created over 32 km of new shoreline. Bahrain has constructed the Durrat Al Bahrain with an area of 20 km^2 off its southeastern shore and the Amwaj Islands with an area of 4.3 km^2 off its northeastern shore. In the United Arab Emirates there is a cluster of spectacular islands produced in the twenty-first century. Dubai has built the Palm Jumeirah, which adds 78 km of extra shoreline, the Palm Jebel Ali, and The World, an artificial archipelago of c. 300 islands in the rough shape of a world map. It adds roughly 232 km of extra shoreline. Further

Figure 3.1 Artificial Islands in the Middle East. Top left, Al Marjan Island (© Digital Globe, 2014, scale bar c. 2.0 km, December 26, 2013); top right, Palm Jebel Ali (© Digital Globe, 2014, scale bar, 3.5 km, December 13, 2010); bottom left, Durrat Al Bahrain (© Digital Globe, 2014, scale bar 1.9 km, April 21, 2013); bottom right, The World, Dubai (© Digital Globe, 2014, scale bar 5.2 km, February 5, 2011).

north, in Ras Al Khaimah Emirate, Al Marjan Island has been constructed off Jazirat Al Hamra. In addition to these islands, large amounts of reclamation have been undertaken along extensive tracts of coastline between Abu Dhabi and the Oman frontier north of Ras Al Khaimah. All these schemes have their environmental implications: they have involved the dredging of prodigious amounts of sediment from the sea floor, the construction of breakwaters requiring intensive quarrying of rock onshore, the destruction of mangrove swamps and natural lagoons, the degradation of corals (Rutz, 2012) and the alteration of wave climates and sediment budgets.

In some parts of the world, such as Hong Kong and Osaka, Japan, artificial islands have been built to accommodate new airports, while in the South China Sea, China is constructing artificial islands for strategic reasons.

3.4 Artificial Reefs

Artificial reefs are submerged structures constructed on the seabed deliberately to mimic some characteristics of a natural reef (Baine, 2001). They are created to provide a habitat for fishes and thus to help artisanal fisheries, give coastal protection, restrict damage to sea grasses by trawlers, act as a substratum upon which corals can grow, rehabilitate degraded coral reefs (Clark and Edwards, 1999; Perkol-Finkel et al., 2006), provide a location for recreational divers, dispose of waste, and generate improved conditions for surfers. They can be constructed from various materials, including sunken ships and barges, old cars and tires, dumped rock and concrete blocks, sand-filled geotextile cages, train carriages, pulverized fuel ash, rope netting, and abandoned oil rigs. Their use has developed since the 1960s and reviews of European examples are provided by Jensen (2002) and Fabi et al. (2011). Artificial surfing reefs are a developing type of structure in the United Kingdom (Rendle and Rodwell, 2014), Portugal (Mendonça et al., 2011), and elsewhere, and experiments are being conducted to ascertain the best shape for a generation of optimal surfing waves (e.g. Black and Mead, 2009).

3.5 Coastal and Lake Reclamation

Deliberate coastal reclamation to produce more available productive land has a long history in Britain (Rippon, 2000). The Romans, for example, started the reclamation of the Fens (Doody and Barnett 1987), Romney Marsh, and the Somerset Levels (Rippon, 2006). Many schemes were also developed in Europe in medieval times (Charlier et al., 2005).

Coastal mega-cities, such as Hong Kong, Macau, Singapore, and Abu Dhabi have now expanded by reclaiming land from the sea. In the case of Singapore this amounts to about 20 percent of its total area (Wang et al., 2015) and has caused loss of coral reefs and other landform types (Lai et al., 2015). Murray et al. (2014) have mapped coastal reclamation around the Yellow Sea, between China and Korea, and their analysis revealed that as a result of this activity 28 percent of tidal flats existing in the 1980s had disappeared by the late 2000s (a loss of 1.2 percent annually). Moreover, reference to historical maps suggests that up to 65 percent of tidal flats were lost over the past five decades. For China as a whole, there has been a reduction of coastal wetland area by slightly over 50 percent (Wang et al., 2014). In Essex, England, 4.5 million tons of clay spoil from tunneling under London to construct the Crossrail project has recently been used to recreate, for nature conservation purposes, a wetland landscape of mudflats, saltmarsh, and lagoons last seen 400 years ago. This Wallasea Island Wild Coast Project will deliver 670 hectares of marshland (www.rspb.org.uk/news/324715-construction-of-europes-largest-manmade-coastal-reserve-starts, accessed July 6, 2015).

Land reclamation and poldering can also cause a reduction in the size of lakes (Zhao et al., 2005). In China, the Dongting and Poyang lakes are the two key examples showing the rapid decrease in lake surface area since the 1950s (Yang and Lu, 2014) as a result of this phenomenon. The surface of Dongting Lake in the Central Yangtze region, formerly the largest freshwater lake in China, decreased by 37 percent, from 2,825 km^2 in 1950s to 1,785 km^2 in 2008 primarily due to anthropogenic activity, including littoral land reclamation. The area of Poyang Lake has also decreased significantly as a result of land reclamation. The total reclaimed land from 1949 to 2007 amounted to 2,300 km^2 and resulted in a decrease in the lake's surface area from 5,200 km^2 to 2,900 km^2 (a 45 percent decrease in the sixty year period).

3.6 Terracing

Hillsides, valley floors, and stream channels have been terraced for centuries, notably in the arid and semiarid highlands of the New World (Donkin, 1979; Denevan, 2001; Whitmore and Turner, 2001), around the Mediterranean basin (Grove and Rackham, 2001), in Portugal (Figure 3.2a), on the mountains of South Asia and Arabia (Figure 3.2b), and in Africa south of the Sahara (Grove and Sutton, 1989), including northern Ethiopia (Nyssen et al., 2000). In Asia, rice cultivation is often associated with extensive terraced hillsides while some of the Pacific Islands such as Hawaii were terraced for taro cultivation (Allen, 1971). However, examples of terracing are also known from southern England, where in

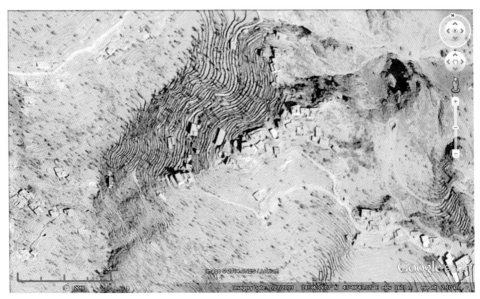

Figure 3.2 Terracing. Upper image: Terraces viewed during field-work on the Douro River, Dow Port Lodge, Portugal. Lower image: Intensively terraced hillside in Yemen. Scale bar 144 m. ©2014 CNES/Astrium.

Roman and medieval times strip lynchets were produced by plowing on steep slopes (Whittington, 1962).

Traditional terraces have a variety of forms that have been characterized by Grove and Rackham (2001, p. 107) as: (1) *step terraces*, which are parallel to the contours; (2) *braided terraces*, which zig-zag up the slope; (3) *pocket terraces*, which consist of crescent-shaped walls producing rootholds for individual crop trees; and (4) *terraced fields*, in which one end is built up above the hillside and the other end is sunk in. Grove and Rackham (2001, p. 110) propose six possible reasons why terraces have been built in the Mediterranean lands: to redistribute sediment, increase root penetration, make a less steep slope on which to cultivate, make a wall out of the stones which would interfere with cultivation, increase absorption of water by the soil in heavy rain events, and control erosion.

Some terrace structures have been built for harvesting and control of flood waters (Beckers et al., 2013). Terraced wadis exist in Jordan and evidence of ancient floodwater farming, possibly beginning in the Neolithic and Bronze Age (McCorriston and Oches, 2001), was found in the extraordinary Wadi Faynan field system, adjacent to the Arava Valley. The great number of ancient agricultural installations in the Negev highlands is outstanding, including tens of thousands of stone terrace walls in ephemeral stream valleys (Evenari et al. 1982; Ore and Bruins, 2012). Terracing in the Negev dates back to the Neolithic and Chalcolithic, but was also practiced in the Bronze Age, Iron Age, Nabataean, Roman, and Byzantine times (Bruins, 2012). Runoff/floodwater-based agricultural systems are also known in North Africa and southeast Spain (Giraldez et al., 1988).

Large tracts of central and southern America were terraced in pre-European times by the Incas (Londoño, 2008) and the Mayans (Dunning and Beach, 1994; Fedick, 1994) to retain soil and water and to facilitate cultivation on steep slopes. However, in Middle America, terracing suffered as a result of the Spanish conquest as the massive loss of the native population reduced the labor required to maintain terraces, and the modern use of terracing is probably a pale shadow of that which existed on the eve of conquest (Whitmore and Turner, 2001).

At the present time in Peru (Inbar and Llerena, 2000) the move to the cities has caused many terracing systems to be abandoned. This is also the case in many Mediterranean countries (Lesschen et al. 2008; Bellin et al., 2009; Tarolli et al., 2014). Some of the terraces have suffered from subsequent erosion processes such as piping (Romero Diaz et al., 2007), though others have seemed more durable (Sole-Benet et al., 2010). On the other hand, check dams have been installed in some areas since the 1970s to control debris flows, flooding, and erosion, and to permit cultivation of vines, olives and other crops in Spain (Boix-Fayos et al. 2007, 2008; Castillo, et al. 2007) and in Italy (Lenzi, 2002). The same is true of Ethiopia (Nyssen et al., 2004).

3.7 Tells and Other Mounds

Long-continued settlement in roughly the same location leads over time to the accumulation of waste, such as dung and ash, and the remains of generations of decaying mud, brick, and stone houses. These accumulations give rise to large mounds (Figures 3.3a and b) that are normally called "tells." They occur in many parts of the world, including Mesoamerica (Hall, 1994), West Africa (Macdonald, 1997), Europe (Bailey et al., 1998; Lubos, 2013; Raczy, 2015), Iran (Maghsoudi et al., 2014), India, and above all, Mesopotamia (Menze and Ur, 2012). Many date back to the beginnings of urban settlement in the early Holocene. Table 3.2 lists the present heights of selected famous tells of the Middle East, some of which are up to 50 m high.

In the Americas there are many mounds, which were constructed for a range of purposes, including for temples, burial, and settlement (Denevan, 1992). In the Midwest and South of the United States there may be as many as several hundred thousand artificial mounds that largely predate European colonization, and some are enormous. Notable is a mound at Cahokia, near St. Louis, Missouri, which is 30.5 m high and covers an area of 6.9 ha.

Burial mounds, also called tumuli, barrows, or kurgan, are a feature of many cultures. One of the most heavily mounded areas is in northern Bahrain, where there are as many as 172,000 Dilmun Mounds (Lowe, 1986). The size of the mounds varies, but the majority of them measure 4.5 by 9 m in diameter and 1–2 m high. These date back to the Bronze Age. Even Neolithic mounds, dating back to c. 4,500 years ago, can be of substantial size, as is the case with Silbury Hill in southern England, which is c. 40 m high and 160 m in diameter (Bayliss et al., 2007).

Table 3.2 *Heights of selected tells in the Fertile Crescent area of the Middle East*

Tell	Height (m)
Sultantepe, Turkey	50
Tell Brak, Syria	40
Tell Barri, Syria	32
Irbil Citadel, Iraq	>30
Tell Arbid, Syria	30
Tell Bazmusian, Iraq	23
Tell Gawra, Iraq	22
Çatal Höyük, Turkey	20
Tell Shemshhara, Iraq	19
Tell Chuera, Syria	18
Göbekli Tepe, Turkey	15
Tell es-Sweyhat, Syria	15
Tell Zeidan Syria	15

Figure 3.3 Upper. Settlement mound, dating back to the Harappan Civilization, in Haryana, northern India. Note the piles of drying buffalo dung on the mound and the luxuriant growth of cannabis plants in the foreground. Lower. A Harappan and later settlement mound in northwest India, rising up above the alluvium of the Haryana plains (background).

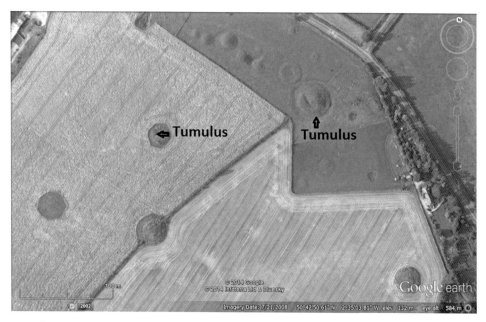

Figure 3.4 Bronze Age round barrows near Kingston Russell, Dorset, southern England. Scale bar 100 m. ©2014 Google, ©2014 Infoterra Ltd & Bluesky.

It is the largest burial mound in Europe. Britain also has various long rampart and ditch structures of Neolithic age, called cursus monuments (Thomas, 2006). In Europe, long barrows were built from the early Neolithic. A major spasm of barrow construction took place in Europe in the Copper and Bronze Ages, with round barrows being one of the most characteristic forms at that time (Figure 3.4). Impressive burial mounds also occur in northwestern China. They were constructed during the Western Xia dynasty (1038–1227 AD; Figure 3.5).

A hill fort is a type of earthwork used as a fortified refuge or defended settlement, located to exploit a rise in elevation for defensive advantage. They are typically European and date to the Bronze and Iron Ages, though some were used in the post-Roman period. The fortification usually follows the contours of a hill, consisting of one or more lines of earthworks, with stockades or defensive walls, and external ditches. A notable example is Maiden Castle in Dorset (Figure 3.6).

There are many other types of artificial landform constructed for military purposes, ranging from the 7,240 km long Great Wall of China (started more than 2,000 years ago), to Roman ditches and ramparts, and various medieval ramparts and fortresses (Ilyés, 2010). Moats and associated ditches are common in Britain (see, e.g., Emery, 1962), with more than 5,000 known from England and 140 from Wales. Many were constructed between the twelfth and fourteenth centuries.

Figure 3.5 Mounds in northwestern China that were constructed during the Western Xia dynasty.

Figure 3.6 Maiden Castle hill fort, Dorset, southern England. Scale bar 266 m. Note the multiple ramparts and ditches. ©2014 Infoterra Ltd. and Bluesky Image, Getmapping plc., Google.

3.8 Embankments and Levees

Embankments (also called levees, flood banks, or dykes) are linear ridges or walls which have been constructed along many rivers, very often for flood protection. Some of the earliest examples were constructed c. 4,600 years ago by the Indus Valley (Harappan) Civilization in Pakistan and North India. They were also constructed over 3,000 years ago in Egypt, where a system of embankments was built along the left bank of the Nile for around 1,000 km, stretching from Aswan to the Nile Delta. The Mesopotamian and ancient Chinese civilizations also built large systems. More recently, prominent levees have been built along the Mississippi and Sacramento rivers (the United States) and in Europe (e.g. along the Po, Rhine, Meuse, Rhone, Loire, Vistula, and Danube).

The Mississippi system represents one of the largest found anywhere in the world. It comprises over 5,600 km of levees. They were begun by French settlers in Louisiana in the 18th century to protect New Orleans. By the mid-1980s, they had reached their present extent and averaged 7.3 m in height, though some are as high as 15 m.

3.9 Mine Spoil Heaps

Waste material excavated from mines is often dumped as spoil heaps at the ground surface. It has been calculated that there are at least 2,000 million tons of shale lying in pit heaps in the coalfields of Britain (Richardson, 1976), and Price (2011) notes that these are among the most significant anthropogenically produced land-forms in the United Kingdom. Some spoil heaps are impressively high. One of the highest in Western Europe is in the former mining area of Pas-de-Calais, northern France. It comprises a range of five cones, of which two reach 180 m, surpassing the highest natural peak in Flanders. Nyssen and Vermeersch (2010) describe the spoil heaps of Belgium and the erosional processes that have occurred on them. Another region of Europe with spoil heaps is the Donbass, in Ukraine, especially around the city of Donetsk, which alone boasts about 130 of them. In Heringen, Hesse, Germany, is Monte Kali, made of spoil from potash mining and rising some 200 m above the surrounding terrain. Spoil heaps from gold mines dominate the landscape of Johannesburg in South Africa (Figure 3.7).

Some waste may be dumped in stream channels, causing subsequent changes in channel morphology (see discussion in Chapter 7) and on coastal foreshores and over cliffs. A classic example of the latter is provided by the disposal of limestone blocks on Portland, Dorset, United Kingdom, where not only have cliff profiles been transformed, but the weight of the dumped material has caused landslides since the seventeenth century (Brunsden et al., 1996; Figure 3.8).

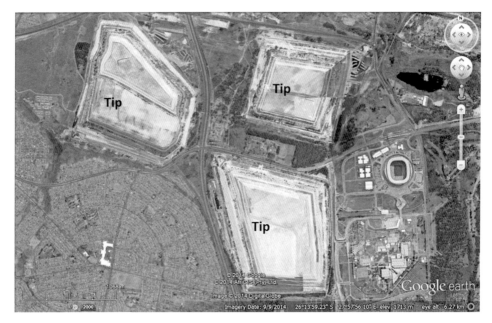

Figure 3.7 Mine dumps in the center of Johannesburg, South Africa. Scale bar c 1 km. ©2014 Google, ©AfriGIS (Pty) Ltd, ©2014 Digital Globe.

Figure 3.8 Quarry waste tipped over the cliff at the West Weares, Isle of Portland, Dorset, United Kingdom.

In the United Kingdom there are many mounds and embankments associated with the coastal salt-making industries, especially in low-lying areas like the dismal Essex estuaries, the Humber marshes, the Wash, the Solway Firth, Morecambe Bay, and the North Kent coast, where some are more than 50 m across and over 4.5 m in height (Holmes, 1980).

3.10 Artificial Lakes: Dams and Reservoirs

Many lakes have been produced by impoundment of water in reservoirs behind dams (Figure 3.9). The first recorded dam was constructed in Egypt some 5,000 years ago, but since that time the adoption of this technique has spread, variously to improve agriculture, prevent floods, generate power, or provide a reliable source of water. There are now some 75,000 dams in the United States. Most are small, but the bulk of the storage of water is associated with a relatively limited number of structures. The 1960s saw the greatest spate of dam construction in American history (18,833 dams were built then). Since the 1980s, however, there have been only relatively minor increases in storage. Globally, the construction of large dams increased markedly, especially between 1945 and the early 1970s, but the need for hydropower means that there may be another spasm of dam construction in the next few decades (Zarfl et al., 2015). Engineers have now built more than 45,000 large dams (i.e. more than 15 m high) around the world, and one of the striking features of dams and reservoirs is that over time they have become increasingly large. Thus in the 1930s the Hoover or Boulder Dam in the United States (221 m high) was by far the tallest in the world and it impounded the biggest reservoir, Lake Mead, which contained just under 38 billion cubic meters of water (Figure 3.10). By the 2000s it was exceeded in height by at least 29 others. Some of the associated reservoirs are vast (see data in Table 3.3), and have become important sites for sediment retention (Vörösmarty et al., 2003; Wisser et al., 2013). This is a theme that will be returned to in Chapter 7.

Some landscapes are dominated by small reservoirs. A striking example is the "tank" landscape of southeast India, where myriads of little streams and areas of overland flow have been dammed by small earth structures to give what Spate and Learmonth (1967: 778) likened to "a surface of vast overlapping fish-scales" (Figure 3.11). In all there are around 1.2–1.5 million tanks ("village ponds") in rural India (Pandey, 2000; Figure 3.12). Some of these are now of declining importance as a result of siltation, privatization of water resources and the mining of groundwater (Gunnell and Krishnamurthy, 2003), but they still remain as an important component of the landscape.

Table 3.3 *The world's largest reservoirs*

Reservoirs with an area $>5,000$ km^2		Reservoirs with a volume >100 km^3	
Volta, Ghana	8,482	Kariba, Zambia/Zimbabwe	181
Smallwood, Canada	6,527	Bratsk, Russia	169
Kuybyshev, Russia	6,450	Nasser, Egypt	162
Kariba, Zambia/Zimbabwe	5,580	Volta, Ghana	150
Bukhtarma, Kazakhstan	5,490	Manicouagan, Canada	142
Bratsk, Russia	5,426	Guri, Venezuela	135
Nasser, Egypt	5,248		

From data provided by the International Commission on Large Dams (www.icold-cigb.org). See also https://en.wikipedia.org/wiki/List_of_reservoirs_by_surface_area, and https://en .wikipedia.org/wiki/List_of_reservoirs_by_volume. (Accessed November 17, 2015).

Figure 3.9 A large pit dug into a sand dune in the United Arab Emirates to provide aggregates for the urban expansion of Dubai.

In China reservoir construction has been massive in recent decades (Yang and Lu, 2014). By 2013, 98,000 reservoirs had been built across the country. The total estimated storage capacity of the reservoirs (794 km^3) is triple that of lakes (268 km^3). The two decades of the 1950s and 1960s saw the addition of nearly 72,000 reservoirs in China and the trends since then are shown in Figure 3.13.

Figure 3.10 Lake Nasser, near Abu Simbel, Egypt.

Figure 3.11 The Hoover Dam and Lake Mead, United States.

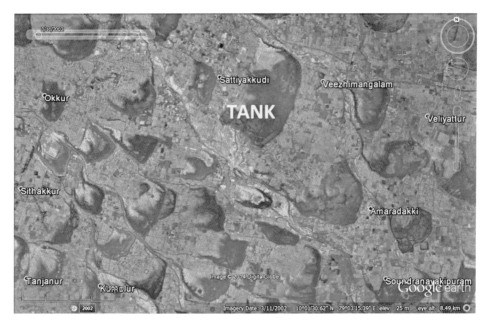

Figure 3.12 The tank landscape of southeast India. Scale bar 1.95 m. ©2014
Digital Globe.

Lakes have also been created by using dykes to cut off low-lying coastal
areas and inlets from the open sea. The classic example is IJsselmeer, a shallow
artificial lake of 1,100 km² in the central Netherlands, with an average depth of
5–6 m. It is now the largest lake in Western Europe (Figure 3.14). It was created
in 1932 when an inland sea, the Zuiderzee, was closed by a 32 km dam, the
Afsluitdijk. Devastating floods in 1953, initiated the damming of most of the
Rhine-Meuse delta in the southwest of the Netherlands to enhance safety levels.
Consequently, the coastline was drastically shortened and former estuarine inlets
were transformed into disengaged and stagnant fresh- and saltwater lakes (van
Wesenbeeck et al., 2014). Wolff (1992), describes the history of draining lakes
and of poldering in the Netherlands.

3.11 Landforms Produced by Excavation

3.11.1 Depressions Associated with Mining

Landforms produced by excavation are widespread, and may be old. For example,
Neolithic peoples in the Breckland of eastern England used antler picks and other
means to dig deep pits in the chalk, as at Grimes Graves. The purpose was to obtain

Figure 3.13 A village tank in northwest India.

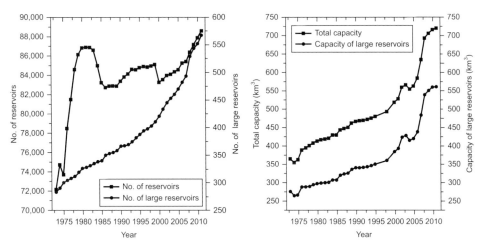

Figure 3.14 Changing number of reservoirs in China from 1972 to 2012 (from Yang and Lu, 2014, figure 4).

good-quality nonfrost-shattered flint to make stone tools. Collapse hollows have developed in such ancient chalk mining areas in various parts of southern England (Bell et al., 2005). Comparable Neolithic flint mines are known from Spiennes in Belgium, from Jurassic limestones at Krzemionki in Poland, and from northern Spain (Tarriño et al., 2014). In parts of Britain, chalk has also been excavated to provide marl for improving acidic, light, sandy soils. Prince (1962, 1964) estimated that 27,000 pits and ponds in Norfolk may have resulted mainly from this activity, which was particularly prevalent in the eighteenth century. Very large numbers of marl pits, dating to the Iron Age and the Roman period, are also found in northeastern France (Etienne et al., 2011).

The Broads are a group of twenty-five freshwater lakes in Norfolk, eastern England. They are of sufficient area, and depth, for early researchers to have precluded a human origin (see Matless, 2014, for a discussion) and Jennings (1952) postulated that they were formed in Romano-British times by a transgression of the sea over earlier valley peats. It is now clear, however, that as Lambert et al. (1970) showed, the Broads result from human work, for some of them have rectilinear and steep boundaries, most of them are not mentioned in early topographic books, and archival records indicate that peat-cutting was widely practiced in the area. It is believed that peat-diggers, before AD 1300, excavated 25.5 million m^3 of peat and so created the depressions in which the lakes have formed. The flooding may have been aided by sea-level change. Comparably extensive peat excavation was also carried on in the Netherlands, notably in the fifteenth century (Van Dam, 2001).

Mineral extraction remains one of the most important causes of excavation (Figure 3.15). The environmental devastation produced by one type of extraction – strip mining – is exceptional and has transformed landscapes in areas such as the lignite fields of central Europe (Hüttl, 1998). This form of mining is also a particular environmental problem in the US coal regions of Pennsylvania, Ohio, West Virginia, Kentucky, and Illinois. Strip mining is the practice of mining a mineral seam by first removing a long strip of overlying soil and rock (the overburden). It is only practical when the material to be excavated is relatively near the surface. This type of mining uses some of the largest machines on Earth, including bucket-wheel excavators which can move as much as 12,000 m^3 of material per hour. There are three forms of strip mining. Firstly, *Area Stripping* is used on fairly flat terrain, to extract deposits over a large area. As each long strip is excavated, the overburden is placed in the excavation produced by the previous strip. Secondly, *Contour Stripping* involves removing the overburden in hilly terrain, where the mineral outcrop usually follows the contour of the land. This method commonly leaves behind terraces on mountain sides. Thirdly, *Mountain-top Removal Mining* uses explosives to blast overburden off the top of some

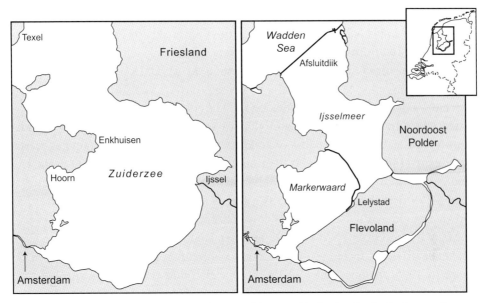

Figure 3.15 The reclamation of the Zuiderzee, Netherlands. Left shows the Zuiderzee before 1918. Right shows the polders and the IJsselmeer in 1990 (redrawn and modified from Wolff, 1992, figure 4).

Appalachian Mountains. It involves the mass restructuring of landscape in order to reach the coal seams as deep as 300 m below the surface. Mountaintop removal replaces the original steep landscape with a rather flatter one (Miller et al., 2014), and large extents of valleys may be filled in. Major changes may occur in water quality and channel forms as a result (Lindberg et al., 2011; Jaeger, 2015).

The depressions associated with some opencast mines are huge (www.mining-technology.com/features/feature-top-ten-deepest-open-pit-mines-world/; accessed November 17, 2015). The Bingham Canyon mine in Utah in the United States is the deepest open-pit mine in the world. It is more than 1.2 km deep and approximately four kilometers wide. Chuquicamata copper mine in Chile is the second deepest open-pit mine in the world. It is 4.3 km long, 3 km wide, and more than 850 m deep. Escondida copper mine located in the Atacama Desert, also in Chile, ranks as the third deepest open-pit operation. It has two pits. The Escondida pit is 3.9 km long, 2.7 km wide, and 645 m deep, while the Escondida Norte pit is 525 m deep. Udachny diamond mine in the Eastern-Siberian Region of Russia is currently the fourth deepest open-pit mine in the world, at 630 m deep. Muruntau gold mine pit in Uzbekistan, ranks as the fifth deepest. It is 3.5 km long, 3 km wide, and more than 600 m deep. Aitik open-pit mine, located in Northern Sweden, is currently 430 m deep.

Some mining involves removal of sediment from river beds. This can cause channel changes, including incision, bank undermining, and knickpoint migration, especially in rivers that are not actively aggrading (Rinaldi et al., 2005; Rovira et al. 2005; Rempel and Church, 2009; Martin-Vide et al., 2010). Similarly, gravels are often mined from floodplain pits, causing lakes to develop (Kondolf, 1997), as along the upper Thames Valley to the west of Oxford, United Kingdom (Figure 3.16).

Features excavated in one generation may be filled in by another, since hollows produced by mineral extraction are both wasteful of land and also suitable locations for waste disposal. The loss of ponds (small lakes) in Great Britain has been a fairly general phenomenon, with their numbers falling from c. 800,000 in the late nineteenth century to only c. 200,000 by the mid-1980s (Jeffries, 2012). Since that time, partly because their biodiversity value is now appreciated, their number has shown a modest increase (Williams et al., 2010).

Many mineral excavation sites are used for landfill and at the present time large quantities of waste, including fly-ash from coal-burning power stations, are sent to them. In 1995, member states of the European Union (EU) landfilled more than 80 percent of their waste, but under the EU Landfill Directive of 1999 this figure has been reduced substantially to around 37 percent in 2010. Nonetheless the amounts of material produced are immense (http://ec.europa.eu/eurostat/statistics-explained/index.php/Waste_statistics, accessed August 6, 2015). In 2012, the total waste generated in the twenty-eight nations of the EU by all economic activities and

Figure 3.16 Lakes created in gravel pits excavated in Late Pleistocene gravels in the Thames floodplain near Oxford, England. Scale bar c. 1.5 km. ©2015 Google, ©Infoterra Ltd. and Bluesky. L = lake.

households amounted to 2,515 million tons. Construction contributed 33 percent of the total (with 821 million tons) and was followed by mining and quarrying (29 percent or 734 million tons), manufacturing (11 percent or 270 million tons), households (8 percent or 213 million tons) and energy (4 percent or 96 million tons); the remaining 15 percent was waste generated from other economic activities.

3.12 Craters Produced by War

Other excavations result from war, especially craters caused by bombs, shell impact, or the action of mines. Recently, Hupy and Schaetzl (2006) and Hupy and Koehler (2012) have introduced the term "bombturbation" to describe the process of land disturbance caused by explosive munitions. They suggested that in the twentieth century soil displacement by munitions amounted to "billions of cubic meters." The First World War battlefields of Flanders were pocked by huge numbers of craters and one, the Lochnagar Crater, which was created by a massive mine explosion, had a diameter of 100 m and a depth of 30 m. Regrettably, human power to create such forms is increasing. It has been calculated (Westing and Pfeiffer, 1972) that between 1965 and 1971, 26 million craters, covering an area of 171,000 ha, were produced by bombing in Indo-China. This represents a total displacement of no less than 2.6 billion m^3 of earth, a figure much greater than calculated as being involved in the peaceable creation of the Netherlands. In Laos, satellite imagery and ground-based surveys indicate that even after nearly four decades, bomb craters remain discernible at densities that commonly exceed 200 km^{-2} and in some cases exceed 800 km^{-2} (Kiernan, 2015). A nuclear test in Nevada in 1962 removed 12 million tons of earth and rock and left behind a crater 390 m in diameter and 97 m in depth.

3.13 Qanat

Qanat, also called *falaj* (Oman), *foggara* (North Africa), *karez* (Afghanistan), and *khettara* (Morocco) are underground, gently sloping tunnel wells dug backward into alluvial fans until the water table is pierced (Wilkinson, 1977). The water is then fed down the tunnels by gravity to be used for irrigation. They are also characterized by vertical shafts which are provided for disposal of spoil and to provide air to the miners who make them. The shafts are surrounded by accumulations of mined debris (Figure 3.17). In Iran, some shafts are 100–120 m deep and the tunnels may run for as far as 40–50 km. These remarkable feats of hydraulic technology originated 2,700–2,500 years ago in Iran, northern Iraq, and eastern Turkey, and subsequently spread to Arabia (Lightfoot, 2000), Central Asia (e.g. the Turpan depression), North Africa, Spain, Mexico, Chile, and Peru. Some are

Figure 3.17 Lines of Qanat craters from Shahrud, northern Iran. Scale bar 150 m.
© 2014 Digital Globe, Google.

still actively used, but elsewhere, partly because of the lowering of the water table by modern systems of pumping, they are in decline, as in Morocco (Lightfoot, 1996 a), Syria (Lightfoot, 1996b), and Algeria (Remini et al., 2011).

3.14 Canals and Other Artificial Channels

Canals are artificial rivers that are used for irrigation, drainage, transport, and travel. Irrigation canals dominate the landscapes and hydrology of areas like Mesopotamia, the Nile Valley, and northern India (Figure 3.18), while land drainage can also require channel construction (Marsha et al., 1978). The oldest known were built for irrigation in Mesopotamia circa 6,000 years ago. In Egypt, canals date back at least to the time of Pepi I Meryre (who reigned from 2332–2283 BC). He ordered a canal to be built to bypass a cataract on the Nile near Aswan. In ancient China, large canals for river transport were established as far back as the Warring States (481–221 BC). The Grand Canal (also known as the Beijing-Hangzhou Grand Canal), a UNESCO World Heritage Site, is the longest canal or artificial river in the world. The oldest parts of the canal date back to the fifth century BC, although the various sections were finally combined during the Sui Dynasty (581–618 AD). Its total length is 1,776 km. (https://en.wikipedia.org/wiki/Canal, accessed November 17, 2015).

Figure 3.18 An irrigation canal in Haryana, northern India.

Canals for carrying freight have been built more intensively in recent centuries, not least since the start of the Industrial Revolution, with some linking up seaways and large lakes. Notable examples are the Kiel, Suez, and Panama Canals, and the St. Lawrence Seaway. The new Suez Canal, opened in August 2015, required the excavation or dredging of around 500 million m^3 of material (www.suezcanal .gov.eg/sc.aspx?show=69, accessed August 6, 2015). Even more impressive is the construction of the new Nicaragua Canal, which started in 2014. This, if completed, will involve the dry excavation of 4,019 million m^3 material of rock and soil, 739 million m^3 of freshwater dredging and 241 million m^3 marine dredging (https://en.wikipedia.org/wiki/Nicaragua_Canal, accessed August 6, 2015).

3.15 Conclusions

Some of the landforms produced by deliberate constructional activity are of considerable antiquity, predate the Industrial Revolution, and may still be dominant features of certain landscapes. The ages of some examples used in this chapter are summarized in Table 3.4. Conversely, however, there has been a great acceleration in earth moving in recent decades, as a result of the need to

Table 3.4 *Examples of the antiquity of artificial constructional and excavational landforms*

Landform	Date
Barrows, long	Neolithic
Barrows, round	Bronze Age
Broads, peat diggings	Medieval
Canals in Mesopotamia	Neolithic
Coastal reclamation in Britain	Romans
Cupules, India	Paleolithic
Dams and reservoirs	Ancient Egypt
Dilmun Mounds, Bahrain	Bronze Age
Embankments on Indus	Harappan Civilization (c. 4,600 years ago)
Hill Forts	Bronze and Iron Ages
Marl pits, France	Iron Age and Roman
Mining pits for flint in Chalk, East Anglia	Neolithic
Mounds in the USA	Precolonization by Europeans
Qanat, Middle East	Achaemenid period (c.2,700–2,500 years ago)
Strip lynchets, UK and Europe	Roman and Medieval
Terraces in Central and South America	Mayan and Incan
Terraces, Middle East	Neolithic and subsequently

acquire new land (e.g. artificial islands and coastal reclamation), or because of our ability to use machinery to move huge quantities of material (e.g. with strip and opencast mining), or because of the availability of powerful explosives (e.g. bombturbation, nuclear craters, etc.). Some features, such as qanat, Indian tanks, and Mediterranean terraces are, however, falling into disuse, and disused excavational features often become filled in for waste disposal, causing them to become less prominent features of the landscape.

4

Subsidence in the Anthropocene

4.1 Introduction

Humans cause ground subsidence in a variety of ways (Figure 4.1): dewatering of karstic formations, mining of minerals such as coal and salt, groundwater, and geothermal fluid abstraction, exploitation of hydrocarbon resources, the draining of organic soils, hydrocompaction, and the disruption of permafrost (thermokarst formation; see Chapter 10). Such subsidence has a suite of geomorphological consequences, including ground fissure formation, coastal inundation, river flooding, and lake development.

It is important to remember, however, that the causes of observed subsidence are often complex and involve both natural and anthropogenic factors. In the case of coastal Louisiana, for example, the natural factors include tectonics (faulting and salt movements), Holocene sediment compaction, the isostatic effects of sediment loading, and glacio-isostatic adjustments. The anthropogenic factors include fluid removal and surface water drainage and management (Yuill et al., 2009). That said, the very rapid rates of subsidence in the Mississippi Delta Plain at the present time, which are much higher than those in the previous 5,000 years, indicate the importance of subsidence caused by hydrocarbon extraction (Morton et al., 2005). However, as the example of Venice shows (Carbognin et al., 2010) it is possible to slow or to reverse rates of subsidence by controls on fluid extraction.

4.2 Karstic Collapse

Williams (1993) provides a survey of the diverse effects of human activities on limestone and dolomite terrains. Among these are the effects of groundwater abstraction. Dangerous and dramatic collapses have occurred because of the dewatering of the rock body.

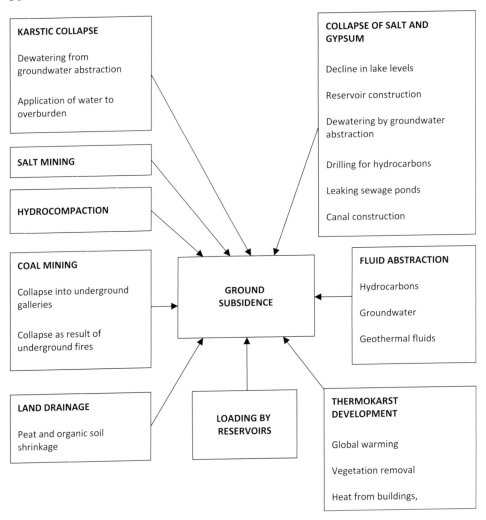

Figure 4.1 Anthropogenic causes of ground subsidence.

As Gutierrez et al. (2014, p. 67) explain, "The main potential effects of lowering the groundwater level include the loss of buoyant support of sediments, the increased groundwater velocity, and the replacement of phreatic flows by downward vadose flows with a greater capacity to induce suffusion." In the dolomite region of the Far West Rand of South Africa (De Bruyn and Bell, 2001), gold mining has required the abstraction of so much water that the local water table has been lowered by >300 m. This has caused clays and other materials filling the roofs of large caves to dry out and shrink so that they have collapsed into the underlying void. One collapse created a depression 30 m deep and 55 m across, killing twenty-nine people. In all some 700 holes have been produced. Some of the

South African collapses may also have been caused by seepage from leaking urban infrastructure and as a result of vibrations associated with industrial processes (Buttrick and Van Schalkwyk, 1998).

Karst collapse due to dewatering has also been observed in parts of China (He et al., 2003; 2004), Majorca, Spain (Garcia-Moreno et al., 2011), and Iran (Taheri et al., 2015). In Alabama in the United States, water-level decline consequent upon pumping has had equally serious consequences; and Newton (1976) estimated that since 1900 about 4,000 induced sinkholes or related features have been formed, while fewer than fifty natural sinkholes have been reported over the same interval. Sinkholes that may result from such human activity are also found in Georgia, Florida, Tennessee, Pennsylvania, and Missouri (Sinclair, 1982). In some limestone areas, however, a reverse process can operate. The application of water to overburden above the limestone may render it more plastic so that the likelihood of collapse is increased. This has occurred beneath reservoirs, such as the May Reservoir in central Turkey, and as a result of the application of wastewater and sewage to the land surface. Ford and Williams (1989, p. 525) drew attention to the focusing of runoff from man-made impermeable surfaces such as roofs, roads, and parking lots, and to the role of leakage from canals, buried water pipes, and so forth.

4.3 Solutional Collapse of Salt and Gypsum

In areas where thick halite (salt) deposits occur in the sedimentary record, solution may cause sinkholes to develop. An example of this is provided by the Dead Sea basin, where the recent decline in levels caused by water abstraction has promoted their formation. More than a thousand potentially dangerous sinkholes have developed along its shorelines since the early 1980s as a result of the flow of undersaturated groundwater dissolving the evaporites. As Yechieli et al. (2006, p. 1075) explain it in the context of the Israeli side of the basin:

The abrupt appearance of the sinkholes, and their accelerated expansion thereafter, reflects a change in the groundwater regime around the shrinking lake and the extreme solubility of halite in water. The eastward retreat of the shoreline and the declining sea level cause an eastward migration of the fresh-saline water interface. As a result the salt layer, which originally was saturated with Dead Sea water over its entire spread, is gradually being invaded by freshwater at its western boundary, which mixes and displaces the original Dead Sea brine.

Similar issues have been encountered on the Jordanian side of the Dead Sea, with karstic collapse creating problems for the chemical plants on the Lisan Peninsula (Closson et al., 2007). Dewatering of gypsum in the Oviedo area of Spain has also promoted sinkhole development (Pando et al., 2013).

Conversely, the application of water may accelerate subsidence. Reservoir construction in areas underlain by gypsum has caused solution holes to develop, as in Siberia (Trzhtsinsky, 2002), while in the Zaragoza district of Spain, gypsum dolines (sinkholes) have resulted from the application of irrigation water (Benito et al., 1995). Leaking sewage ponds have caused subsidence to occur in a gypsum area of South Dakota in the United States (Johnson, 1997), and elsewhere drilling through gypsum for hydrocarbon development has also led to accelerated solution as a result of the ingress of unsaturated water. In 1930, a canal was excavated to connect the Lesina Lagoon in southern Italy with the Adriatic Sea (Fidelibus et al., 2011). It exposed highly cavernous gypsum bedrock. During the last two decades, a large number of cover collapse and cover suffosion sinkholes have formed, and the area affected by subsidence increased exponentially from 1999 to 2009. There is a tight spatial correlation between the sinkholes and the canal.

4.4 Coal Mining Subsidence

The importance of subsidence caused by coal mining varies according to the thickness of seam removed, its depth, the width of working, the degree of filling with solid waste after extraction, the geological structure, and the method of working adopted (Wallwork, 1974).

Studies in the Ruhr, Germany, indicate amounts of subsidence that can be as high as 25 m, though the mean is 1.6 m (Harnischmacher, 2010; Harnischmacher and Zepp, 2014). In Limburg, in the Netherlands, the amounts of subsidence locally exceed 10 m (Bekendam and Pöttgens, 1995). Subsidence may disrupt surface drainage (Sidle et al. 2000) and the resultant depressions then become permanently flooded. In the Ruhr, Lake Lanstrop formed between 1963 and 1967 in response to up to 9 m of subsidence (Bell et al., 2000). In the United Kingdom (Price et al., 2011), it is not uncommon for between 1 and 3 m of subsidence to have occurred as a result of coal-mining operations, though near Wigan in Greater Manchester up to 13 m of subsidence has taken place. The huge expansion of coal mining in China has also led to widespread subsidence in recent years, as in Shandong Province (Wu et al., 2009). Subsidence may also cause fault reactivation, international examples of which are given by Donnelly (2009). Not all subsidence results directly from collapse into mine workings. In some cases, as in China, underground fires in coal seams can also have the same effect (Kuenzer and Stracher, 2012). On the cessation of mining, flooding of underground workings may occur, and this can lead to modest vertical uplift (Bekendam and Pöttgens, 1995).

4.5 Salt Mining

Coal-mining regions are not the only areas with serious subsidence. In Cheshire, northwest England, rock salt is extracted from thick, shallow seams in the Triassic

Figure 4.2 Late Nineteenth Century Subsidence in Northwich, Cheshire, Northeast England. Source: www.visitnorthwich.co.uk/heritage/. Accessed August 5, 2015.

beds (Cooper, 2002). This material is highly soluble, so the flooding of mines may cause additional collapse. These three conditions – thick seams, shallow depth, and high solubility – have produced optimum conditions for subsidence and subsidence lakes called "flashes" have developed (Wallwork, 1956). Buildings were also destroyed (Figure 4.2). Brine abstraction and collapse also affect land in Lancashire and Teeside in northeast England (Price et al., 2011). Collapses have also occurred in the Carrickfergus area of Northern Ireland (Bell et al., 2005). In the Perm district of Russia huge sinkholes (up to 200 m deep) have developed in association with phosphate mines and similar problems have been encountered with potash mines in Spain (Lucha et al., 2008). Up to 40 m of subsidence, operating at rates of up to 40 cm per year, has also resulted from brine pumping in Bosnia-Herzegovina (Mancini et al., 2009).

4.6 Hydrocarbon Abstraction

Subsidence produced by oil abstraction occurs in some parts of the world (Table 10.1; Nagel, 2001). Subsidence around the Goose Creek oil field near Houston, Texas, United States, was the first evidence that rapid, large volume extraction of hydrocarbons was capable of causing the ground to sink around the producing wells. The induced subsidence, which was discovered shortly after field development began in 1917, indicated that accelerated withdrawal of oil, gas,

Table 4.1 *Amount and rates of ground subsidence resulting from hydrocarbon extraction.*

Location	Amount (m)	Date	Rate (mm yr^{-1})
Azerbaydzhan, former USSR	2.5	1912–62	50
Atravopol, former USSR	1.5	1956–62	125
Wilmington, United States	9.3	1928–71	216
Inglewood, United States	2.9	1917–63	63
Maracaibo, Venezuela	5.03	1929–90	84

and associated water from shallow unconsolidated reservoirs could lower the land elevation, cause minor earthquakes, and activate faults around the periphery of producing fields (Pratt and Johnson, 1926). However, the classic area for this phenomenon is Los Angeles, where 9.3 m of subsidence occurred as a result of exploitation of the Wilmington oilfield between 1928 and 1971. The neighboring Inglewood oilfield displayed 2.9 m of subsidence between 1917 and 1963 and some associated coastal flooding problems occurred at Long Beach. Similar subsidence has been recorded from the Ekofisk field in the North Sea, Lake Maracaibo in Venezuela, and from some Russian fields. Subsidence in the Mississippi delta and elsewhere in the Gulf Coast region of the United States may also be in part the result of hydrocarbon extraction (Morton et al., 2005; 2006). Gas abstraction can also lead to subsidence, as in northeast Italy (Gambolati et al., 1991; Bau et al., 2000) and Indonesia (Chaussard et al., 2013).

4.7 Groundwater Abstraction

A more widespread problem is posed by groundwater abstraction for industrial, domestic, and agricultural purposes. The methods for detecting such subsidence have been reviewed by Galloway and Burbey (2011). They present data on current maximum rates for various parts of the world (Table 4.2). Additional long-term data for amounts and rates of subsidence resulting from both groundwater and geothermal water abstraction are presented in Table 4.3.

The ratios of subsidence to water-level decline are strongly dependent on the nature of the sediment composing the aquifer. Ratios range from 1:7 for Mexico City to less than 1:400 for London, England (Rosepiler and Reilinger, 1977). In the United States, particularly notable has been the extent of subsidence in the San Joaquin Valley of California (Figure 4.3; Chi and Reilinger, 1984). In recent years of drought, the rates have accelerated, amounting to over 30 cm in an eight month period in 2014 (http://water.ca.gov/groundwater/docs/NASA_REPORT.pdf, accessed September 2, 2015).

Table 4.2 *Measured subsidence rates. These represent the local maximum measured rate for the specified period (modified from Galloway and Burbey, 2011, table 1).*

Location	Rate (mm yr^{-1})	Period	Measurement method
Bangkok, Thailand	30	2006	Leveling
Bologna, Italy	40	2002–6	Differential interferometry
Changzhou, People's Republic of China	10	2002	Leveling
Coachella Valley, California, United States	70	2003–9	Differential interferometry
Datong, People's Republic of China	20	2004–8	Differential interferometry
Houston-Galveston, Texas, United States	40	1996–8	Differential interferometry
Jakarta, Indonesia	250	1997–2008	GPS
Kolkata, India	6	1992–8	Differential interferometry
Mashhad Valley, Iran	280	2003–5	Differential interferometry
Mexico City, Mexico	300	2004–6	Differential interferometry
Murcia, Spain	35	2008–9	Differential interferometry
Phoenix-Scottsdale, Arizona, United States	15	2004–10	Differential interferometry
Saga Plain, Japan	160	1994	Leveling
Semarang, Indonesia	80	2007–9	Differential interferometry
Tehran Basin, Iran	205–50	2004–8	Differential interferometry
Toluca Valley, Mexico	90	2003–8	Differential interferometry
Yunlin, Republic of China	100	2002–7	Leveling

Subsidence has also taken place in Houston, Texas, but the rate has slowed in recent years because of controls on groundwater withdrawals (Kearns et al., 2015). In Japan subsidence has also now emerged as a major problem (Nakano and Matsuda, 1976). In 1960 only 35.2 km^2 of the Tokyo lowland was below sea level, but continuing subsidence meant that by 1974 this had increased to 67.6 km^2, exposing a total of 1.5 million people to major flood hazard. Since then, controls on groundwater exploitation have reduced the threat. Groundwater-induced subsidence is widespread in China and many cities have been affected (Figure 4.4). It was first found in Shanghai and Tianjin in the 1920s and became

Table 4.3 *Ground subsidence resulting from fluid abstraction*

Location	Amount (m)	Date	Rate (mm yr^{-1})
Ground subsidence produced by **groundwater abstraction**			
London, England	0.06–0.08	1865–1931	0.91–1.21
Savannah, Georgia (United States)	0.1	1918–55	2.7
Mexico City	7.5	–	250–300
Houston, Galveston, Texas	1.52	1943–64	60–76
Central Valley, California	8.53	–	–
Tokyo, Japan	4	1892–1972	500
Osaka, Japan	>2.8	1935–72	76
Niigata, Japan	>1.5	–	–
Pecos, Texas	0.2	1935–66	6.5
South-central Arizona	2.9	1934–77	96
Bangkok, Thailand	2.05	1933–2002	Up to 120
Shanghai, China	2.62	1921–65	60
Bandung, Indonesia	–	2000–12	80
Ground subsidence caused by **abstraction of geothermal water**			
New Zealand	15	50 years	330
Baja California (Mexico)	–	–	70

serious in the 1950s–60s in both cities. By 2005, land subsidence had been found in more than ninety cities. It is mainly distributed in three regions: the Yangtze River Delta, North China, and the Fenwei plains (Xu and Shen, 2008).

Shanghai and neighboring parts of the Yangtze Delta have suffered up to 2–3 m of subsidence since 1921 (Chai et al. 2004). Similar amounts of subsidence are also known from Tianjin and Xi'an (Xue et al., 2005), and as China develops subsidence is becoming more widespread (Hu, 2006; Yin et al., 2006). Groundwater abstraction has also caused severe subsidence in several areas in Taiwan, and the role of smectite dehydration in this process has been investigated (Liu et al., 2001). On the alluvial fan of the Choshui the subsidence rate reaches 14.3 cm yr^{-1} (Liu et al., 2004).

Subsidence occurs elsewhere in Asia. In Bangkok, subsidence in the early 1980s proceeded at up to 12 cm yr^{-1}, and by 2002, 2.05 m of subsidence had taken place (Phien-wej et al., 2006), causing flooding and building damage. Subsidence is also occurring in the Mekong Delta of Vietnam (Erban et al., 2014). Flood risk has been accelerated by subsidence in Jakarta, Indonesia (Abidin et al., 2015), while surveys in the Bandung basin of West Java, which were conducted within the period 2000–2012, showed an average subsidence rate of about 8 cm yr^{-1}, but reaching 16.9 cm yr^{-1}.

Figure 4.3 The extent of ground subsidence resulting from groundwater abstraction in the San Joaquin Valley, California, United States, between the 1920s and 1977. Courtesy of the US Geological Survey.

Subsidence caused by groundwater abstraction has also been a problem in Venice, where it has greatly increased flood risk, but the rate of subsidence has fallen since c. 1970 because of controls on this abstraction (Carbognin et al., 2010).

4.8 Geothermal Fluid Abstraction

In some volcanic regions, geothermal water is abstracted for power generation, as in New Zealand, where up to 15 m of land subsidence has occurred in 50 years in the Wairakei area (Allis, 2000; Allis et al., 2009). Subsidence from this cause has also been identified in southern California (Massonnet et al., 1997) and in Baja

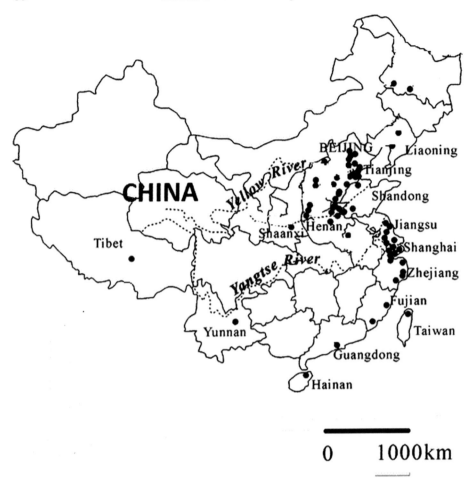

Figure 4.4 Cities with land subsidence in China (black dots) (simplified from Xue et al., 2005, figure 1).

California, Mexico (Carnec and Fabriol, 1999), where mean rates of subsidence are around 7 cm per year (Glowacka et al., 2000).

4.9 Hydrocompaction

Hydrocompaction is the compaction and reduction in volume of soils and sediments that occurs when their moisture content is increased. It causes subsidence when unconsolidated sediments (collapsible soils) of low density are wetted, as for example by the application of irrigation water, the disposal of wastewater, or runoff from urban surfaces. It is a feature of arid and semiarid lands (Houston et al., 2001) where materials such as loess, debris flow deposits or alluvial sediments above the water table are not normally wetted below the root zone and have high void ratios. When dry, such materials may have sufficient strength to support considerable effective stresses without compacting. However, when they are wetted, their intergranular strength is weakened because of the rearrangement of their particles. The associated subsidence creates fissures in the ground and is a process that needs to be considered during the construction of canals, pipelines, dams, and irrigation schemes (Shlemon, 1995; Al-Harthi and Bankher, 1999). It can cause structural deterioration of buildings.

4.10 Land Drainage

Land drainage promotes subsidence, notably of organic soils. The lowering of the water table makes peat susceptible to oxidation, shrinkage, consolidation, and deflation so that its volume decreases. One of the longest records of this process has been provided by the measurements at Holme Fen in the English Fenlands (Dawson et al., 2010). Approximately 3.8 m of subsidence occurred between 1848 and 1957 (Fillenham, 1963), with the fastest rate occurring soon after drainage had been initiated. The present rate averages about 14 mm per year (Richardson and Smith, 1977). At its maximum natural extent the peat of the English Fenland covered around 1,750 km². Now only about one-quarter (430 km²) remains. One consequence of peat wastage in the Fens was the formation of "roddons." These are fossilized silt- and sand-filled tidal creek systems of mid- to late-Holocene age incised into contemporaneous clay deposits (Figure 4.5). However, anthropogenic change (drainage and agriculture) has caused the former channels to become positive topographical features (Smith et al., 2010). Similar features are found in Romney Marsh in southern England.

Subsidence following drainage has also occurred in the peat soils of the Netherlands, with rates at around 0.2–0.8 cm per year (Nieuwenhuis and Schokking 1997; Schothorst, 1977; Hoogland et al., 2013). Peatland ecosystems once covered a major proportion (40 percent) of the Dutch land surface, but the area of peat soils has been reduced to less than 10 percent since drainage started in the 11th century. It has been predicted that the peat areas in Friesland will subside by a further 40–60 cm between 1999 and 2050 (Brouns et al. 2015). Subsidence has also occurred in the vicinity of Venice, Italy (Gambolati et al., 2005; Gambolati et al, 2006), where rates of subsidence of histosols as high as 2–3 cm per year have been recorded (Fornasiero et al., 2003), causing 1.5–2.0 m of subsidence over a period of seventy years (Zanella et al., 2011). Comparable rates of subsidence of organic soils (2.3 cm yr^{-1} between 1910 and 1988) have been recorded from the Sacramento–San Joaquin delta area of California (Rojstaczer and Deverel, 1995). In tropical Southeast Asia rates are rather higher, averaging around 4.0–5.5 cm yr^{-1} (Hooijer et al., 2012). However, studies over a few decades suggest that in Malaysia such rates decline through time to about 2 cm yr^{-1} (Wosten et al., 1997).

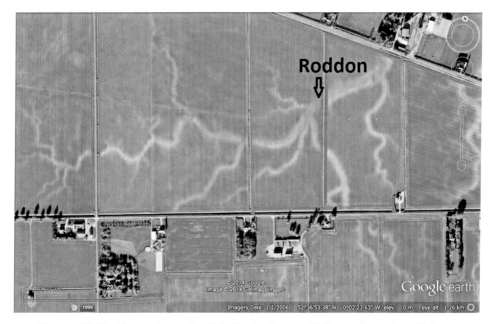

Figure 4.5 Roddons of the English Fenland near Tollney. Scale bar 291 m. ©2014Google, ©2014 GetMapping plc.

Alluvial floodplains may store large amounts of carbon as partially decomposed organic matter. In these systems, the sediments are preserved by the anoxic conditions, which are maintained by near constant saturation and/or inundation, similar to processes that facilitate organic carbon storage in peat bogs. If river channelization increases the channel depth, these conditions may change, and the organic material may decay, inducing floodplain subsidence (Kroes and Hpp, 2010).

The drying of peat as a result of climate warming and human activity (e.g. drainage, forestry, agriculture, peat harvesting, and road construction) lowers the water table in peatlands and thus increases the frequency and extent of peat fires. This is important because although peatlands cover only about 2–3 percent of the Earth's land surface they store around 25 percent of the world's soil carbon. The loss of peat through burning will thus release substantial amounts of carbon, much of it previously stored for millennia, into the atmosphere, thereby affecting the whole Earth system (Turetsky et al., 2015).

4.11 Induced Seismic Activity

Human activities can cause seismic activity (Davies et al., 2013), sometimes in areas where it is naturally rare. These activities include loading of the crust by reservoirs, fluid abstraction, and injection, underground waste disposal, hydro-fracturing (fracking), carbon sequestration, mining, tunnel construction, and exploitation of geothermal resources. While some of the seismic activity is minor, higher magnitude events have occurred.

Earthquakes can result from the effects on the Earth's crust of large masses of water impounded behind reservoirs (Guha, 2000; Gupta, 2002; Durá-Gómez and Talwani, 2010; El Hariri et al., 2010; Liu et al., 2011). Large reservoirs impose stresses of significant magnitude on crustal rocks, especially those located on areas with normal or strike-slip faults (Qiu and Fenton, 2015). With the ever-increasing number and size of reservoirs the threat rises. Detailed monitoring has shown that earthquake clusters occur in the vicinity of some dams after their reservoirs have been filled, whereas before construction activity was less clustered and less frequent. Similarly, there is evidence from some locations of a linear correlation between the storage level in the reservoir and the logarithm of the frequency of shocks. It is seldom easy, however, to say definitively that seismic activity in the vicinity of a reservoir is necessarily caused by the presence of the reservoir, and in China there is ongoing debate as to whether the magnitude 7.9 Wenchuan earthquake of 2008, which killed 80,000 people, was triggered by the reservoir behind the 156 m tall Zipingpu Dam (Qiu, 2012).

Cases where seismicity and faulting can be attributed to fluid extraction come from the oilfields of Texas and California and the gas fields of the Po Valley in Italy, the Netherlands, and Uzbekistan (Cypser and Davis, 1998; Donnelly, 2009; Suckale, 2009; Kraaijpoel and Dost, 2013). Equally, fluids may be injected to aid exploitation of oil, leading to earthquakes (Keranen, et al., 2013; Improta et al., 2015). Some induced seismic activity may be related to "hydrofracturing" (fracking), whereby water is injected into the rock to create distinct fractures that increase permeability and thus facilitate the extraction of fluids and the production of geothermal energy (Majer et al., 2007). Fracking for shale-gas exploitation can trigger seismicity, as in Canada (Atkinson et al., 2015), because it can cause an increase in the fluid pressure in a fault zone (Rutquist et al., 2013). Enhanced geothermal systems, where water is injected into rocks for the purposes of extracting geothermal energy, are another source of induced seismicity (Baisch et al., 2010; Grünthal, 2014), as has been documented in Indonesia, central Europe, the Philippines, Japan, Kenya, North and South America, Australia, and New Zealand (Zang et al., 2014). Water injection has also been employed to dispose of saline water from river systems in Texas and the American West and has generated small earthquakes (Hornbach et al., 2015; Yeck et al., 2015). Other types of wastewater disposal have the same effects (Ellsworth, 2013; Stabile et al., 2014). Fears have recently been expressed that carbon capture that involves long-term subsurface storage of large volumes of aqueous solutions of CO_2 and supercritical CO_2 under high pressure, may cause earthquakes (Mazzoldi et al.,

2012; Yarushira and Bercovia, 2013). Coal mining (Bischoff et al., 2010) also causes earthquakes, as do construction of tunnels for roads and railways (Husen et al., 2012).

All these different mechanisms may contribute to what appears to be an increasing incidence of earthquakes. Ellsworth (2013) reported that within the central and eastern United States, the earthquake count has increased dramatically over the past few years. More than 300 earthquakes with $M \geq 3$ occurred in the three years from 2010 through 2012, compared with an average rate of twenty-one events per year observed from 1967 to 2000. The greatest rise in activity occurred in 2014 when 659 $M \geq 3$ earthquakes occurred. States experiencing elevated levels of seismic activity included Arkansas, Colorado, New Mexico, Ohio, Texas, and Virginia, but the greatest concentration was in Oklahoma (Rubinstein and Mahani, 2015). McGarr et al. (2015) have gone so far as to refer to "a seismic surge."

4.12 Conclusions

Humans have caused ground subsidence by dewatering of karstic formations, mining of minerals such as coal and salt, groundwater, and geothermal fluid abstraction, exploitation of hydrocarbon resources, the draining of organic soils, and hydrocompaction. In the future, ongoing exploitation of energy resources may cause further subsidence and a surge in seismic activity.

5

Weathering Processes in the Anthropocene

Weathering is a key geomorphological process which shapes natural landscapes, causes deterioration of the built environment, and links the land surface to other parts of the Earth System through the contribution chemical weathering makes to biogeochemical cycling. Like other geomorphic processes, weathering has been complexly influenced by human activities in the Anthropocene, although unlike many other processes there are few datasets available with which to quantify changing rates of weathering and ascribe their causes. One of the complexities surrounding weathering is that it comprises a range of individual processes, causing chemical decomposition and/or physical breakdown of rocks and minerals at or near the Earth's surface on bare rock surfaces, and in soils and regolith. A further source of complexity is that many weathering processes are influenced by biology, with plants, animals, and microorganisms contributing directly and indirectly to many types of chemical and physical weathering. Weathering is also complexly linked to erosion, with steady state and eroding landscapes presenting very different situations (see hypotheses two and three in Brantley et al., 2011). Ultimately, weathering rates are controlled by climatic and tectonic factors which together control the availability of water, the biology, the rock types, topography, and other important factors influencing the weathering system (see Goudie and Viles, 2012, figure 3). Through the influence that chemical weathering has on the global carbon cycle, changes in weathering rates also feed back to the global climate system (see Chapter 11). Over tens of millions of years changes in climate, biology, and weathering rates have been tightly interlinked with, for example, evidence that the evolution of trees and seed plants in the Devonian led to enhanced chemical weathering and soil formation, resulting in a drawdown of atmospheric CO_2 and concomitant global cooling (Berner, 1998).

Over the different phases of the Anthropocene human activities have potentially influenced weathering processes and rates in a number of direct and indirect ways, and in different settings, through impacts on climate, biology, hydrology, and erosion (see also Pope and Rubenstein, 1999, tables 1 and 2). In terms of impacts on biology and erosion, changes in agriculture have been suggested as causing notable "spikes" in chemical weathering records during the paleoanthropocene in several different parts of the world. More recently, agricultural influences on groundwater hydrology have led to considerable problems of soil salinization with knock-on effects on the weathering of archaeological sites. In terms of impacts on climate, changes in urban and regional air quality have been particularly notable controls on weathering processes in the Industrial Era, while during the Great Acceleration increasing land use change and global change have been seen to be starting to affect weathering systems. Increasing atmospheric CO_2 levels are hypothesized to exert a major control over mineral weathering rates in future (Brantley et al., 2011), and changes in rainfall and temperature likely to exert a strong influence on building stone deterioration. Proposals have also been made for future geoengineering responses to climate change which exploit the negative feedback that silicate weathering exerts on atmospheric CO_2 contents.

5.1 Weathering in the Paleoanthropocene

A number of recent studies have utilized new sources of evidence and methods to identify the impacts of land use change on chemical weathering rates in the paleoanthropocene. For example, Bayon et al. (2012) use isotopic and Al/K ratio data from a 40,000 year marine sediment core from the mouth of the Congo River to identify an unprecedentedly high intensification of chemical weathering around 3,000 years ago, contemporaneous with the arrival of Bantu agricultural groups. Bayon et al. (2012) infer that land use change and associated soil erosion led to increased surface area of rock exposed to weathering. Their interpretation of the causality of this spike in weathering has been criticized (Maley et al., 2012), but human activity should not be ruled out as a contributing factor. Similar inferences have been made using evidence from a high resolution sediment core from the South China Sea by Wan et al. (2015). Here, an increase in chemical weathering and erosion has been picked up using isotopic and other proxies between 6400 and 4000 cal year BP followed by a decrease from 4000 to 1800 cal year BP. Both can be clearly related to climatic changes. However, a further intensification in chemical weathering and erosion has occurred from 1800 cal year BP decoupled from climatic changes, and has been ascribed by Wan et al. (2015) to deforestation, cultivation, and mining.

Evidence from a range of rather different studies helps to strengthen these putative links between agricultural and other human activities and chemical weathering rates over the paleoanthropocene. Pope and Rubenstein (1999), for example, used Scanning Electron Microscopy (SEM) observations of porosity of mineral grains from soils to illustrate that those close to an archaeological site in Arizona are more weathered than those further away from the site. Heitkamp et al. (2014) used a comparison between comparable grazed and inaccessible soil sites in the Andes to provide further evidence for the impact of agriculture on weathering rates. Using a chemical index of alteration (based on the ratio of labile elements to aluminum) they were able to illustrate significantly higher rates of weathering on grazed sites. They hypothesized that vegetation change resulting from grazing leads to lower vegetation cover, higher soil surface temperatures, and soil acidification which, in turn, produce higher rates of chemical weathering.

Chemical weathering releases ions which can become reprecipitated and dating the build-up of these precipitates can be used to infer changing weathering rates. Deposits such as tufa and travertine (freshwater accumulations of reprecipitated carbonates) are ideal for such studies. During the 1970s and 1980s an increasing body of isotopic dates became available for deposits of tufa (also sometimes known as travertine). Some of these dates suggested that over large parts of Europe, from Britain to the Mediterranean basin and from Spain to Poland, rates of tufa formation were high in the early and middle Holocene, but declined markedly thereafter (Weisrock, 1986, pp. 165–7). Many studies have tended to support this concept. For example, in northern France at Daours, tufa formation ceased shortly after c. 5,000 years ago (Limondin-Lozouet et al., 2013). Similarly, Luzon et al. (2011) found that the reduction in tufa deposition in the Añamaza system of central Spain also occurred at that time, as did Wehrli et al. (2010) working in Italy and Gradiński et al. (2013) working in Slovakia. A Holocene tufa decline has also been proposed for further afield, as in Iran (Mokhtari et al., 2011).

Vaudour (1986) maintains that since around 3000 BP humans have been responsible for the decline in tufa accumulation, but he gives no clear indication of either the basis of this point of view or of the precise mechanism(s) that might be involved. If the late Holocene reduction in tufa deposition is indeed a reality (which is contested by, e.g., Baker and Sims, 1998), then it is necessary to consider a whole range of possible mechanisms (Table 5.1), both natural and anthropogenic (Nicod, 1986: 71–80). As yet the case for an anthropogenic role is unproven (Goudie et al., 1993) and it is possible that climate changes, such as the ending of the middle Holocene "climatic optimum," were a more important determinant of the tufa decline (Wehrli et al., 2010). In some parts of Europe, tufa deposition is occurring still and the climatic factor may have been dominant in controlling tufa nature and deposition (Dabkowski et al., 2015). Nonetheless, in the Trabaque region of central Spain (Dominguez-Villar et al., 2012) there was a drastic diminution in the amount of tufa precipitation after ~4 ka BP, probably caused by anthropogenic deforestation. Also working in Spain, González-Amuchastegui and Serrano (2015) found that in the Holocene, tufa deposition

Table 5.1 *Possible mechanisms to account for the alleged Holocene tufa decline.*

Climatic/natural	Anthropogenic
Discharge reduction following rainfall decline leading to less turbulence	Discharge reduction due to over-pumping, diversions, etc.
Degassing leads to less deposition	Increased flood scour and runoff variability of channels due to deforestation, urbanization, ditching, etc.
Increased rainfall causing more flood scour	Channel shifting due to deforestation of floodplains leads to tufa erosion
Decreasing temperature leads to less evaporation and more CO_2 solubility	Reduced CO_2 flux in system after deforestation
Progressive Holocene peat and podzol development through time leads to more acidic surface waters	Introduction of domestic stock causes breakdown of fragile tufa structures
	Deforestation means there are less fallen trees to act as foci for tufa barrages
	Increased stream turbidity following deforestation reduces algal productivity

and erosion had responded to climatic fluctuations until c. 6.5 ka, but subsequently human activity had determined landscape evolution and especially from 4.5 ka.

5.2 Weathering in the Industrial Era

Although relatively few data are available and the effects are generally not immediately obvious or easy to measure, there is evidence that human activities have produced changes in both the nature and rate of weathering over the past few hundred years (Winkler, 1970). The factors involved include acidification of precipitation by sulfates, nitrates, and carbon dioxide; lateritization by vegetation removal; dry deposition of sulfates, etc.; rises in groundwater levels which promote the "wick effect" and salt weathering; and higher temperatures which accelerate rates of silica dissolution, reduce rates of carbonate solution and the incidence of frost weathering.

Air pollution. One of the prime causes of increasing rates of weathering since the start of the Industrial Revolution is air pollution. As a result of emissions of sulfur dioxide through the burning of fossil fuels, there are high levels of sulfuric acid in precipitation over many industrial areas. This produces acid rain, which has a pH of less than 5.65, this being the pH which is produced by carbonic acid in equilibrium with atmospheric CO_2. In many parts of the world, rain may be markedly more acidic than this normal, natural background level. The seriousness of acid rain caused by sulfur dioxide emissions in the Western industrialized nations peaked in the mid 1970s or early 1980s. In China, however, from

2000 to 2006 sulfate emissions increased by 53 percent, an annual growth rate of 7.3 percent (Lu et al., 2010) and precipitation became increasingly acidic. Acid precipitation may cause accelerated dissolution of carbonate rocks and lead to increasing calcium loads in streams, as has been observed in the northeastern United States (Kaushal et al., 2013), though empirical studies are sparse, so that there is little direct evidence that acidification significantly affects chemical weathering rates. Moreover, experimental studies have demonstrated that dissolution rates exhibit very little change as a function of pH in the near neutral range (White and Blum, 1995). The impact of acidification will also depend on the buffering capacity of catchments and the thickness and composition of the materials in the critical zone (see the work of April et al. [1986] on rising chemical weathering rates in the Adirondacks of the eastern United States).

Chemical reactions involving sulfur dioxide can also generate salts such as calcium and magnesium sulfates, which may be effective in causing the physical breakdown of rock through salt weathering. They also lead to the formation of crusts on exposed surfaces (Marszalek et al., 2014) as, for example, on natural sandstone outcrops in Central Europe (Vařilová et al., 2015), and on many important examples of built heritage (Barca et al., 2014). Figure 5.1 shows examples of severe weathering and black crust development from the limestone walls of Oxford.

Changes in industrial technology, in the nature of economic activity, and in legislation caused the output of SO_2 in Britain to decrease by 35 percent between 1974 and 1990. In 2009, SO_2 emissions were 89 percent lower than 1990 levels. This was also the case in many other industrialized countries, including the United States (Malm et al., 2002). When controls on sulfate emissions take place and atmospheric sulfur dioxide levels diminish, rates of weathering may also decline, as has been found for St. Paul's Cathedral, London (Inkpen et al., 2012a and b). On the other hand, stone decay in cities can in addition be accelerated by nitrates derived from air pollution, particularly in proximity to vehicle exhausts (Massey, 1999; Schiavon et al., 2004, Maguregui et al., 2008).

Salt weathering. Salt weathering can be accelerated by changes in groundwater levels resulting from irrigation, vegetation clearance, and urbanization (Goudie and Viles, 1997; Figure 5.2). This occurred in the Indus Plain in Pakistan (Goudie, 1977), where irrigation caused the water table to be raised by about 6 m since 1922. This produced increased evaporation and salinization. The salts that are precipitated by evaporation above the capillary fringe include sodium sulfate, a very effective cause of stone decay at the great archaeological site of Mohenjo-Daro (Figure 5.3). Similarly, in Uzbekistan the ancient towns of Kiva, Bukhara, and Samarkand (all World Heritage Sites) have suffered from irrigation-induced groundwater rise and salinization, causing deterioration to the lower courses of the buildings (Akiner et al., 1992).

Figure 5.1 Weathering of buildings and walls along New College Lane, Oxford.
Note the black crust formation.

There are also many threats to Egypt's ancient cultural heritage sites, with salt damage and other problems relating to rising groundwater being among the most serious, as identified by Keatings et al. (2007) at the mudbrick ruins of Hawara Pyramid. Studies by Wüst and Schlüchter (2000) illustrated the widespread threat of groundwater rise and salt crystallization processes around Thebes (Luxor and Karnak; Figure 5.4), with damage caused by sodium chloride the major problem. Weathering problems are also notable and alarming in Cairo (Fitzner et al., 2002). Kamh et al. (2008), in a study of Islamic archaeological sites in the city, noted the

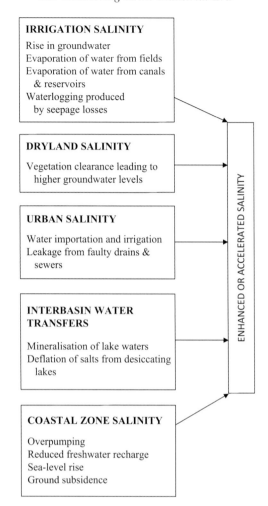

Figure 5.2 Causes of enhanced or accelerated salinization.

synergy between salt weathering and the 1992 earthquake. Some of the sites most badly damaged by the earthquake were those previously weakened by intense salt attack. In the Kharga Oasis, there are temples and cemeteries that are being damaged by salt weathering, and as elsewhere in Egypt the problem is being exacerbated by changes in groundwater levels brought about by irrigation (Salman et al., 2010).

Another potential cause of accelerated salt weathering is the increased salinity of stream waters and of surface materials that can be associated with ground disturbance. In a semiarid landscape, substantial concentrations of salts may occur at relatively shallow depths. Construction of dirt roads, well pads, and pipelines necessarily disturb soils, thus exposing or bringing subsurface salts closer to the

Efflorescence of
sodium sulphate

Figure 5.3 Sodium sulfate has attacked the Harappan site at Mohenjo-Daro, Pakistan, causing rapid decay since groundwater levels were raised by irrigation. Upper image shows the main stupa area. The lower image shows the Great Bath. Note the white efflorescence and the decayed bricks.

Figure 5.4 Rising damp from rising groundwater levels, as here in Karnak, Egypt, has caused increasing decay of major heritage sites.

surface. Runoff can carry those salts to streams, and the wicking process that produces efflorescent salts can successively bring more salt to the surface (Bern et al., 2015).

5.2.1 Wetting and Drying Cycles

Ta Keo in Cambodia is a 1,000 year old temple which was cleared of forest in the early twentieth century. Studies using GIS-based analysis of historic imagery indicate an average tenfold increase in stone loss rates at the site (0.2 instead of 0.02 percent per year). André et al. (2014) attributed this to the climatic stress provoked by the exposure of fragile ornamented sandstones to the harsh impact of tropical sunshine and monsoon rains on clearing the surrounding forest. The resulting threefold increase in daily temperature and humidity ranges was crucial, for pronounced wetting–drying cycles are conducive to enhanced swelling–shrinking movements that provoke sandstone delamination and contour scaling.

5.2.2 Laterite

In some parts of the tropics are extensive sheets of laterite, an iron and/or aluminum-rich duricrust. These develop naturally, either because of a preferential removal of silica during the course of extensive weathering (leading to a *relative* accumulation of the sesquioxides of iron and aluminum), or because of an *absolute* accumulation of these compounds. One of the properties of laterites is that they harden on exposure to air and through desiccation. Exposure may take place through accelerated erosion, while forest removal may cause such a change in microclimate that desiccation of the laterite surface takes place. There are records from many parts of the tropics of accelerated induration brought about by forest removal (Gourou, 1961; Goudie, 1973). In West Africa, the process of laterite hardening is called *bowalization* (Padonou et al., 2013; 2014).

5.3 Weathering in the Great Acceleration

Atmospheric carbon dioxide levels have been rising steadily because of deforestation and the burning of fossil fuels. Increasing CO_2 levels have the potential to accelerate chemical weathering of carbonate and silicate rocks. Calmels et al. (2014), for example, have shown from a study across a climate transect in the Jura mountains that carbonate weathering is strongly sensitive to vegetation type because of its control on soil CO_2 concentrations. They suggest that carbonate weathering and associated CO_2 consumption will quickly respond to any land-use change or future global change. Land-use change may also affect the delivery of dissolved silica and other solutes to the oceans (Conley et al., 2008; Struyf et al., 2010; Li et al., 2011; Meunier et al., 2011; Clymans et al., 2011; Chen et al., 2014). As Carey and Fulweiler (2012) have explained, the influence of this on riverine silicon (Si) export is hypothesized to be due in large part to release and uptake of Si by terrestrial vegetation. In addition, it is well known that plants play a major role in both the weathering of silicate minerals and the solubility of Si in soil. The application of nitrogenous fertilizers can accelerate rates of limestone dissolution, for during the nitrification of the NH_4^+ supplied by fertilization, nitrate and H^+ ions are produced in the soil and so increase acidification (Brunet et al., 2011).

5.3.1 *Mining and River Solutes.*

In mining areas, increased weathering leads to increased export of ions in stream-flows. This is the case, for example, in the Appalachian coal mining area of the eastern United States. Here, excess overburden from mountaintop mining is disposed of in constructed fills in small valleys adjacent to the mining site. Leachate persistently increases the downstream concentrations of major ions (Hopkins et al., 2013; Lindberg et al., 2011; Griffith et al., 2012). In areas with underground mining, acidic metal-rich mine drainage can develop as a result of accelerated oxidation of iron pyrite (FeS_2) and other sulfidic minerals resulting from the exposure of these minerals to both oxygen and water, as a consequence of the mining and processing of metal ores and coals (Raymond and Oh, 2009).

5.3.2 Carbon Export in Rivers.

Carbon budgets are an important issue, given our need to understand the carbon cycle and related climate changes. There is, therefore, an increasing interest in how environmental changes are impacting upon soil organic carbon and its transport in streams to the oceans. Weathering within soils may play a key role in carbon cycling.

Bauer et al. (2013) argued that inputs of carbon to the global ocean can be altered through changes in the water balance (precipitation and evapotranspiration) and in carbon stocks and flows in river basins. Hydrological alterations include the effects of climate change on the amount and frequency of rainfall events and temperature regulation of evapotranspiration. Land management practices that alter rates of evapotranspiration, such as irrigation and clearance of vegetation, can also be important. In addition, inputs of sulfuric acid, agricultural practices, peatland disturbance, the thawing of permafrost, wetland removal, and reservoir construction can alter carbon stocks and flows at the drainage-network level. In the UK uplands, an increase in heather burning has coincided with an increase in dissolved organic carbon (DOC) concentrations. Räike et al. (2012) reported that over the past twenty to thirty years these have increased in freshwater systems in mid- to high-latitudes of the Northern Hemisphere and that changes in land-use practices have caused increases in DOC export. They reported that in Finland, for instance, draining of peatlands and forest management practices (such as clear-cutting and ditching) have temporarily enhanced leaching of DOC to surface waters. Clark et al. (2010) indicated that in streams DOC has shown strong increases in areas which have experienced strong downward trends in pollutant sulfur and/or sea salt deposition. Monteith et al. (2007) explained this by the fact that atmospheric deposition can affect soil organic matter (SOM) solubility through at least two mechanisms: by changing either the acidity of soils or the ionic strength of soil solutions, or both.

5.4 Future Weathering

Remarkably little work has been done on how weathering rates will respond to future changes in climate. Lack of data, as well as the complexity of the weathering system, makes confident predictions very difficult. Beaulieu et al. (2012) have indicated that the reaction of continental weathering to anthropogenic forcing is difficult to determine because chemical weathering is a complex process that depends on many factors. These variables, which are often tightly intertwined, include, *inter alia*, climate, the mineralogy of outcropping rocks, physical erosion, rainfall pH, soil temperature, and vegetation cover (see also Goudie and Viles, 2012).

As Beaulieu et al. (2012) explain, in addition to being affected by the melting of ice in the soil zone (increasing the liquid water availability for weathering), weathering systems will also be sensitive to climate warming, through perturbations in net primary production and heterotrophic soil respiration. These perturbations will induce changes in soil acidity (organic acid release and below-ground CO_2 accumulation linked to microbial respiration) and impact on the chemical weathering rate. Temperature rise should also promote dissolution of silicate minerals, but should, conversely, reduce carbonate dissolution, the rate of which is inversely related to temperature. In addition, vegetation productivity changes, together with modifications in rainfall pattern, should impact on weathering processes through perturbations of below-ground hydrologic behavior and changing rates of leaching of dissolved material. There has been very little work on testing of these ideas, though Gislason et al. (2009) found that in Iceland, rates of basalt weathering, as determined from river discharge and solute data, have increased over the last four decades of warming. Cotton et al. (2013) argued that terrestrial silicate weathering, which is influenced by ecosystem productivity, should respond to increasing atmospheric CO_2 concentrations, as silicate weathering drives carbon sequestration by converting atmospheric CO_2 to CO_3 and moving it to terrestrial and oceanic reservoirs. They also suggested that a large proportion of chemical weathering occurs during the reaction of carbonic acid with silicate minerals in soils and that the speed of these reactions would be affected by future temperatures and precipitation amounts.

Viles (2002) attempted to conceptualize the impact of future climate change on building stone deterioration in the United Kingdom (Table 5.2). There she contrasted the likely impacts of climate change on building stones between the northwest and the southeast of the country under twenty-first-century climate change scenarios. She argued that in the northwest, which would likely have warmer and wetter winters, chemical weathering's importance might be increased, whereas in the southeast, which would have warmer, drier summers, processes like salt weathering might become more important. Viles (2003) has also tried to model the impact of climate change on limestone weathering and karstic processes in the UK, but the conceptual model has wider significance and could be applied to karst areas worldwide. Further work on future stone deterioration has been done by the NOAH's Ark project which focused on the impact of climate change on European cultural heritage. Bonazza et al. (2009a) believed that as carbon dioxide levels

Table 5.2 *Possible consequences of future climate change for building stone decay in the United Kingdom (from Viles, 2002, table 3).*

	Dominant building stone type	Process responses	Other threats	Overall response
Northwest: warmer, wetter winters	Siliceous sandstones and granites	Increased chemical weathering. Less freeze–thaw weathering. More organic growths contributing to soiling.	Increased storm activity may cause episodic damage. Increased wave heights may encourage faster weathering in coastal areas. Increased flooding may encourage decay.	Enhanced chemical decay processes and biological soiling, reduced physical decay processes.
Southeast: warmer, drier summers	Limestones	Less freeze–thaw weathering. Reduced chemical weathering as a result of less available water. Increased salt crystallization in summer. More deteriorating organic growths.	Increased drying of soils (especially clay-rich soils) will encourage subsidence and building damage. Low-lying coastal areas will be particularly prone to marine encroachment. Increased drought frequency may encourage decay.	Enhanced physical and biological weathering, more dust for soiling, reduced chemical weathering.

climb, and rainfall becomes more acidic, karstic weathering will speed up. Smith et al. (2010, 2011) suggested that higher winter rainfall amounts in the UK could cause the "greening" of buildings by algal films, while deeper penetration of moisture might affect the operation of processes like salt weathering. Viles and Cutler (2012) discussed the ways in which pollution levels, and changes in temperature and precipitation, might affect organisms on rock surfaces and thus modify rates of bioerosion and bioprotection. Hall et al. (2011) suggested that in southern England, predicted increases in potential evaporation for the period 2070–2100 would cause substantial increases in water flux in walls, from which one could expect increased damage rates from salt crystallization. As the climate warms, the number of freeze–thaw cycles to produce frost weathering in susceptible rocks in areas like central England are likely to decrease (Grossi et al., 2007), as they did during the warming of the nineteenth and twentieth centuries (Brimblecombe and Camuffo, 2003). There may also be increasing thermal stresses on marbles (Bonazza et al., 2009b), and more frequent wildfires to cause rock disintegration.

In areas such as the Nile Delta, where salt weathering is prevalent, future sea-level rise and over-pumping of groundwater will cause accelerating salt water intrusion to occur (Susnik et al., 2015). This may lead to more widespread salt weathering. Figure 5.5a shows the present situation with regard to groundwater salinity in the Nile Delta, and demonstrates the sharp gradient that occurs in salt concentrations as one moves inland. Figure 5.5b shows the effects of sea-level rise on the Delta, and suggests that it is highly probable that the location of areas of high groundwater salinity will occur in areas that are now relatively fresh.

With regard to organic carbon export in rivers, Mattsson et al. (2015) suggest that in northern boreal regions there will be increasing temperature and precipitation, and that the changes will be larger in winter than in summer. They believe that these precipitation and discharge patterns, coupled with shorter ice cover/soil frost periods in the future, would be expected to contribute significantly to changing flow paths of organic matter over a range of land-use patterns. The projected increase in precipitation in northern Europe is expected to result in increased export of DOC to lakes and rivers from catchments (Räike et al., 2012).

Weathering and geoengineering. Enhanced silicate weathering has been proposed as one potential geoengineering solution to climate change, as silicate weathering has been shown to contribute to drawdown of atmospheric CO_2. However, how to achieve measurable impacts in desirable timeframes remains the source of some debate. Moosdorf et al. (2014) present the results of a modeling study to assess the efficiency of terrestrial enhanced weathering schemes, which apply ultramafic silicate rock floor to increase natural weathering rates. Their study investigated source rocks, transport pathways, and potential application areas in

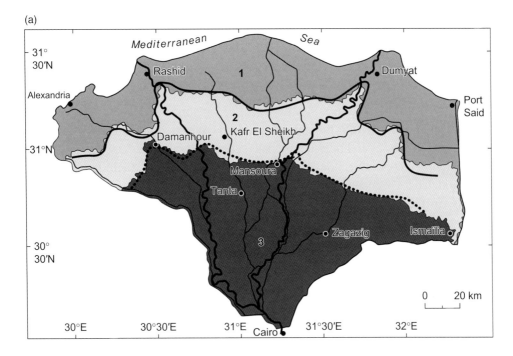

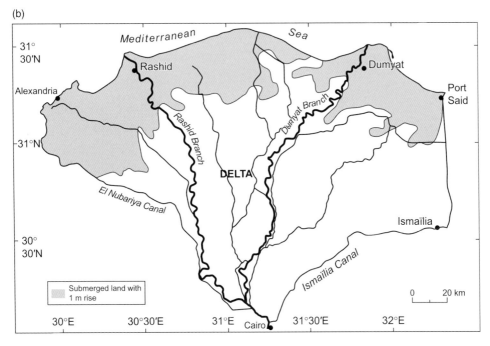

Figure 5.5 The Nile Delta. (a) Present groundwater salinity concentrations.
1 = Saltwater zone (>35,000 ppm), 2 = Mixing zone (1,000–35,000 ppm, and
3 = Freshwater zone (<1,000 ppm). Solid line is observed concentration of
35,000 ppm, and dotted line is observed concentration of 1,000 ppm. (b) Area
likely to be submerged by a 1 m rise in sea level (modified from Sefelsnasr and
Sherif [2014, figures 3 and 4]).

order to model the balance between emission and drawdown of CO_2. Many uncertainties remain about the practicability of such schemes, but the study implies that they could potentially make a worthwhile contribute to CO_2 drawdown.

5.5 Conclusions

As yet we know relatively little about the extent to which humans have modified weathering processes and rates across paleoanthoprocene, Industrial Era, and Great Acceleration timescales. Locally, however, it is clear that such factors as air pollution, land-use change, and changes in groundwater levels have had significant effects. In the future changes in atmospheric carbon dioxide, temperature, and precipitation will affect weathering in complex ways because of the interdependencies of weathering, biological, soil, and other geomorphological processes.

6

Hillslope Processes in the Anthropocene

6.1 Modification of Infiltration Capacities and Other Soil Properties

Soil erosion is a major and serious aspect of the human role in environmental change (Sauer, 1938; Montgomery, 2007). There is a long history of weighty books on the subject (Chapter 1; see e.g. Bennett, 1938; Jacks and Whyte, 1939; Morgan, 2005) and there are many examples of severe erosion phenomena and rates associated with human activities. Lowdermilk (1935, p. 409) wrote:

The great enemy of the human race is soil erosion which has been associated with the habitations of man since before the dawn of history. It is no new land disease, but has only recently been diagnosed and named for what it is. The removal of vegetation . . . exposes soils to the dash of rain or the blast of wind, against which they had been protected for thousands of years. Topsoils blow away or wash away, or both. Unprotected sloping lands are usually bared to hard and tight subsoils, which drain off the water as from a tiled roof.

Dotterweich (2013) provides a good global review of the history of soil erosion; while a discussion of rates of erosion in Tennessee, United States, in pre-European and post-European times is provided by Dotterweich et al. (2015). McNeill (2000, p. 35) identified three pulses of accelerated soil erosion. The first occurred with the initial development of agriculture, the second with the spread of European land practices to other parts of the world, and the third with the spread and intensification of agriculture from the 1950s. In some parts of the world, soil erosion has been so severe that bare rock surfaces (especially on limestones) have developed – a process that Chinese workers call "rocky desertification" or "karst rocky desertification"(Jiang et al., 2014). The impact of humans on karstic features has a long history. For example, Drew (1983) argued that the Burren, a plateau karst on the west coast of Ireland, characterized by thin soils, patchy vegetation, and large areas of bare rock, supposedly a legacy of glacial erosion, may have been stripped by human activities. Archaeological and palynological evidence suggested that the area was well populated and forested for part of prehistoric time. Investigation of paleosols and of karren forms on ancient structures supported the idea that forest clearance removed an extensive cover of mineral soil in the later Bronze Age. More recent research suggests that there may have been subsequent episodes of erosion as well (Feeser and O'Connell, 2009). Intense human activities in modern times may exacerbate the situation.

Much surface erosion, referred to by Montgomery (2007, p. 3) as "the skinning of our planet," is caused by surface runoff. This in turn is often controlled by the infiltration capacities of surface materials. These can be greatly affected by grazing by domestic stock, the replacement of grassland with shrubland, and deforestation. In addition, there are other structural properties of soils that affect their erodibility, including the nature of their aggregates and their humus contents.

6.2 Grazing

The number of domestic stock in the world (Table 6.1) has greatly increased in recent decades. Cattle numbers, for example, have more than tripled since 1890, with severe implications for trampling, soil compaction and erosion, and removal of riparian vegetation. Heavily grazed lands tend to have considerably lower infiltration capacities than those found in ungrazed areas. The relationships

Figure 6.1 Exposed tree roots show the extent of surface lowering that has taken place on a gentle slope near Lake Baringo, Kenya.

Table 6.1 *Number of domestic stock in the world, from 1890 to 2000 (x 1000).*

	1890	1900	1910	1920	1930	1940	1950	1960	1970	1980	1990	1998	2000
Asses	8,062	10,016	13,306	16,815	23,165	25,761	31,181	36,271	38,563	38,038	43,709	43,344	40,577
Buffalo	43,444	47,965	52,857	58,228	64,446	71,243	78,886	87,505	107,307	121,649	148,355	162,586	164,297
Cattle	410,084	448,447	517,030	592,207	665,313	692,221	771,453	919,910	1,080,115	1,214,862	1,294,841	1,287,154	1,336,941
Goats	137,465	148,987	175,210	186,361	229,367	236,279	265,719	344,626	375,966	462,184	580,150	705,152	722,224
Horses	97,794	102,260	109,524	108,331	106,138	82,678	72,392	65,857	60,793	59,456	60,976	60,929	56,722
Mules	2,075	2,313	2,899	3,414	3,509	4,429	6,366	9,228	11,350	12,917	14,784	14,146	13,196
Pigs	177,060	194,579	215,882	208,028	245,506	257,167	249,610	393,619	545,705	797,145	857,260	955,121	906,066
Sheep	625,291	603,974	683,873	677,925	779,258	770,687	778,929	998,052	1,061,006	1,096,201	1,203,644	1,063,720	1,049,503

Source: http://themasites.pbl.nl/tridion/en/themasites/hyde/landusedata/livestock/index-2.html (accessed March 9, 2015).

between grazing pressures and soil infiltration capacities are, however, complex. On the one hand, moderate stocking levels may increase infiltration capacities by breaking down surface biological or rain-beat crusts, while on the other, high stocking levels may remove all vegetation cover, cause breakdown of soil aggregates, and produce severe trampling and soil compaction, thereby decreasing soil infiltration rates (du Toit et al., 2009). Trimble and Mendel (1995) demonstrated why it is that cows have the ability to cause soil compaction. Given their large mass, small hoof area, and the stress that may be imposed on the ground when they are scrambling up a slope, they are probably remarkably effective in causing soil compaction. Particular fears have been expressed that the replacement of Amazonian rainforest by cattle-trampled pasture could lead to great increases in the frequency and volume of stormflow. One study showed that the frequency of storm flow in such grazed areas increased twofold, while its volume increased seventeenfold (Germer et al., 2010). In northern Australia, large numbers of feral water buffaloes have had adverse effects on soil compaction, erosion rates, and the state of stream banks (Skeat et al., 2012).

6.3 Replacement of Grassland by Shrubland

The infiltration capacities of dryland surfaces are highly variable, and one reason for this is the nature of vegetation cover. In areas like the western United States, humans often caused grassland to be replaced by shrubland (Bhark and Small, 2003). The differences between grass and shrub surfaces have been the subject of an extensive literature (Ravi et al., 2009), and Parsons et al. (1996) found from rainfall simulation experiments that compared to grasslands, the interrill portions of shrubland hill slopes had higher runoff rates, overland flow velocities, and rates of erosion. In general, it can be argued that under a shrub canopy, infiltration capacities will be relatively high because the addition of organic matter and the activity of roots increase soil porosity. On the other hand, the decreased infiltration capacity in intercanopy areas may more than offset higher infiltration capacities under the shrub canopies, with the net result that runoff from shrub-dominated hill slopes may be many times greater than those dominated by grassland. Furthermore, as Turnbull et al. (2010, p. 410) explain on the basis of their erosion and runoff plot studies in New Mexico:

Over the grassland to shrubland transition, the connectivity of bare areas where runoff tends to be preferentially generated increases. Therefore, from the grass, grass-shrub, shrub-grass to shrub plots, flow lines become increasingly well connected, which increases the capacity for flow to entrain and transport sediment, leading to the greater sediment yields monitored over shrubland.

6.4 Deforestation

Forests (Figure 6.2) protect underlying soil from the direct effects of rainfall. The canopy shortens the distance over which raindrops fall, decreasing their velocity, and their kinetic energy.

Although there are examples of certain tree types in certain environments creating large raindrops, in general most canopies reduce the erosional effects of rainfall. Also important is the presence of humus in forest soils (Trimble, 1988), for this both absorbs the impact of raindrops and has an extremely high permeability. Forest soils also have many macropores produced by roots and their rich soil fauna. Thus they have high infiltration capacities. They also tend to be well aggregated, making them resistant to both wetting and water drop impact. This superior degree of aggregation results from the presence of considerable amounts of organic material, which is an important cementing agent in the formation of large water-stable aggregates. Furthermore, earthworms and other soil-dwelling organisms produce aggregates. Finally, deep-rooted trees, by increasing the total shear strength of the soils, help to stabilize steep slopes and reduce mass movements.

Figure 6.2 Luxuriant rainforest, as here in Sabah, Malaysia, provides an environment for low rates of erosion.

Figure 6.3 Badland and gully (donga) landscape developed in Swaziland as a result of forest removal. The association of some of these dongas with the remains of Iron Age smelters suggests that deforestation for fuel may have been the trigger for their development. Note the check dam composed of gabions.

It is therefore to be anticipated that with the removal of forest, rates of soil loss will rise, rock will be exposed to create "rocky desertification" (Yang et al., 2011), mass movements will increase in magnitude and frequency, and gullies will form in susceptible materials, as with the case of the *dongas* of southern Africa (Figure 6.3) and the *calanchi* of Italy (Cocco et al., 2015).

6.5 Soil Compaction by Agriculture, Vehicular Activity, and Ski Resorts

Soil compaction, which involves the compression of a mass of soil into a smaller volume, reduces the rate of water infiltration, which may change the soil moisture status and accelerate surface runoff and erosion (Hamza and Anderson, 2005; Nawaz et al., 2013). Plowing produces a compacted soil layer – the "plow sole" – at the base of the zone of plowing. This is because the normal action of the plow is to leave behind a loose surface layer and a dense subsoil where the soil aggregates have been pressed together by the sole of the plow. The compacting action can be especially injurious when the depth of plowing is both constant and long term, and when heavy machinery is used on wet ground. Indeed, excessive use of heavy agricultural machinery is perhaps the major cause of soil compaction. Infiltration capacities may also be modified by off-road vehicular and motorcycle movements (Crozier et al., 1978), by grazing animals, and by forestry operations (Goutal, et al., 2013). There is also evidence that intensive utilization of ski slopes can also produce soil compaction and associated erosion (de Jong et al., 2015).

6.6 Driving Forces

6.6.1 Deforestation and Afforestation

In areas with high rainfall intensities and soils with low infiltration capacities, moderately high rates of erosion may occur under forest (Zimmermann et al., 2012), but deforestation normally leads to increases in soil erosion because of its effects on soil properties and on runoff processes. This is made evident when one considers the results of plot studies that have been undertaken in Africa (Table 6.2) and in the humid tropics (Table 6.3). The latter table also shows the importance of the land-use treatments that follow deforestation. It also demonstrates that median rates of erosion on clean-weeded crop land are sixty times higher than where the tree crops have a mulch or cover crop, and 160 times higher than under natural forest. Labrière et al. (2015) compiled and analyzed more than 3,600 measurements of soil loss from twenty-one countries in the humid tropics. They established that the erosion rate in forests was ca. 1/10th and 1/150th than that of croplands and bare soils, respectively, while mean soil loss values for grasslands and shrublands were about half that of croplands.

The replacement of savanna woodland with cultivation is also prone to cause a jump in runoff and erosion rates. This was demonstrated in the *cerrado* of Brazil by Oliveira et al. (2015), where on average the runoff coefficient was ~20 percent for the plots under bare soil and less than 1 percent under native *cerrado* vegetation. The mean annual soil losses in the plots under bare soil and *cerrado* were 1,240 t km^{-2} yr^{-1} and 1.0 t km^{-2} yr^{-1}, respectively.

The replacement of forest with perennial plantation crops, such as oil palm, rubber, tea, cocoa, and coffee, may be associated with low rates of erosion, but this depends on the nature of the cover crop under the canopy, its management and maintenance, the density of planting, the exposure of access routes, and the time taken for a full canopy to develop (Hartemink, 2006). It has often been maintained that erosion rates may be high under teak plantations which is usually attributed to: (1) reduction in understory vegetation due to excessive light reduction and/or allelopathy; (2) low organic matter accumulation due to low litter production; and (3) increase in raindrop erosivity because the large leaves of the teak induce an increase in raindrop size. However, not all studies show this

Table 6.2 *Runoff and erosion under various covers of vegetation in parts of Africa.*

Locality	Average annual rainfall (mm)	Slope (%)	Annual runoff (%)			Erosion (t km^{-2} yr^{-1})		
			A	B	C	A	B	C
Ouagadougou (Burkina Faso)	850	0.5	2.5	2–32	40–60	10	60–80	1000–2000
Sefa (Senegal)	1,300	1.2	1.0	21.2	39.5	20	730	2,130
Bouake (Ivory Coast)	1,200	4.0	0.3	0.1–26	15–30	0.1	100–2600	1800–3000
Abidjan (Ivory Coast)	2,100	7.0	0.1	0.5–20	38	0.03	10–9000	10800–17000
Mpwapwa (Tanzania)	c.570	6.0	0.4	26.0	50.4	0	7,800	14,600

Note: A = forest or ungrazed thicket; B = crop; C = barren soil
Source: After various sources in Goudie (2013b).

Table 6.3 *Median rates of erosion (in tons km^{-2} yr^{-1}) under natural forest and different land use types in the humid tropics (modified from Walsh and Blake, 2009, table 8.4).*

Natural forest	30
Shifting cultivation: cropping phase	280
Shifting cultivation: fallow regrowth phase	20
Tree crops: clean weeded	4,800
Tree crops: with a cover crop or mulch	80
Multistory tea gardens	10
Plantations	60
Forest plantations: litter removed or burned	5,300
Forest plantations: young with agricultural intercropping	520

to be the case and many of the studies reporting high erosion rates were conducted in places where prescription fires are a common management practice or where management is poor (Fernández-Moya et al., 2014). Betel nut plantations have increased greatly in some Asian countries, including Taiwan, and it has been found that with crown height usually exceeding 10 m and wide spaces between trees, most betel nut plantations are less effective than forests in protecting soil from raindrop impact (Cheng et al., 2008). Eucalyptus has been widely planted to control erosion, but sometimes it appears to lead to greater fire frequencies and associated erosion.

The clearance of forest can lead to severe gully formation, soil stripping, colluviation, and alluviation, as occurred in the United States when agriculture was introduced to previously forested lands in the nineteenth century. This has been graphically described in the context of the Upper Mississippi Valley Hill Country by Trimble (2013).

The change in erosion rates following deforestation, however, depends on the harvesting methods used (Edeso et al., 1999) and the degree of clearance and disturbance that occurs. Mechanized harvesting may seriously damage forest soils, causing compaction, rutting, and accelerated runoff (Cambi et al., 2015). Conversely, the replacement of intensively cultivated land with forest or shrubland, a process called deintensification, may lead to a reduction in onsite soil erosion and sediment export to rivers and lakes (Bakker et al., 2008), unless soil protection structures such as terraces are abandoned and collapse in the process.

One interesting issue is the impact that temporary deforestation associated with shifting cultivation in the tropics has on rates of erosion. This ancient form of agriculture is currently practiced by 200–500 million people around the world.

Its two key features are the use of fire to prepare fields for cultivation and the subsequent abandonment of those fields as productivity declines as a result of the depletion of soil nutrients and the invasion of pests and weeds (Cornell and Miller, 2007). Once abandoned, however, fields are allowed to return to a more natural state as native plant and tree species reclaim them. If long intervals occur between clearance phases, slash and burn agriculture appears to cause less soil erosion than more intensive forms of agriculture (Ziegler et al., 2009). Douglas (1996) reported that in Borneo shifting cultivation does not increase sediment yields significantly, largely because so much organic debris is left on the ground. Data such as these suggest that governmental attempts to restrict shifting cultivation may be misguided (Lestrelin et al., 2012).

6.6.2 Tillage Erosion

One type of erosion associated with agriculture is "tillage erosion" (Baartman et al., 2012; Su et al., 2012; Logsdon, 2013; Meijer and Heitman, 2013). Tillage is responsible for the movement of soil material, particularly on slopes, and leads to a net soil loss from convex landscape positions and a net soil gain in concave landscape positions. Figure 6.4 shows a widely accepted model of long-term evolution of a complex slope profile as predicted by a simple simulation using a fixed soil transport rate by tillage operation (De Alba et al., 2004). This shows that tillage results in soil loss from shoulder and ridge top positions. It causes smoothing of the landscape and reduces slope angles by moving soil from convexities to concavities. Water erosion, on the other hand, is most severe at mid-slope positions, which are usually steeper and suffer from accelerated water movement from upslope. Field borders, fences, and vegetated strips that interrupt soil fluxes lead to the creation of topographic discontinuities or lynchets in many parts of the world (Vieira and Dabney, 2011; Nyssen et al., 2014). In Laos, weed invasion

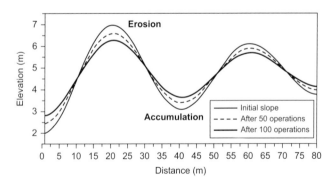

Figure 6.4 A model of slope evolution associated with tillage erosion (modified after De Alba et al., 2004, figure 1).

means that tillage has become a more important feature of cultivation in areas of shifting agriculture (Dupin et al., 2009), causing erosion rates to climb exponentially. Based on a global review, Van Oost et al. (2006) have indicated that erosion rates for mechanized agriculture are often of the same order of magnitude or larger than water erosion rates.

In England the land surface was altered by the development of ridge and furrow patterns created by tillage in the Middle Ages. Such patterns are a characteristic feature of many of the heavy soils of lowland England where, especially in the Midland lowlands, large areas are striped by long, narrow ridges of soil, lying more or less parallel to each other and usually arranged in blocks of approximately rectangular shape. They were formed by heavy plows pulled by teams of oxen; the precise mechanism has been described by Coones and Patten (1986, p. 154). Brown (1970) was of the opinion that the formation of such features was the most important geomorphological event in Buckinghamshire since the Pleistocene.

6.6.3 Overgrazing

About 40 percent of Earth's land surface is used for grazing by domestic stock, and overgrazing is a major cause of soil degradation (Evans, 1998). A distinction needs to be drawn, however, between overgrazing and sustainable grazing. The latter may enable a satisfactory vegetation cover to be maintained, which can control erosion rates (Kairis et al., 2015).

Domestic stock reduce the vegetation cover, and trample and compact the ground surface (thereby reducing infiltration capacities; Gifford and Hawkins, 1978), and physically dislodge particles so that they move down slope (Ries et al., 2014). Working in China, Zhou et al. (2010) found significant differences in soil physical properties in the semiarid grasslands of the northern Loess Plateau with different levels of grazing and trampling by goats and sheep. The soil in the ungrazed area had a significantly lower bulk density and significantly higher water content, proportion of stable aggregates, and infiltration rate than that in the grazed area. Recent increases in sheep numbers in parts of highland Britain have exacerbated the problem, especially in areas covered in peat (Evans, 1997). However, this is a problem in many parts of the world, as in East Africa (Murray-Rust, 1972), where it causes siltation of reservoirs, and in South Africa where it has probably contributed to the formation of erosional scars called *dongas* (Boardman, 2014). In Australia (Figure 6.5), Olley and Wasson (2003) found that since European settlement began 180 years ago, the sediment flux in the upper Murrumbidgee River has changed as a result of grazing, climate variations, and dam closures. They believed that of these, the introduction of grazing stock, which triggered widespread gully erosion, had the largest effect,

Figure 6.5 Exposed roots showing the degree of soil erosion caused by cattle in the West Kimberley area, Australia.

increasing the sediment flux by a factor of more than 150. The sediment flux out of this 10,500 km^2 catchment prior to settlement was estimated to have been ~2,400 t yr^{-1}. Degradation of the vegetated valley bottoms by introduced stock in the 1840s and 1850s triggered a massive phase of gully erosion, so that over the next forty to fifty years the erosion rates in the headwater areas increased by a factor of nearly 245, and ~43,000,000 t of sediment was generated. However, as the gully networks reached maximum extension, sediment yield from the headwater areas declined by a factor of ~40 and are now estimated to be about six times the pre-European rates.

In Colorado, United States, Carrara and Carroll (1979), used dendrogeomorphological techniques to show that erosion rates over the last 100 years have been about 1.8 mm yr^{-1}, whereas in the previous 300 years they were between 0.2 and 0.5 mm yr^{-1}, indicating an acceleration of about six-fold. This they attributed to the introduction of large numbers of cattle.

On the other hand, some assertions about the role of heavy grazing in promoting long-term land degradation in the Mediterranean lands are not necessarily based on sound data (Perevolotsky and Seligman, 1998), and Rowntree et al (2004) have suggested that the role of overgrazing in communal lands in parts of South Africa may have been exaggerated.

6.6.4 Urbanization and Erosion Rates

The ground-breaking studies of Wolman and Schick (1967) and Wolman (1967) have shown that the equivalent of many decades of natural or even agricultural erosion may take place during a single year in areas cleared for construction (Table 6.4). For instance, in Maryland, United States, they found that sediment yields during construction reached 55,000 t km^2 yr^{-1}, while in the same area rates under forests were only c. 80–200 t km^2 yr^{-1} and those under farms c. 400 t km^2 yr^{-1}. New road cuttings in Georgia were found to have sediment yields up to 20,000–50,000 t km^2 yr^{-1}. In Virginia, Vice et al. (1969) noted equally high rates of erosion during construction and reported that they were ten times those from agricultural land, 200 times those from grassland, and 2,000 times those from forest in the same area. Likewise, in Devon, England, Walling and Gregory (1970) found that suspended sediment concentrations in streams draining construction

Table 6.4 *Rates of erosion associated with construction and urbanization.*

Location	Land use	Source	Rate (t km^{-2} yr^{-1})
Maryland, United States	Forest	Wolman (1967)	39
	Agriculture		116–309
	Construction		38,610
	Urban		19–39
Virginia, United States	Forest	Vice et al. (1969)	9
	Grassland		94
	Cultivation		1,876
	Construction		18,764
Detroit, Michigan, United States	General nonurban	Thompson (1970)	642
	Construction		17,000
	Urban		741
Maryland, United States	Rural	Fox (1976)	22
	Construction		37
	Urban		337
Maryland, United States	Forest and grassland	Yorke and Herb (1978)	7–45
	Cultivated land		150–960
	Construction		1,600–22,400
	Urban		830
Wisconsin, United States	Agricultural	Daniel et al. (1979)	<1
	Construction		19.2
Tama New Town, Japan	Construction	Kadomura (1983)	c. 40,000
Okinawa, Japan	Construction	Kadomura (1983)	25,000–125,000

areas were two to ten times (occasionally up to 100 times) higher than those in undisturbed areas. Rates of sediment production may be especially high in humid tropical urban areas, where there is intense rainfall (Chin, 2006). It is therefore not surprising that great care needs to be exercised to minimize erosion from construction sites (Harbor, 1999). Even small rural villages can produce substantial effects. For example, De Meyer et al. (2011) have shown that in Uganda, village compounds and associated unpaved footpaths and roads generate sediment that is a significant source of pollution in neighboring Lake Victoria.

6.6.5 *Unpaved Roads, Footpaths, Erosion Rates, and Erosional Phenomena*

Construction of roads constitutes a damaging facet of forestry activities: the forest has to be cleared for them and they are thus a cause of deforestation. Unpaved forest roads and skid trails (Sidle et al., 2004) change soil properties and the hydrogeomorphic behavior of hillslopes (Reid and Dunne, 1984; Megahan et al., 2001; Arnáez et al., 2004). Roads increase the sediment yield, especially in tropical areas with intense rainfall, as a result of mass movements on steep embankments (Wemple et al., 2001; Cerdà, 2007; Sidle et al., 2014) or as a consequence of the direct impact of raindrops and turbulent runoff. The alteration of hillside profiles, with consequent disruption of surface and subsurface flows, the construction of cutslopes with steep gradients (Luce and Black, 1999), the lack of plant cover to protect the soil, the highly compacted surface of the roadbed (Croke et al, 2001), and the low infiltration capacities of unpaved road surfaces (Ziegler and Giambelluca, 1997) largely explain the variety and intensity of erosion processes.

High rates of sediment production occur after the construction of forest roads (Clarke and Walsh, 2006) when they are used for frequent transport of logs or when no upkeep is carried out. The connection of road ditches and culverts with stream networks facilitates the movement of runoff that quickly reaches channels (Wemple and Jones, 2003). Consequently, it may produce faster flow peaks and higher total discharges that can lead to gully formation (Nyssen et al., 2002). Forest roads often act as linearly connected systems, so that surface runoff flow may travel downslope toward the stream network. Roads are therefore potentially susceptible to hydraulic erosion processes, and may contribute substantially to stream sedimentation (Sheridan et al., 2006), even during low-magnitude rainfall events.

Not all accelerated erosion associated with roads and paths is related to forestry. In southern England, especially in areas with sandy lithologies, sunken lanes have developed as a consequence of long-term use (Boardman, 2013). Similar "hollow ways" are common in the Middle East around Bronze Age and Chalcolithic mounds (tells), where they form networks of roughly straight, shallow valleys that either radiate from the tells or, alternatively, trend across country, linking those sites as cross-country routes (Wilkinson et al., 2010). Footpaths used for recreation in highland areas such as the Lake District in the United Kingdom (Coleman, 1981), the Mourne Mountains in Ireland (Ferris et al., 1993), Norway (Gellatly et al., 1986), and the Drakensbergs in South Africa (Garland et al., 1985), are sources of erosion. Not surprisingly, the rates of this tend to increase with greater visitor pressures, but long-term studies of rates are few (Rodway-Dyer and Walling, 2010).

6.6.6 Fires and Soil Erosion

Fires may start naturally (Figure 6.6), especially in dry areas and in dry seasons, and it is often difficult to disentangle the effects of climatic changes from the effects of human actions (Pierce et al., 2004; Valese et al., 2014). Nevertheless,

Figure 6.6 A recently burned area in Mesa Verde National Park, Colorado, United States.

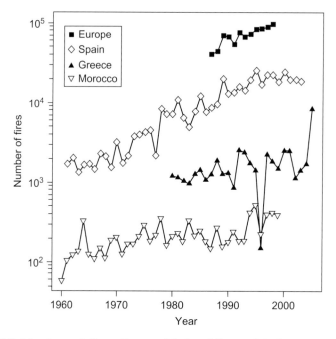

Figure 6.7 Number of fires (log scale) in different Mediterranean regions since 1960 (from Pausas, et al., 2009, figure 2).

many fires are started by humans, either deliberately or nondeliberately, and this is one of the longest and most powerful of human influences on the environment, dating back to the Paleolithic (see Chapter 2). Human influences may explain the observed increase in fires in the Mediterranean lands (Figure 6.7) in recent decades. Fox et al. (2015) report that of the 30,000–60,000 fires that occur there annually, more than 90 percent are of human origin. However, the effects of drier conditions related to global warming have been implicated in such upward trends in areas like the southwest United States. Increases in fire occurrence in the western United States have also been created because millions of hectares of rangeland have been invaded by annual and woody plants. These increase the role of wild-land fire (Pierson et al., 2011). The severity of burning can, paradoxically, be increased by fire suppression policies, which have been adopted widely in Europe and North America.

Because fires remove vegetation, cause loss of leaf litter, and expose the ground, they tend to increase erosion (Morris and Moses, 1987; Table 6.5). Overland flow is magnified in volume and speed both as a result of reduced rainfall interception and the increased percentage of rainfall available and also because of the reduction in surface roughness caused by the removal of the vegetation and litter. The resulting exposure of the soil leaves it prone to rain-splash detachment (Shakesby

Table 6.5 *Examples of fire-induced soil erosion studies.*

Source	Location
Shakesby (2011)	Mediterranean region
Smith and Dragovich (2008)	Eastern Australia
Moody and Martin (2001)	Colorado Front Range, United States
Hyde et al. (2014)	Montana and Idaho, United States
Pierson et al. (2009)	Idaho, United States
Martin et al. (2011)	Rockies, Canada
Warrick et al. (2013)	California, United States
Pierson et al. (2008)	Nevada, United States
Neris et al. (2013)	Canary Islands
Pardini et al. (2004)	Spain
Wilson (1999)	Tasmania, Australia
Inbar et al. (1998)	Israel
Spigl and Robichaud (2007)	Montana, United States

et al., 2007). Most, though not all, studies indicate that fires can lead to a decrease in soil resistance because of a reduction in soil organic content and in aggregate stability. If the fire does not succeed in destroying the organic-rich surface layer (sometimes called "duff"), then the effect on soil erosion may be limited, as in parts of the Canadian Rockies (Martin et al., 2011). The amount of ash produced in a fire is also significant, for an ash cover can dampen down surface erosion (Cerda and Doerr, 2008).

Watershed experiments in the chaparral scrub of the southwest United States have shown marked increases in runoff and erosion after burning (Warrick et al., 2013). This is because there is normally a distinctive "nonwettable" layer in the soils in which the chaparral grows. This is composed of soil particles coated by hydrophobic substances leached from the shrubs or their litter, and is normally associated with the upper part of the soil profile. It builds up through time in the unburned chaparral. The high temperatures which accompany chaparral fires cause these hydrophobic substances to be distilled so that they condense on lower soil layers. This process results in a shallow layer of wettable soil overlying a non-wettable layer. This condition, especially on steeply inclined slopes, can create the conditions for severe surface erosion (Figure 6.8). (De Bano, 2000; Shakesby et al., 2000; Letey, 2001; Ravi et al., 2009; Shakesby, 2011). Water repellency has also been found in the Rocky Mountains (Robichaud, 2000). However, its importance varies between different vegetation types so that observations made in American chaparral may not necessarily apply to Australian eucalyptus forest (Shakesby et al., 2007). Similarly, the time taken for forests to recover from a severe burn also

Figure 6.8 The Santa Monica Mountains near Los Angeles. A seasonally dry climate, makes it prone to fires that burn its resinous vegetation and cause severe erosion in denuded areas.

varies between different plant types, and eucalypts, for example, may quickly resprout, whereas conifers may rely upon seed germination to recover.

However, in general, burning can, especially in the first years after the fire event, lead to high rates of soil loss (Connaughton, 1935), though the controls are highly complex (Moody et al., 2013). Burned forests often have rates of soil erosion a whole order of magnitude higher than those of protected areas, though this depends on the severity of the fire (Lavee et al., 1995) and the intensity of subsequent rain storms (Prosser and Williams, 1998). The eroded sediment can rejuvenate alluvial fans and lead to the formation of stream terraces (Benda et al., 2003). Fires also generate debris flows and slides (Wondzell and King, 2003) and can trigger gully incision in mountain landscapes (Hyde et al., 2014).

6.6.7 Irrigation and Erosion

Although there has been little research, there is evidence that irrigation can lead to soil erosion. One reason for this is that irrigation water may cause the dispersion of clay minerals in soils, which changes their structure so that they become erodible and susceptible to piping (García-Ruiz et al., 1997). Another reason is that erosion may occur along irrigation furrows (Berg et al., 1980). Indeed, Koluvek et al. (1993) reported that of the 15 million hectares of irrigated land in the United States, 21 percent is affected by soil erosion to some extent and that it is often excessive on slopes greater than 2 percent. Irrigation-induced furrow erosion has also been reported from the Pamirs (Golosov et al., 2015). Water erosion can also be associated with center pivot systems and with sprinklers. With sprinkler irrigation, water drop energy detaches particles, some of which may be transported down-slope by shallow interrill flow if the water application rate exceeds the soil infiltration rate (Trout and Neibling, 1993).

6.6.8 Mediterranean Land Uses

Soil erosion has a long history in the Mediterranean lands. The classical authors, including Plato and Aristotle, were fully aware of the problem (see discussion in Montgomery, 2007, pp. 50–2). Casana (2008) found evidence for severe erosion in southern Turkey in Roman times, while Fuchs et al. (2004), working in southern Greece, used optical dating of colluvium to suggest that there were various phases of anthropogenic erosion and associated sedimentation: the Neolithic, the middle and late Bronze Age, Roman times, and since the sixteenth century AD. Changes in land use, such as the introduction of orchard crops, terracing, and cultivation, have had a profound and variable influence on the Mediterranean environment throughout the Holocene (Butzer, 2005). However, it has not always been possible to confirm that changes have resulted from human activities rather than as a result of climatic variability (e.g. Miller Rosen, 1997).

García-Ruiz (2010) has provided a full analysis of how land use changes in Spain have affected soil erosion rates. One change that has taken place in recent decades is the abandonment of farmland because of rural depopulation and the problems of mechanization on small packets of land on steep slopes. On the one hand, one would expect vegetation recolonization to cause erosion rates to be reduced, but on the other the lack of maintenance to field terraces on steep slopes can cause gully erosion (Koulouri and Giourga, 2007; Lesschen et al, 2007; Arnáez et al., 2015). Soil erosion is also severe in olive plantations, vineyards, and almond orchards, some of which have been subsidized by the European Union (Faulkner, 1995; Zuazo and Pleguezuelo, 2008).

Olive growing is often designated as one of the most critical land uses contributing to soil erosion (Figure 6.9). Farmers tend to till intensively to avoid competition of weeds with trees for water and nutrient uptake, since most orchards are rain fed. The combination of this human-induced low vegetation cover with the

Figure 6.9 Cultivation of olives, as here in the Douro Valley of Portugal, often involves a very limited vegetation cover between trees. This can lead to accelerated erosion, though terracing may reduce the problem.

steep slope gradients on which these orchards are located, and with the high-intensity rainfall events that characterize the Mediterranean climate, make the environment susceptible. Vanwalleghem et al. (2010) used the presence of mounds under olive trees to show that in their Spanish field area the average profile truncation varied between 0.33 and 1.18 m per site, with an overall average of 0.57 m. For all surveyed sites, the estimated age of the trees was between 55 and 100 years. However, many published rates are based on soil plot studies or models based on the Universal Soil Loss Equation (USLE), and some of these approaches may overestimate erosion rates. Also, rates depend greatly on land treatment. For example, on steeper slopes, terracing may be employed and this leads to lower erosion rates, as does the presence of a stone cover on many steeper slopes (Fleskens and Stroonijder, 2007). Nontillage cultivation, the use of cover crops, and vegetation strips can also dampen down erosion (Gomez et al. 2005; Hernandez et al., 2005; Francia Martinez, et al. 2006).

Vineyards are extensive in lower and middle latitudes and in 2009 covered a total of c. 7.7 million ha worldwide (Corti et al., 2011). However, there has been a

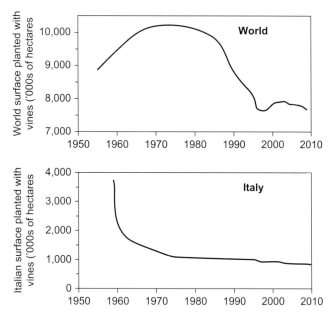

Figure 6.10 Trends in vine cultivation for the world (top) and for Italy (bottom). (based on data from, the World Viticulture Organization, in Corti et al, 2011).

decline in the areas planted to grapes since a peak in the 1970s, with a particularly sharp decline being evident in Italy (Figure 6.10). Vineyards are often reported as having high rates of erosion (Tropeano, 1984; Tarolli et al., 2015) because of rill and gully formation, landsliding, surface wash during high-intensity storms, and runoff from access roads. Different means of measurement give different rates of erosion in the same area, with rates from small field plots being larger than those from larger catchments (Raclot et al., 2009).

One novel means of estimating relatively long-term rates of denudation was undertaken by Casali et al. (2009). They found that in Navarre, Spain, grafting in vines was made until the 1990s directly in the field, and almost at soil surface level, so that the quantification of erosion/sedimentation rates around a single plant can be performed by using the large, identifiable callus forming around the graft as a paleosurface marker. They found that overall erosion rates in vineyards were c. 3,000 t km^{-2} yr^{-1} or 2 mm yr^{-1}, which, on average, greatly exceeded even the most conservative soil loss tolerance thresholds. Moreover, their results suggested that tillage erosion was the leading soil loss process in their vineyards rather than water erosion. A similar method was used in Haraut, France, by Paroissien et al. (2010), who found that the average soil loss was 1,050 t km^{-2} yr^{-1}. However, a compilation of rates from different areas by Quiquerez et al. (2008; see Table 6.6), showed that the variability between locations is high. Recent

Table 6.6 *Compilation of erosion rates in different vineyards in Europe (modified from various sources in Quiquerez et al., 2008).*

Location	Time scale (yr)	Mean soil loss (t km^{-2} yr^{-1})
Aisne, France	3	3,500
Aloxe-Corton, France	50	1,490
Alsace, France	> 5	3,200
Champagne, France	> 1	116
Champagne, France	3	150
Champagne, France	3	60
Douro, Portugal	10	39
Douro, Portugal	10	110–280
Languedoc, France	40	550
Monthélie, France	32	1350
Switzerland	> 1	1,500–2,500
Vosne-Romanée, France	54	760

developments in mechanization may have exacerbated erosion rates (Ramos and Martinez-Casasnovas, 2007). Field preparation for maximum mechanization has required intensive leveling in order to achieve larger fields with low to moderate slopes. In many cases, topsoil layers have been removed, mixed, or placed in deeper layers. The resulting soils have a low organic matter content, weak structural properties, and high susceptibility to sealing. These have negative effects on erosion processes. Conversely, cover crops can be effective at reducing erosion, and Ruiz-Colmonero et al. (2011) found that erosion plots under traditional tillage in central Spain yielded substantially more erosion (588 t km^{-2} yr^{-1}) than under two cover crops, *Brachypodium* (78 t km^{-2} yr^{-1}) or *Secale* (127 t km^{-2} yr^{-1}). The erosion rates found in Europe are similar to those found in the Napa Valley of California, United States (Battany and Grismer, 2000).

6.7 Case Study: Eroding Peat in Britain

Globally, peatlands cover around 400 million ha and store between one-third and one-half of the world's soil carbon (C) pool. Because of their long-term ability to sequester and store C as peat, they play a major role in global carbon cycling (see also Chapter 11). Additionally, peatlands provide other ecosystem services, including water retention and biogeochemical cycling. The formation of erosive gully systems is a major concern as it leads to drainage of the peat and is a large source of particulate organic carbon to peatland streams (Clay et al., 2012). Globally, Gallego-Sala and Prentice (2012) have argued that blanket bog regions are at risk of progressive peat erosion and vegetation changes as a direct consequence of climate change

Peat degradation is a problem in the highlands of Britain (Bragg and Tallis, 2001; Evans and Warburton, 2007), where, paradoxically, human-induced hydrological changes associated with deforestation in the Holocene may have contributed to the initiation of peat bogs and blanket mires (Moore, 1987). Over many areas, including the Pennines and the Brecon Beacons, blanket peats are being severely eroded to produce pool and hummock topography, areas of bare peat, and incised gullies (*haggs*), morphological details of which are given by Evans and Lindsay (2010). Many rivers draining such areas are discolored by the presence of eroded peat particles, and sediment yields of organic material are appreciable (Labadz et al., 1991). Bare areas can also generate higher amounts of runoff (Grayson et al., 2010) and so increase flood risk.

UK peats have been subject to a range of natural and anthropogenic disturbances over the last few hundreds of years, including climate change, sulfur deposition, drainage, prescribed burning, overgrazing, and wildfires (Radley, 1962; Parry et al., 2014).

Natural processes. Some of the observed peat erosion may be an essentially natural process, for the high water content and low cohesion of undrained peat masses make them inherently unstable. Moreover, the instability must normally become more pronounced as peat continues to accumulate, leading to bog slides and bursts round the margins of expanded peat blankets (Conway, 1954). Blanket bogs are sensitive to climate because their existence depends on a permanently high water table and because the characteristic *Sphagnum* moss rapidly suffers damage at temperatures greater than 15 °C (Moore, 2002). Higher temperatures will also lead to faster decomposition rates. In the future it is possible that peat bog erosion may be accelerated by a combination of worsening summer droughts and increasingly erosive winter rainstorms (Evans et al., 2006).

Pollution. Many blanket peatlands in the United Kingdom are located near heavily industrialized cities that have emitted large amounts of heavy metals, sulfur dioxide (SO_2) and nitrous oxides (NO_x) into the atmosphere. Much of this pollution has been deposited on blanket peatland surfaces. Peatland vegetation, particularly *Sphagnum*, is sensitive to this pollution which has thus been linked to the exposure of large areas of bare peat and the initiation of gullies. On the other hand, in Scotland lake core studies indicate that severe peat erosion was initiated between AD 1500 and 1700, prior to air pollution associated with industrial growth, and Stevenson et al. (1990) suggested that this erosion may have been initiated either by the adverse climatic conditions of the Little Ice Age or by a greater intensity of burning as land use pressures increased.

Peat cutting. Subsistence hand cutting of peat has provided fuel, but the mechanization of peat cutting in the 1980s fueled the growth of the horticultural and peat fuel industries, thereby causing a rapid expansion in the areas of peat being cut. Artificial drainage has been carried out on blanket peatlands to lower the water tables in preparation for peat cutting. This lowering of the water table is unfavorable for peat growth.

Grazing. Peat is susceptible to damage from sheep grazing, which can initiate and exacerbate erosion, result in vegetation change, and prevent the colonization of later successional vegetation species such as birch.

Burning. Prescribed rotational patch burning of heather is used as an ecological management tool to produce heather stands of different ages, and this can impact on the peat itself. Wildfires can also result in peat ignition and the exposure of large areas to erosion.

Draining and afforestation. To prepare peatlands for commercial forestry, narrowly spaced artificial drainage ditches are dug and fertilizer is applied. Planting is often very dense, which results in increased transpiration losses and interception rates. Since the 1950s, large areas of upland peat have been afforested in northern European countries. In the United Kingdom, between the 1950s and 1980s, forests were planted on about 500,000 ha of peatland.

Trampling and other disturbances. Installation of access tracks and buildings for forestry, windfarms, and estate management can disturb considerable quantities of peat and alter their ecological and hydrological characteristics. Other causes may include footpath erosion (Wishart and Warlenston, 2001) and human trampling (Robroek et al., 2010).

Restoration. Restoration of eroding UK peatlands has been a conservation concern for several decades. There has been a move to actively restore areas of peat using techniques such as the revegetation of gully walls and interfluves through stabilization with textiles and seeding (Shuttleworth et al., 2015). Blocking of drainage ditches is another practice used in blanket peatland restoration projects in the United Kingdom and Ireland (Wilson et al., 2011). To prevent further expansion of the gully network, gullies are often dammed from the most headward portion downward in order to trap sediment, slow down water flows, and raise local water tables. Reprofiling of the sides of gullies also occurs in many places in an attempt to reduce gully side slope steepness to reduce erosion rates.

6.8 Soil Erosion Rates: Introduction

Many techniques have evolved for the study of soil erosion rates, some of them relating to short time scales and small spatial scales (e.g. plot studies), some of them to medium spatial and temporal scales (e.g. reservoir sedimentation data [Rapp et al., 1972]), but others taking a longer or larger perspective (or both). With respect to the last category, one way of obtaining rates of soil erosion during the Holocene is to estimate rates of sedimentation on continental shelves. This method was employed by Milliman et al. (1987) to evaluate sediment removal down the Yellow River in China during the Holocene. They found that, because of accelerated erosion, rates of sediment accumulation on the shelf over the last 2,300 years have been ten times higher than those for the rest of the Holocene (i.e. since around 10,000 BP). In the Orange River basin of South Africa, work by Compton et al. (2010) using current sediment loads and the amount of sediment that has accumulated offshore have indicated that before the building of large dams there was a tenfold increase in the flux of sediment down the river compared to the mean Holocene flux because of land degradation associated with grazing and cultivation.

Long-term rates of erosion can also be obtained using cosmogenic radionuclides. A study in the Piedmont of the eastern United States by Reusser et al. (2015) showed that background, landscape-scale erosion rates, calculated from the concentration of ^{10}Be in river sediment, were much lower than rates of both soil erosion and sediment transport during peak land-use periods. During the early 1900s, aerially averaged rates of hillslope erosion exceeded ^{10}Be-derived background erosion rates by more than one-hundredfold.

Pimentel (1976) estimated that in the United States, soil erosion on agricultural land operates at an average rate of about 3,000 t km^{-2} yr^{-1}, which is approximately eight times quicker than topsoil is formed. He calculated that water runoff delivers around 4 billion tons of soil each year to the rivers of the forty-eight contiguous states, and that three-quarters of this comes from agricultural land. He estimated that another billion tons of soil is eroded by the wind, a process which created the Dust Bowl of the 1930s (see Chapter 8). More recently, Pimentel et al. (1995) argued that in the United States about 90 percent of cropland is losing soil above the sustainable rate, and that about 54 percent of pasture land is overgrazed and subject to high rates of erosion. Similarly, in Europe soil erosion rates exceed sustainable levels (Verheijen et al., 2009). However, as Boardman (1998), Trimble and Crosson (2000), and Verheijen et al. (2009) sagely point out, determination of general rates of soil erosion is fraught with uncertainties, with different techniques of measurement giving very different results.

Cerdan et al. (2010) evaluated available plot-based erosion data for Europe (Table 6.7). This confirmed the dominant influence of land use and cover on erosion rates. Sheet and rill erosion rates were highest on bare soil; vineyards showed the second highest soil losses, followed by other arable lands (spring crops, orchards, and winter crops). Land with a permanent vegetation cover (shrubs, grassland, and forest) was characterized by soil losses which were generally more than an order of magnitude lower than those on arable land. Disturbance of permanent vegetation by fire leads to momentarily higher erosion rates but rates are still lower than those measured on arable land. They also noticed important regional differences in erosion rates, which were generally much lower in the

Table 6.7 *Comparison of plot erosion rates (t km^{-2} yr^{-1}) for different land uses between the Mediterranean lands and the rest of Europe (simplified from Cerdan et al., 2010, table 5).*

Cover	Other regions in Europe	Mediterranean lands
Arable	633	84
Bare	1,712	905
Forest	0.3	18
Grassland	29	32
Orchard	2,060	167
Shrub	13	54
Vineyard	2,364	862

Table 6.8 *Examples of the evaluation of methods to control water erosion on slopes in drylands.*

Techniques	Source
Addition of fertilizer	Lasanta et al. (2000)
Afforestation	Romero-Diaz et al. (2010)
Bench terracing	Ternan et al. (1996)
Blade plowing and exclosure	Eldridge and Robson (1997)
Control of trampling by sheep	Eldridge (1998)
Crop residue management	Unger et al. (1991)
Engineering trenches	Marston and Dolan (1999)
Fallow cropping	Valentin et al. (2004)
Geotextiles	Rickson (2006)
Hedges	Poesen et al. (2003)
Juniper control	Belski (1996)
Mulching	Bautista et al. (1996)
No or minimum tillage	Kabakci et al. (1993)
Plant Strips	Martinéz Raya et al. (2006)
Rock mulches	Poesen et al. (1994)
Soil compaction	Poesen et al. (2003)
Synthetic polymers and biopolymers	Orts et al. (2007)
Trash and stone lines	Wakindiki and Ben Hur (2002),
Vegetation cover	Durán Zuazo et al. (2004)

Mediterranean as compared to other areas in Europe; this could mainly be attributed to the high soil stoniness in the Mediterranean area.

6.9 Erosion Management

Numerous techniques have been applied to try and reduce erosion (Morgan, 2005) (Table 6.8). These include the establishment of a good vegetation cover (e.g. by afforestation, fertilization, and the replacement of bush with grass), careful land management (e.g. suppression of fires, control of overgrazing, and trampling), modification of soil structures by the addition of gypsum and/or polymers (Graber et al., 2006), and control of slope runoff (e.g. by terracing, transverse hillside ditches, contour plowing, and vegetation strips). Since the Dust Bowl era, the modern concept of no-till farming – now generally known as Conservation Agriculture (CA) – has developed. This involves the simultaneous application of three interlinked principles (Kassam et al., 2014): (i) permanently minimizing or avoiding mechanical soil disturbance (no-till seeding); (ii) maintaining a continuous soil cover of organic mulch with plants (crop residue, stubble, and green manure/cover crops, including legumes); and (iii) growing diverse plant species in associations, sequences, or rotations.

Even in the face of increasing population growth, such techniques can be successful in restricting land degradation, as demonstrated by the important study undertaken in the Machakos region of Kenya (Tiffen et al., 1994). However, they are not invariably successful. For example, in semiarid Tunisia, where approximately one million ha of agricultural land have been installed with antierosive contour benches, Baccari et al. (2008) have shown that the life of the benches has proved to be limited, with those on gypsum clays being especially prone to failure. It was also found that soil conservation was in part achieved at the cost of a reduction in runoff and in available water resources for irrigation. Indeed, soil erosion countermeasures can be counterproductive, as has been made evident by reviews of large-scale afforestation schemes in China (Cao et al., 2010) and Spain (Romero-Diaz et al., 2010). This latter study showed that aggressive land sculpting and the subsequent planting of inappropriate trees on scrublands, actually accelerated rates of erosion on marl slopes by between 1 and 2 orders of magnitude. Bulldozing can be a curse in Mediterranean environments. One also needs to be aware that some species of exotic plant that have been introduced for control of water erosion can prove to be highly invasive, as with mesquite, *Prosopis juliflora*, in Kenya (Muturi et al., 2009), with potential knock-on effects on geomorphology and ecology.

6.10 Lake Sedimentation Rates

Changes in erosion rates brought about in catchments as a result of land cover changes have been shown to have had an impact on sedimentation accumulation rates (SARs) in lakes from the Neolithic onward (Edwards and Whittington, 2001; Tylmann, 2005). A core from Llangorse Lake in the Brecon Beacons of Wales (Jones et al., 1985) provides excellent long-term data on changing SARs (Table 6.9):

Table 6.9 *Holocene sedimentation rates from*
Llangorse Lake, Wales, United Kingdom
(from data in Jones et al., 1985)

Period	Sedimentation rate (cm 100 yr^{-1})
9000–7500 BP	3.5
7500–5000 BP	1.0
5000–2800 BP	13.2
2800–AD 1840	14.1
c. AD 1840–present	59.0

Table 6.10 *Data on rates of erosion and sedimentation in Loe Pool,*
Cornwall, United Kingdom, since 1860 (from O'Sullivan et al., 1982)

Dates	Activity	Rates of erosion in catchment (R. Cober) as determined from lake sedimentation rates (t km^{-2} yr^{-1})
1860–1920	Mining and agriculture	174
1930–6	Intensive mining and agriculture	421
1937–8	Intensive mining and agriculture	361
1938–81	Agriculture	12

The thirteenfold increase in rates after 5000 BP seems to have occurred rapidly and is attributed to initial forest clearance. The second dramatic increase of more than fourfold, which has taken place since c. 1840, is a result of agricultural intensification.

In the last two centuries, SARs in British lake basins have changed in different ways in different basins according to the differing nature of economic activities in catchments. In the case of the Loe Pool (Cornwall), SARs were high while the mining industry was active, but fell dramatically when it was curtailed (Table 6.10). At Seeswood Pool (Warwickshire; Table 6.11), which has a dominantly agricultural catchment area, the highest SARs have occurred since 1978 in response to land management changes, including larger fields, continuous cropping, and increased dairy herd size. In Cumbria, cores from Blelham Tarn showed an exponential increase in SAR since 1950 (Van der Post et al., 1997) and this was attributed to increased stocking densities of sheep in its catchment. In other catchments, preafforestation plowing may have caused sufficient disturbance to cause accelerated sedimentation. For example, Battarbee et al. (1985) looked at

Table 6.11 *Data on sediment yield from catchment of Seeswood Pool, Warwickshire, United Kingdom (from data in Foster et al., 1986).*

Dates (AD)	Sediment yield $(t \ km^{-2} \ yr^{-1})$
1765–1853	7.0
1854–80	12.2
1881–1902	8.1
1903–19	9.6
1920–5	21.6
1926–33	16.1
1934–47	12.7
1948–64	12.0
1965–72	13.9
1973–7	18.3
1978–82	36.2

sediment cores in the Galloway area of southwest Scotland, and found that in Loch Grannoch the introduction of plowing in the catchment caused an increase in SAR from 0.2 cm yr^{-1} to 2.2 cm yr^{-1}.

Rose et al. (2011) compiled SARs for 207 European lakes derived from ^{210}Pb-dated cores to assess how rates have changed through time. They found that 71 percent of these cores showed surface SARs higher than "basal" (mainly nineteenth century) rates, 11 percent showed no change while 18 percent showed a decline. Little change in SAR occurred prior to 1900 and most increases occurred in more recent periods, in particular 1950–1975 and post-1975. They argued that while increases in SAR in many lowland areas may be due to changes in agriculture, the same could not be said of upland and mountain lake-types where, apart from possible changes in rough grazing pressures, land use has remained largely unchanged over the period covered by the chronologies considered here. One proposed explanation for SAR increases in these lakes is climate warming which could elevate both nitrogen and phosphorus inputs to lakes, lengthen periods of biological productivity, and hence elevate autochthonous inputs to basins.

Schiefer et al. (2013) compiled similar data for lakes in western Canada. They found that the median change in SAR in recent decades has been about 50 percent greater than background (first half of twentieth century), and they attribute this in many cases to forestry activities, including the construction of forestry roads.

In Minnesota and Wisconsin, United States, Mulla and Sekely (2009) and Engstrom et al. (2009) have demonstrated significant increases in the accumulation of sediment and phosphorus in Lake Pepin over the last 140 years (Figure 6.11). They showed that the SAR has increased almost six-fold from

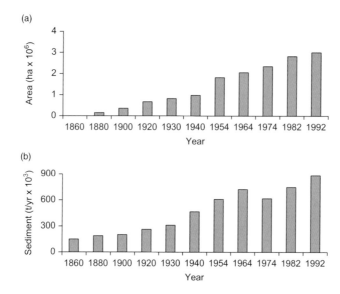

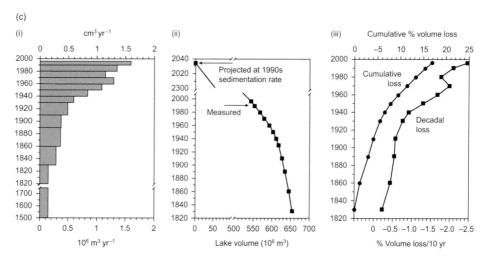

Figure 6.11 Changes in Lake Pepin since 1860. Top: Historical changes in the area of land under row cropping in selected years for the Upper Mississippi, Minnesota, and St. Croix River basins. Middle: Historical changes in decadal average sediment accumulations in Lake Pepin (from Mulla and Sekel, 2009, figures 2 and 8). Bottom: Linear and volumetric sedimentation rates for Lake Pepin (i), projected lake volume from 1990 datum (ii), and percent volume loss relative to 1830 datum (iii) (from Engstrom et al., 2011, figure 14).

151,625 metric tons yr^{-1} in 1860 to present day rates of 875,576 tons yr^{-1}. However, the rate of infilling has clearly accelerated in the 170 years since European settlement, greatly shortening the projected life of the lake. Today about 2.5 percent of the 1830 lake volume is lost every decade, while just prior to

European settlement the decadal loss was 0.23 percent. These studies found strong evidence relating historical increases in sediment accumulation in Lake Pepin to increases in the areal extent of agricultural row crop production systems and increases in river flows, especially from the Minnesota River basin. During pre-European settlement, rates of upland erosion were negligible in the Minnesota, Upper Mississippi, and St. Croix River basins because the landscape was covered with prairie grass and forest. In recent times, 92 percent of the Minnesota River basin has been converted to row crop agriculture, which generates moderate to high rates of soil erosion when practised on steep slopes.

Another lacustrine environment in the United States that has been modified as a result of land use change is the small closed playa, which is so characteristic of the Great Plains. Intense cultivation has resulted in the loss or degradation of approximately 95 percent of playas in the southern High Plains (Daniel et al., 2015).

Exponential growth in SARs since the 1980s as a result of erosion have also been found in Brazil (Godoy et al., 2002). In Madagascar, the island's largest lake, Alaotra, has almost disappeared because of siltation caused by rapid soil erosion in its catchment (Bakoariniaina, et al. 2006), while in the East African Rift, due to sedimentation the depth of Lake Baringo declined from an average 8 m in 1972 to the current 3 m (Lwenya and Yongo, 2010).

In China, accelerating human activities during the past three centuries means that the sediment load entering Poyang Lake, Jiangxi Province, has undergone drastic changes (Gao et al., 2015). Because of intensified soil loss, the total sediment load entering this lake during AD 1800–1950 increased by 58.7 percent, compared with AD 1000–1700. However, after AD 1950, the sediment load variation is mainly impacted upon by the combined influence of dam emplacement and soil erosion. As sediment interception by dams has increased, the total sediment load entering Poyang Lake during AD 1990–2000 has been only 60.9 percent of that of the highest riverine sediment flux during AD 1951–1980, which is almost equal to that of the lowest level during AD 1000–1700. Similarly, working in Qionghai Lake watershed, located on the southeastern edge of the Qinghai–Tibetan Plateau, Chen et al. (2015) established that the mean denudation rate of the watershed was about 0.82 mm per year during the Holocene, but that since 1952 the rate has increased to 1.82 mm per year, accompanied by a greatly increased sedimentation rate.

The work of Binford et al. (1987) on the lakes of the Peten region of Guatemala, an area of tropical lowland dry forest, is instructive with respect to the effects of early agricultural colonization. Combining studies of archaeology and lake sediment stratigraphy, they were able to reconstruct the diverse environmental consequences of the Mayan civilization. The period of Mayan success saw a marked reduction in vegetation cover, more catchment soil erosion, an increase in lake

SARs and the supply of inorganic silts and clays to the lakes, a pulse of phosphorus derived from human wastes, and a decrease in lacustrine productivity caused by high levels of turbidity. In parts of Guatemala, a spasm of soil erosion occurred during the Maya Late Classic phase (AD 550–830), and here depressions record 1–3 m of aggradation in two centuries (Beach et al., 2006). Beach et al. (2015) refer drolly to the "Mayacene," a microcosm of the Early Anthropocene that occurred from c. 3000 to 1000 BP. They synthesized the evidence for Mayan impacts on climate, vegetation, hydrology, and the lithosphere, from studies of soils, lakes, floodplains, wetlands, and other ecosystems. The Mayans altered ecosystems with vast urban and rural infrastructure that included thousands of reservoirs, wetland fields and canals, terraces, field ridges, and temples.

The impact of land-use changes since European colonization in southeastern Australia has been discussed by Sloss et al. (2011). They found that SARs over the last 100 years within two lakes, Illawarra and St. Georges Basin, showed the effects of catchment land-use change and native vegetation clearance. Both catchments had similar lake and catchment areas but have experienced different degrees of modification. In the heavily modified catchment of Lake Illawarra sedimentation rates close to fluvial deltas can exceed 16 mm per year, and be between 2 and 4 mm per year in the adjacent central basin. This is approximately an order of magnitude greater than the pre-European rates. In contrast, at St. Georges Basin, where the catchment has experienced much less modification, SARs in the central basin appear to have remained close to those prior to European settlement.

Just as many rivers are currently delivering reduced amounts of sediment to the oceans, because of dam and reservoir construction (see Chapter 7), so there are examples of reduced sediment inputs to lakes. In the case of Lake Baikal, sediment inputs since the mid-1970s have declined by 49–82 percent (Potemkina and Potemkin, 2014). This is attributed partly to climatic change altering vegetation cover and thus erosion rates in the headwater streams that feed the lake, and partly due to anthropogenic changes in the catchment area, including land-use change.

Other anthropogenic processes can also influence SARs. In particular, elevated nutrient inputs associated with pollution can increase the productivity of organisms, and as they die this can lead to accelerated accumulation on the lake floor. However, erosion-derived nutrient inputs can also cause a change in organic productivity. This was demonstrated for the oligotrophic Fayetteville Green Lake, New York, United States, where in the 1800s deforestation led to erosion which led to nutrient inputs and higher rates of organic deposition. The SAR increased sevenfold (Hilfinger et al., 2001).

Sometimes the threat of excessive sedimentation in lakes has been exaggerated. This was demonstrated for the Tonle Sap Lake in Cambodia (a World Heritage Site and the largest freshwater lake in Southeast Asia) by Kummu et al. (2008). Using long-term core-derived data, they were able to show that current rates were not abnormally high and were considerably lower than those in the early Holocene and that therefore the lake was not at risk from this particular threat.

6.11 Acceleration of Mass Movements

There are many examples of humans accelerating mass movements by vegetation removal, loading of debris onto slopes, changing groundwater conditions, excavating the bases of slopes, and cutting away the toes of debris flows.

Deforestation is hugely important, especially in areas with intense rainfall events and steep relief. Forests help to prevent shallow landslides in two ways: (1) by modifying the soil moisture regime via evapotranspiration; and (2) by providing root reinforcement within the soil mantle or regolith. In shallow soils, tree roots may penetrate the entire soil mantle and anchor the soil into a more stable substrate. Dense lateral root systems in the upper horizons stabilize the soil, while larger tree roots provide reinforcement across planes of weakness along the flanks of potential slope failures. Although woody roots significantly reduce shallow ($<$1–2 m deep) landslide potential on steep slopes, deeper soil mantles ($>$5 m) benefit much less from such reinforcement, as root density decreases dramatically with depth and few large roots are able to anchor across the basal failure plane. The only benefit of root strength to the stabilization of deep-seated landslides is when very large lateral woody roots cross planes of weakness along the flanks of potential failures (Stokes et al., 2009). Forests are also important in controlling rock-falls. In some cases, initiation of rock-falls is increased by roots as they may promote detachment while penetrating cracks and fissures in the rock. Rocks may be slowly forced apart by root growth thereby forming a pathway for water to further enter the rock mass. Conversely, however, roots have also been found to reduce rockfall activity by binding rock particles together (Sakals et al., 2006).

Some examples of the effects of deforestation on landslides are listed in Table 6.12. In New Zealand, the clearance of indigenous forest caused actively eroding gullies to develop, particularly from the 1880s to the 1920s (Marden et al., 2012). It also led to a surge in landslide activity (Glade, 2003). In Japan, Imaizumi et al. (2008) found that landslide volumes in forested areas that had been clear-cut zero to twenty-five years previously, were fourfold those that had been clear-cut more than twenty-five years previously.

Table 6.13 provides data on landslide erosion rates for uncut and clear-cut forests from unstable, mountainous terrain in the Pacific Northwest of the United States. The average rate for uncut forests is 63 t km^{-2} yr^{-1} while that of clear-cut forests is 239 t km^{-2} yr^{-1}, an increase of c. 3.8 times. The logging of old-growth

Table 6.12 *Examples of landsliding associated with deforestation.*

Location	Source
Andes	Guns and Vanacker (2014)
Costa Rica	Miller (2011)
Ethiopia	Broothaerts et al. (2012)
Japan	Imaizumi et al. (2008)
Mount Elgon, Uganda	Muggaga et al. (2013)
New Zealand	Marden et al. (2012)
Nilgiris, India	Kumar and Bhagavanulu (2008)
Pyrenees, Spain	García-Ruiz et al. (2010)
Sierra Norte, Mexico	Alcántara-Ayala et al. (2006)
Taiwan, Western Foothills	Chang and Slaymaker (2002)

Table 6.13 *Landslide erosion rates (in t km^{-2} yr^{-1}) from temperate forests in the Pacific Northwest, North America (modified from Sidle and Burt, 2009, table 12.2).*

Area	Uncut	Clear-cut
Capilano River basin, British Columbia	130	590
Chicagof and Prince of Wales Islands, Alaska	110	200
Coastal Southwest British Columbia	14	32
Idaho Batholith, Zena Creek	7	–
Klamath Mountains, Oregon	56	370
Olympic Peninsula, Washington	70	0.0
Oregon Cascades, Blue River	50	420
Oregon Cascades, H. J. Andrews Experimental Forest	110	320
Oregon Cascades, Maple Creek	60	150
Oregon Coast Ranges, Mapleton area	19	700
Average	63	239

forests from California to Alaska took place in an environment with a combination of steep slopes, fractured bedrock, cohesionless colluvium, high rainfall, and snowmelt. Mass movements ensued, but as Orme (2002, p. 467) has cautioned, it is important to maintain a sense of perspective, for the impacts of logging tend to diminish over time as new forests arise from seed and stump sprouting. Moreover, many such regions have long been prone to mass movements, resulting from heavy rains and rapid snowmelt, especially after natural fires and wind-throws.

Road construction can destabilize slopes, especially in mountainous regions (Figure 6.12). Sidle et al. (2011) suggested that landslides occurred because roads excavated into steep mountain slopes create instability by the following means: (1) undercutting steep slopes, thus removing support; (2) over-loading and over-steepening fill slopes; and (3) altering natural hydrologic pathways

Figure 6.12 The road down the Nile Gorge in central Ethiopia has generated and become subject to landslides.

and concentrating water onto unstable portions of the hillslope. In southeastern France, humans have accelerated landslide activity by building excavations for roads and by loading slopes with construction material (Julian and Anthony, 1996). The undercutting and removal of the trees of slopes for the construction of roads and paths has also led to landsliding in the Himalayas (Barnard et al., 2001). This is also the case in mountainous Nepal, where the number of fatal landslides shot up during the 1990s (Petley et al., 2007; Figure 6.13). This seems to be correlated with the rapid development of the road network after about 1990. Similarly, landslides that were triggered by a great earthquake in Kashmir in 2005 occurred preferentially in areas where road construction had taken place (Owen et al., 2008). Sidle et al. (2011) found that in Yunnan, South China, landslide erosion (including dry ravel) exceeded 3,300,000 t km^{-2} yr^{-1} along the most severely eroded sections of a road and averaged more than 960,000 t km^{-2} yr^{-1} along a surveyed 23.5 km of road. They stated that these values are the highest ever reported for road-related landslides. However, many other landslides in China appear to have been triggered by surface excavation and mining activity (Huang and Chan, 2004).

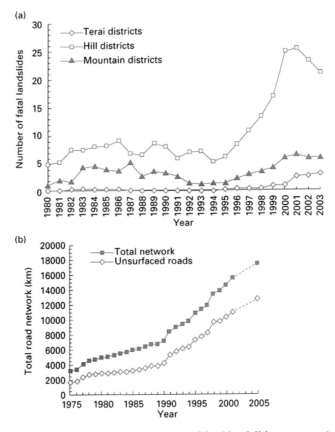

Figure 6.13 Landslides in Nepal. (a) Number of fatal landslides per year (smoothed by five-year running mean), and (b) the growth of the road network (modified from Petley et al., 2007, figures 12b and 13b, in Goudie, 2013, fig. 6.16).

The construction of dams and the impoundment of reservoirs change groundwater conditions, which can impact on slope stability (Schuster, 1979). Examples include the disastrous Vaiont Dam landslide which occurred in Italy in 1963 (Genevois and Ghirotti, 2005). Heavy antecedent rainfall and the presence of young, highly folded sedimentary rocks provided the necessary conditions for a slip to take place, but it was the construction of the dam itself which changed groundwater conditions sufficiently to affect the stability of a rock mass on the margins of the reservoir: 240 million m^3 of ground slipped, causing a rise in water level which overtopped the dam and caused flooding and loss of life downstream. Two-thousand six hundred people were killed. Slope instability also resulted when the Franklin D. Roosevelt Lake was impounded by the Columbia River in the United States (Coates, 1977). More recently, concerns have been expressed about the development of slope failures around the reservoir impounded by the Three Gorges Dam in China (Wang et al., 2008).

In northern Spain, changes in landslide occurrence in the Upper Pleistocene and Holocene can partly be attributed to periods of intense precipitation or seismic activity, whereas the sharp increase in landslide numbers since the early 1980s appears to be related to increasing human pressures (Remondo et al., 2005). In an alpine area in Switzerland, where the area affected by landslides increased by 92 percent from 1959 to 2004, Meusberger and Alewell (2008) indicated that the causes were a combination of an increase in the occurrence of torrential rain events and an increase in cattle stocking. However, on a longer time span of 3,600 years, pulses of increased landslide activity in the Swiss Alps have been linked to phases of deforestation (Dapples et al., 2002). In northwest England, Chiverrell et al. (2007) related phases of gullying, alluvial fan formation, and valley slope deposition to land use pressures in Iron Age/Roman and medieval times, while lichenometric studies by Innes (1983) suggested that there had been an acceleration in debris flow activity in the Scottish Highlands in the last two centuries. The timing of the increase suggests that it was related to burning and overgrazing.

Fire, whether natural or man-induced, can be a major cause of slope instability and debris flow generation: by removing or reducing protective vegetation, increasing peak stream flows, and creating larger soil moisture contents and soil-water pore pressures (because of reduced interception of rainfall and decreased moisture loss by transpiration). Examples of fire-related debris flow generation are known from sites in the United States, including Colorado (Cannon et al., 2001 a), New Mexico (Cannon et al., 2001b), the Rocky Mountains, and the Pacific Northwest (Wondzell and King, 2003).

6.12 Mass Movement Management

Humans have developed techniques to try and control mass movements (Table 6.14).

Methods for landslide mitigation can be classified into three main types: (1) geometric methods, in which the slope of the hillside is changed; (2) hydrogeological methods, in which an attempt is made to lower the groundwater level or to reduce the water content of the material; and (3) chemical and mechanical methods to increase the shear strength of the unstable mass, and to counter destabilizing forces (e.g. anchors, rock or ground nailing, structural wells, piles, geotextiles, and application of shotcrete).

Figure 6.14 shows the Pacific Palisades area of Los Angeles. Here the Pacific Coastal Highway has been constructed on a narrow piece of land between the steep coastal cliffs and the ocean. This has involved some steepening of the base of

Table 6.14 *Methods used to control mass movements on slopes.*

Type of movement	Method of control
Falls	Benching the slope
	Covering of slopes with steel mesh, geotextiles, etc.
	Drainage
	Flattening the slope
	Reinforcement of rock walls by grouting with cement, anchor bolts, etc.
Slides and flows	Drainage of surface water with ditches
	Grading or benching to flatten the slope
	Installation of piles through the potential slide mass
	Retaining walls at the foot of the slope
	Rock and earth buttresses at the foot of the slope
	Sealing of surface cracks to prevent infiltration
	Subsurface drainage

Figure 6.14 Examples of slope stabilization techniques at Pacific Palisades, Los Angeles, California, United States. The top image is at the Villa de Leon, and the lower image is at Pacific Palisades just to the northwest of Sunset Boulevard.

Figure 6.14 (*cont.*)

the slope. In addition, large numbers of buildings have been built on the slopes. Given that the cliffs are composed of largely unconsolidated and friable siltstones and sandstones of Tertiary age, one has all the ingredients for slope collapse. To deal with this, the engineers have had to employ hard engineering methods, including deep piles, rock anchors and bolts, steel sheeting, and the emplacement of geotextiles.

6.13 Future Rates of Soil Erosion and Mass Movements

6.13.1 Rainfall Intensity

Rainfall intensity is a major factor controlling rates of soil erosion and mass movements (Sidle and Dhakal, 2002), and many studies have explored the role of past climatic fluctuations in controlling rates of erosion and of gully formation (see, e.g., Lyons et al. (2013)).

Under increased greenhouse gas concentrations some General Circulation Models (GCMs) exhibit enhanced mid-latitude and global precipitation intensity and shortened return periods of extreme events (New et al. 2001). Over recent warming decades, heavy precipitation events appear to have been intensified over large parts of Northern Hemisphere land areas (Min et al., 2011). Higher intensity storms increase the risk of soil erosion (Hoomehr et al., 2015).

6.13.2 Increasing Fire Activity

Sediment delivery by rivers may be impacted by more fire activity in a warmer and drier environment (Tang et al., 2015), and this could cause an increase in suspended loads as slopes are subjected to greater erosion and debris flow generation (Goode et al., 2012). Forest fire frequency in Mediterranean countries is generally expected to increase with land cover and changes in climate as temperatures rise and rainfall patterns are altered. Although the cause of many Mediterranean fires remains poorly defined, most fires are of anthropogenic origin and are located in the wildland urban interface (WUI), so fire ignition risk depends on both weather and land-cover characteristics. In coming years, changes in fire-fighting strategy and in technical support in the form of improved radio communication and helicopters may contribute to reducing fire frequency and burned area (Fox et al., 2015).

6.13.3 Modeled Future Soil Erosion Rates

Geomorphologists have started to model changes in soil erosion that may occur as a consequence of changes in rainfall amounts and intensity (Nearing, 2001; Yang et al. 2003; Mullan et al., 2013), though it is difficult to determine the likely effects of climate change compared to future land-use management practices (Wilby et al., 1997). In addition, wholesale changes in vegetation may result from future changes in fire frequency and the length of fire seasons (Westerling et al., 2011; Flannigan et al., 2013) and this may also affect runoff and soil erosion.

Sun et al. (2002) modeled runoff erosivity changes for China and produced a map of rainfall erosivity for 2061–99. They suggested that for China as a whole, assuming current land cover and land-management conditions, soil erosion rates will increase by 37–93 percent. For Brazil, the model of Favis-Mortlock and Guerra (1999) indicated that by 2050 the increase in mean annual sediment yield in the Mato Grosso would be 27 percent. For the southeast of the United Kingdom, where winter rainfall is predicted to increase modestly, Favis-Mortlock and Boardman (1995) recognized that changes in rainfall not only impacted upon erosion rates directly, but also through their effects on rates of crop growth and on soil properties. Nonetheless, they showed that erosion rates were likely to rise, particularly in wet years.

In the USA, after analyzing a range of different models, Nearing et al. (2005, p. 151) estimated that if the increase in rainfall would be on the order of 10 percent, with greater than 50 percent of that increase due to an increase in storm intensity, and if no changes in land cover occurred, erosion could increase by something on the order of 25–55 percent over the next century. More recently, modeling by Hoomehr et al. (2015) for Tennessee, confirmed these predictions, suggesting a 7–49 percent increase in annual rainfall erosivity from 2010 to 2099.

It is, of course, likely that climate change will cause farmers to change the ways in which they manage their crops and, in particular, to change the crops they plant in any particular area. This, in turn, may alter runoff and erosion. O'Neal et al. (2005) suggested, for example, that in the Midwest of the United States there might be a shift, because of price and yield advantages, from maize and wheat to soybean cultivation. Rates of runoff and erosion under soybeans would tend to be substantially greater than under the crops they replace.

6.13.4 Future Slope Instability

Slope stability will be impacted upon by climate change in a multitude of different ways (Crozier, 2010; Table 6.15).

Increases in rainfall could be one driving force. Collison et al. (2000) modeled the impact of climate change on an escarpment in southeast England. They found that because increases in rainfall would largely be matched by increases in evapotranspiration, the frequency of large landslides would be unchanged over the next eighty years. They argued that other factors, such as land-use change and human activity would be likely to have a greater impact than climate change.

Table 6.15 *Potential responses of slopes to climate change (partly based on Crozier, 2010, table 1).*

Increase in precipitation and precipitation intensity
Higher water tables leading to reduction in shear strength
Increased weight (surcharge) promoting a reduction in shear strength Reduction in friction
Increased river bank scour and undermining of slopes
Increase in temperature
Reduction in interstitial ice and permafrost leading to reduced cohesion
Rapid snow melt leading to build up of pore water pressure and strength reduction
Reduction in glacier volume causing reduced buttressing of valley side slopes
Thickening of active layer and promotion of active layer detachments
Higher sea levels leading to coastal slope failures

In lower latitudes, a possible increase in hurricane (typhoon) activity could cause an increase in debris flows in places like Taiwan (Chiang and Chang, 2011).

The situation with regard to predictions for the mountains of Europe is confused, with some workers suggesting that there will be an increase in frequency of mass movements, and others suggesting the opposite. Because of differences between GCMs and problems of downscaling, there are still great problems in modeling future landslide activity (Dehn and Buma, 1999). Also, slope stability is affected by a number of different driving forces.

One of these is the nature of precipitation. In the Italian Dolomites, Dehn et al. (2000) suggested that future landslide activity would be reduced because there would be less storage of precipitation as snow. Therefore, the release of meltwater, which under present conditions contributes to high groundwater levels and strong landslide displacement in early spring, would be significantly diminished. In the Swiss Alps, there is evidence that debris flow activity has declined since the Little Ice Age, and if summer precipitation events occur less frequently in the future, their incidence could decline still further (Stoffel and Beniston, 2006). Comparable predictions have been made for the French Alps (Jomelli et al., 2009), where it is proposed that debris flow frequency could be reduced by 30 percent in the late twenty-first century. On the other hand, analysis of mass movement failures in high mountains during recent warm decades by Huggel et al. (2012) has provided no statistically rigorous evidence of a trend in activity.

A second driving force is active layer thickness (ALT). In recent warming decades there has been a widespread, though not universal, increase in this as permafrost has melted (IPCC, 2013, p.364). For example, Sobota and Nowak (2014) investigated changes in the active layer of the high Arctic coastal area in northwest Spitsbergen (Svalbard). The study was conducted on the Kaffiøyra Plain. In 1996–2012 on a beach and on a moraine the trend of annual ALT increase was 1.3 cm and 2.5 cm, respectively. Wu et al. (2015), working in the Tibetan Plateau, showed an increase in ALT between 2002 and 2012 that averaged ~4.26 cm per year, while Peng et al. (2015), working along the Qinghai-Tibet highway found that between 1995 and 2011 the ALT increased about 0.44 m, with an average increasing annual rate of 3.42 cm in cold permafrost regions, while in warm permafrost regions the ALT increased about 0.68 m, with an average increasing annual rate of 5.72 cm. These trends are likely to persist in coming decades, so that surface soils, sediments and rock masses will become more mobile (Arenson and Jakob, 2015; Blais-Stevens et al., 2015; Ravanel and Deline, 2015; Shan, 2015). Mass movements of soil and surface sediment occur along the base of the active layer. These slides, called active layer detachments (ALDs), take place when saturated overburden slides over the frozen substrate, and result in the downslope travel of material up to hundreds of meters over low slope angles

(Lamoureux and Lafrenière, 2009). Rockfall activity has already increased in the Mont Blanc area as a consequence of this (Ravanel and Deline, 2011).

The disappearance of glaciers (see Chapter 10) is another mechanism that can cause slope movements by reducing the stability of glacially undercut and over-steepened slopes (Holm et al., 2004; Vilímek et al. 2005). This generates a risk of increased landsliding and debris avalanches (Kirkbride and Warren, 1999; Haeberli and Burn, 2002). It was one of the reasons for an apparent increase in landslides in the Austrian Alps (Kellerer-Pirklbauer et al., 2012), where substantial surface lowering and retreat of glacier tongues since the Little Ice Age have changed the stress and strain fields in the slopes. This has occurred along with permafrost degradation, which has increased the ALT and the occurrence of unfrozen zones, thereby adding to the likelihood of slope failures. This combination of glacier retreat and permafrost degradation was also championed by Deline et al. (2012) and by Stoffel et al. (2014), who argued that destabilized rock glaciers could lead to debris flows without historic precedents. That recent warming has caused some rock glaciers and associated debris flows to become more active has been confirmed in Alaska by Simpson et al. (2015) and in high Asia by Sorg et al. (2015). Geertsema et al. (2006) showed a recent increase in mass movement activity in British Columbia, Canada, and related this to the same sorts of mechanisms. Conversely, in the Tyrol of Italy, Sattler et al. (2011) failed to find a correlation between debris flows and degrading permafrost at the present time. In New Zealand, Allen et al. (2011) noted the prevalence of recent failures occurring from glacier-proximal slopes and from slopes near the lower permafrost limit, but uttered a word of caution to the effect that demonstrating any influence of atmospheric warming, permafrost degradation, and perennial ice melt on recent higher elevation slope failures and more importantly, distinguishing these influences from tectonic and other climatic forcing, is not yet possible.

6.14 Conclusions

Through changes in land cover, the adoption of agriculture and pastoralism, urbanization, the use of fire, and the development of transport links, rates of soil erosion, and mass movements have been greatly modified. As the study of peatlands shows, multiple factors may be involved in any one situation. However, humans have adopted management strategies to mitigate the effects of increased erosion and slope instability. In some regions, including those from which glaciers are retreating or in which permafrost is melting, the future may bring a further acceleration in landslide and debris flow activity.

7

Fluvial Processes and Forms in the Anthropocene

Humans have had a colossal impact on river systems over centuries (Williams et al., 2014), disturbing the connectivity of flows and forms. In this chapter, we concentrate on three aspects of fluvial geomorphology: changes in flows, sediment loads, and channels. Other aspects of rivers have been discussed in the context of coastal sediment supply (see Chapter 9), lake sedimentation, and soil erosion (see Chapter 6).

7.1 Modifications of River Flow

Humans have deliberately altered fluvial landscapes (Nilsson et al., 2005; Hudson and Middelkoop, 2015) by the building of embankments and levees, the construction of dams and reservoirs, and channelization (see also Chapter 3). As Lynas (2011, p. 139) has graphically remarked:

Whole natural drainage basins, which once responded to the grand seasonal cycles of winter flood and summer drought, now react meekly to the whims of water managers seated in the control rooms that govern sluice gates in tens of thousands of large dams. The Colorado River may have gouged out the most spectacular cutting in the world – the Grand Canyon – but today the flow of this powerful torrent is as much a product of human hydrological engineering as it is of any natural force.

In the Negev Desert of the Middle East, stones were deliberately removed from desert surfaces by Nabatean people (i.e. in Roman times) to generate swifter runoff – a process called water-harvesting – so that crops could be grown on irrigated plots (Evenari et al., 1982; Lavee et al., 1997). In addition, however, all over the world humans have caused many changes to occur inadvertently as a result of land cover and land-use changes, including vegetation removal, urbanization, mining, the facilitation of alien plant invasions, and the like (Figure 7.1).

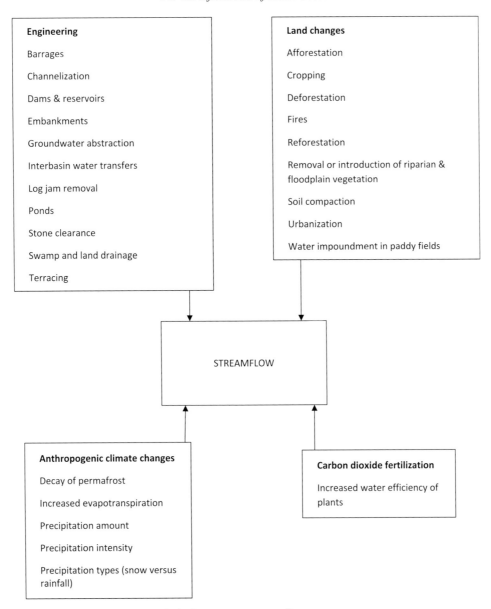

Figure 7.1 Anthropogenic influences on streamflow.

The multiple ways in which dams and reservoirs impact on geomorphology are shown in Figure 7.2. In the United States, Fitzhugh and Vogel (2011) mapped the impacts of dams on flood flows. The percent of rivers with greater than a 25 percent reduction in the median annual flood was 55 percent for large rivers, 25 percent for medium rivers, and 10 percent for small rivers. The greatest reductions have taken place to the west of the Mississippi and especially in the Great Plains, the deserts of

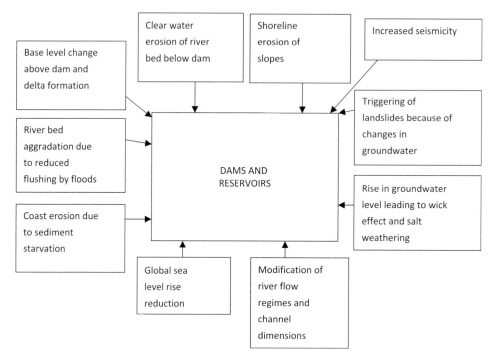

Figure 7.2 The multiple geomorphological effects of dams and reservoirs.

the southwest and in northern California. Flows of the Colorado River at the mouth have steadily declined since the beginning of the twentieth century, and in most years after 1960 it has run dry before reaching the sea. Its average flow rate at the northernmost point of the Mexico–United States border is now less than a tenth of the natural flow – due to upstream water use. Below here, all of the remaining flow is diverted to irrigate the Mexicali Valley, leaving a dry riverbed from Morelos Dam to the sea.

Interbasin water transfers for municipal supply and irrigation are another major reason for changes in river flows (Shiklomanov, 1986), notably in Canada, the United States, India, and the former Soviet Union. In Australia, it has been calculated that the total flow of the Murray at its mouth has been reduced by 61 percent due to humans and that the river now ceases to flow through the mouth 40 percent of the time. This figure would be 1 percent in the absence of water resource development.

In many parts of the world humans obtain water by pumping from groundwater (Wada et al., 2010), especially since the first center-pivot systems were introduced by Frank Zybach in 1949. Now about 38 percent of the world's irrigated areas are groundwater-based and about 43 percent of irrigation water is derived from underground aquifers (Siebert et al., 2010). Groundwater is also important as a source of water for municipal consumption. Increasing population levels and the adoption of new exploitation techniques (e.g. the replacement of irrigation methods involving animal or human power by electric and diesel pumps) has increased these problems. Global groundwater abstraction increased from c. 312 km^3 per year in 1960, to c. 734 km^3 per year in 2000, and this abstraction exceeded recharge, so that over the same period groundwater reserves became depleted. Moreover, the rate of global groundwater depletion has probably more than doubled since the period 1960–2000 (Aeschbach-Hertig and Gleeson, 2012). Flows, especially in streams in sedimentary rocks where groundwater makes a major contribution to stream discharges, may be reduced as a result of this drawdown (Hiscock et al., 2001; Kirk and Hebert, 2002; Weber and Perry, 2006). Zektser et al. (2005) and Wen and Chen (2006) established that extensive irrigational pumping has caused streamflow depletion in the High Plains of the United States. Rugel et al. (2012) found that karstic groundwater exploitation in Georgia had also led to flow declines.

Changes in land use are important, as has been demonstrated by paired catchment studies (Brown et al., 2005). Pliny the Elder was probably the first to allude to the hydrological role of forests in his *Natural History* (written in the first century AD; see Marsh, 1864, and Andréassian, 2004, for a review of early studies). Forest removal and its replacement with pasture, crops, or bare ground effects stream flow in many ways. First, forests modify the accumulation and ablation of snow (Schelker et al., 2013). Thus, a study by Lin and Wei (2008) of the Willow watershed in British Columbia, Canada, suggested that forest harvesting significantly increased peak flows in spring due to an increase of snow accumulation and acceleration of the snowmelt process as a result of removal of forest vegetation. Second, a mature forest probably has a higher rainfall interception rate, has a tendency to reduce rates of overland flow, and generates soils with a higher infiltration capacity and better general structure. All these factors tend to produce both a reduction in overall runoff levels and less extreme flood peaks. However, the response of catchments will depend on other factors, including the nature of the underlying soils and the amount of groundwater contribution to flow (Cosandey et al., 2005). The effects of forests on base-flows remain controversial. Many studies associate higher watershed forest cover with lower base-flows, attributed to the high evapotranspiration rates of forests, while other studies indicate increased base-flow with higher watershed forest cover due to higher infiltration and recharge of subsurface storage (Price, 2011).

Early studies of land-use effects on stream flows were reviewed by Pereira (1973). He believed that the first experiment in which a planned land-use change was executed to enable observation of the effects on flow, began at Wagon Wheel Gap, Colorado, United States, in 1910. Here stream flows from two similar watersheds of about 80 ha each were compared for eight years. One valley was then clear-felled and the records were continued. After the clear-felling the annual water yield was 17 percent above that predicted from the flows of the unchanged control valley.

The substitution of one forest type for another may also affect flow. This can be exemplified from the Coweeta catchments in North Carolina, United States, where paired catchment studies have been conducted since 1934. Here two experimental catchments were converted from a mature deciduous hardwood forest cover to a cover of white pine (*Pinus strobus*). Fifteen years after the conversion, annual stream flow was found to be reduced by about 20 percent (Swank and Douglass, 1974). The reason for this notable change is that the interception and subsequent evaporation of rainfall is greater for pine than it is for hardwoods during the dormant season.

Studies have also been undertaken in other environments. For example, in central Queensland, Australia, Siriwardena et al. (2006) investigated the effects that clearance of Brigalow (*Acacia harpophylla*) forest in the 1960s and its replacement with grass and cropland had had on runoff. They found that runoff in the postclearing period was greater by 58 percent than if clearing had not occurred. In Amazonia, the replacement of forest with soy cultivation also has had major consequences for stream flow (Hayhoe et al., 2011). One type of deforestation that has an impact on stream flow is salvage logging that takes place following tree infestation by beetles and other insects (see, e.g., Pomeroy et al., 2012). Humans have contributed to some outbreaks of infestation by introducing such organisms (Orwig, 2002; Aukema et al., 2010). Extreme fire events (see, e.g., Smith et al., 2011) and die-off following anthropogenic drought may also affect flows (Adams et al., 2012).

In some cases, the greater runoff of surface water following deforestation can lead to the conversion of ephemeral swamps into lakes. After complete deforestation, Woodward et al., (2014), working in Australia, demonstrated that water available to wetlands increases by up to 15 percent of annual precipitation.

Currently a phenomenon called the "forest transition" is taking place in some locations. This is a shift from net deforestation to net reforestation (Meyfroidt and Lambin, 2009). Around the Mediterranean Sea, for instance, farmland abandonment has been widely reported, caused by national and international migrations and because market conditions made farming unprofitable in areas with steep slopes, small field sizes, and with difficulties for access and mechanization (García-Ruiz et al., 2011). In the United States the forested area has increased substantially since the 1930s and 1940s, though it is possible that some authors have exaggerated the extent of abandoned agricultural land and the amount of forest regrowth that has occurred (Ramankutty et al., 2010).

Afforestation and reforestation tend to reverse the effects of deforestation. For example, the work of Lopez-Moreno et al. (2006) in the Ebro basin of Spain, showed a general negative trend in flood intensity in recent decades, together with an increase in the importance of low flows in the total annual contribution. They attributed this to the increase in vegetation cover that is a consequence of the farmland abandonment and reforestation that occurred during the twentieth century. Comparable trends have been reported from the Pyrenees (Buendia et al., 2015). Increasing forest cover in Uruguay has had a similar set of consequences (Silveira and Alonso, 2009). Fears have also been expressed that the replacement of tall natural forests by eucalyptus will produce a decline in stream flow. However, most research fails to support this contention, for transpiration rates from eucalyptus trees are similar to those from other tree species (except in situations with a shallow groundwater table) while their interception losses tend if anything to be generally rather less than those from other tree species of similar height and planting density (Bruijnzeel, 1990).

In northern England, Birkinshaw et al. (2014) analyzed data from the Coalburn research catchment (1.5 km^2). The site was instrumented in 1967, plowed and planted in 1972/73, and the trees have now reached maturity. Their results show that after plowing there was an increase of around 50–100 mm in annual streamflow compared with the original upland grassland vegetation. However, the mature trees now show a decrease of around 250–300 mm in the annual streamflow compared with the original vegetation and a decrease of around 350 mm compared with when the site was plowed.

Farley et al., (2005) analyzed available global data and found that annual runoff was reduced on average by 44 percent and 31 percent when grasslands and shrublands were afforested, respectively. They also established that eucalypts had a larger impact than other tree species in afforested grasslands, reducing runoff by 75 percent compared with a 40 percent average decrease with pines. In addition, they found that runoff losses increased significantly with plantation age for at least twenty years after planting. For grasslands, absolute reductions in annual runoff were greatest at wetter sites, but proportional reductions were significantly larger in drier sites. They suggested that reductions in runoff may be most severe in drier regions, and that in a region where natural runoff is less than 10 percent of mean annual precipitation, afforestation should result in a complete loss of runoff. Where natural runoff is 30 percent of precipitation, it will likely be cut by half or more when trees are planted.

The nature of cropping is another important land-use change to affect stream flow. In many studies, the runoff from clean-tilled land tends to be greater than that from areas under a dense crop cover. In the Upper Mississippi basin in the United States, the tendency for the discharge of the river since the 1940s to increase more than can be explained by increasing precipitation alone, has been attributed to the great spread in soybean cultivation (Schilling et al., 2010). In Costa Rica, plot studies (Algeet-Albarquero et al., 2015) demonstrated that an oil palm plantation plot presented the highest runoff coefficient (RC; mean RC = 32.6 percent), twice that measured under grasslands (mean RC = 15.3 percent), and twentyfold greater than in secondary forest (mean RC = 1.7 percent). Conversely, in southeast Asia, where many paddy fields are terraced pond systems, rice cultivation may cause a reduction in runoff to occur (Wu et al., 2001).

In some parts of the world, such as southern Africa, the spread of invasive exotic plants causes greater loss of water than the native vegetation and so reduces stream flow. The incremental water use of alien plants is estimated to be 3,300 million m^3 yr^{-1}, equivalent to a 190 mm reduction in rainfall, and equivalent to almost three-quarters of the virgin mean annual runoff of the huge Vaal River (Le Maitre et al., 2000). Changes in riverbank vegetation may also have had a strong influence on river flow in the southwest United States, where many streams are lined by the introduced salt cedar (*Tamarix pentandra*). With roots either in the water table or freely supplied by the capillary fringe, these shrubs have full potential transpiration opportunity. Thus, their removal can cause large increases in flow. Some recent studies, however, suggest that the hydrological impact of this nonnative riparian vegetation may not be as serious as has sometimes been proposed (e.g. Hultine and Bush, 2011).

Human encroachment on papyrus swamps has been identified as an influence on streams flowing into Lake Victoria in Uganda. Ryken et al. (2015) indicated that, due to their strong buffering capacity, papyrus wetlands had a first-order control on runoff and sediment discharge. Subcatchments with intact wetlands had a slower rainfall–runoff response, smaller peak runoff discharges, lower rainfall–runoff ratios, and significantly smaller suspended sediment concentrations. Subcatchments with papyrus swamps upon which humans had encroached, had sediment yield values that were about three times larger compared to catchments with intact papyrus vegetation (respectively 106–137 t km^{-2} yr^{-1} versus 34–37 t km^{-2} yr^{-1}).

Land drainage, including that of swamps, peaty areas, and agricultural land, also impacts upon downstream flood incidence, though this has long been a source of controversy. Much depends on the scale of study, the nature of land management and the character of the soil that has been drained. After a review of experience in the United Kingdom, Robinson (1990) found that the drainage of heavy clay soils that are prone to prolonged surface saturation in their undrained state generally led to a reduction of large and medium flow peaks. He attributed this to the fact that their natural response is "flashy" (with limited soil water storage available), whereas the drainage of permeable soils, which are less prone to such surface saturation, improves the speed of subsurface flow, thereby tending to increase peak flow levels.

In the United States, Frans et al. (2013) found that land drainage had caused some increase in flows in the Mississippi basin. Also working in the United States, Schottler et al. (2014) compared changes in hydrology for twenty-one southern Minnesota watersheds and showed that artificial drainage was a major driver of increased river flow over the last 70 years, exceeding the effects of precipitation changes and crop conversion. The majority of the increase in flow was attributed to changes in water residence time on the landscape and subsequent reductions in evapotranspiration resulting from installation of artificial drainage networks. Rivers with altered hydrology also exhibited significant channel widening since the mid-twentieth century, supporting the hypothesis that agricultural land-use changes have created more erosive rivers. Rivers that had significant increases in annual flow volume experienced channel widening of 10–40 percent, whereas rivers with no flow increase had no change in channel width.

Urbanization affects flood runoff characteristics and sedimentation (Hupp et al., 2013; Miller et al., 2014; Hopkins et al., 2015) because of the construction of impermeable surfaces and the installation of sewers and storm drains (see, e.g., Hollis, 1975; Graf, 1977; Sheng and Wilson, 2009). Urbanization generates rapid water runoff with an increase of flash flood frequency at the river basin outlet. In addition, in semiarid southern California, which relies heavily on imported water for domestic use, a synthesis of river discharge data reveals that summer (June, July, and August) river discharge in catchments that have at least 50 percent urban, suburban, and/or commercial land cover has increased by 250 percent or more over the past half century, without any substantial precipitation change during these months. Total annual discharge in the Los Angeles River has also increased at levels up to several hundred percent. According to Townsend-Small et al., (2014), three factors probably account for these trends: (1) increased groundwater recharge rates from leaking water pipelines; (2) inputs of treated wastewater into streams and rivers; and (3) increased runoff or recharge due to overirrigation of ornamental landscaping.

Recently, techniques have been developed to reduce and delay urban storm runoff. These include Sustainable Urban Drainage Systems (SUDS), which tackle urban surface runoff problems at the source using features such as soakaways or infiltration basins (Brander et al. 2004), permeable pavements (Brattebo and Booth, 2003), grassed swales or vegetated filter strips, infiltration trenches, ponds (detention and retention basins), and wetlands to attenuate flood peak flows.

Many rivers across the world have experienced significant streamflow changes over the last decades (Salmoral et al., 2015). Drivers of these are multiple, including climate change, land-use and land-cover changes, water transfers, and river impoundment. Many of these drivers interact simultaneously, sometimes making it difficult to discern the impact of each driver individually.

7.2 Sediment Transport: The Impact of Dams and Soil Erosion

Fluvial sediment transport has been modified by humans in two main ways (Wasson, 2012). On the one hand, dam construction has caused much sediment to be trapped in reservoirs (Wisser et al. 2013; Poeppl et al., 2015), though a suite of techniques (e.g. sluicing) is available, though seldom used, to mitigate this problem (Kondolf, 2013). On the other, sediment delivery to rivers has been increased as a result of accelerated rates of soil erosion. In the southeast United States, the results of these two tendencies were analyzed by McCartney-Castle et al. (2010), who found that under pre-European conditions (1680–1700) the mean annual suspended sediment load transfer rate was 6.2 Mt per year, under predam conditions (1905–1925) the rate was 15.04 Mt per year, whereas now, following dam construction, the rate is 5.2 Mt per year.

Large dams act as barriers to flow, sediment, and aquatic organisms, and so can create major discontinuities in river geomorphological and ecological conditions (Marren et al., 2015). The impounding of water leads to the deposition of sediment within the upstream reservoir, while the release of sediment-starved flow from the reservoir typically leads to erosion, scour, and channel incision downstream. Also, over time, net decreases in flow downstream often result in channel narrowing and expansion of the floodplain, onto which vegetation encroaches. However, many dams are small run-of-river structures, also known as weirs or overflow dams, which typically have heights that do not surpass those of the channel banks. They confine water between the channel banks upstream and allow flow to pass over the crest of the structure unimpeded. These dams do not act as major sediment traps, nor do these structures produce substantial downstream channel erosion (Csiki and Rhoads, 2014). Nonetheless, they occur in large numbers. Over 2,000,000 low-head dams (<7.6 m high) fragment United States rivers (Fencl et al., 2014).

One class of dam, called a Sediment Retention Structure, is constructed specifically to stop slugs of sediment moving downstream. The most famous example of this is an earthen dam, 575 m long and 56 m high, built on the North Fork Toutle River in the US state of Washington by the United States Army Corps of Engineers in 1989. It was designed to prevent sediment from the 1980 eruption of Mount St. Helens from increasing flood risks along the Toutle and Cowlitz rivers. The basin behind the dam could accommodate 197 million m^3 of sediment, but is now effectively full.

Some estimates of the impact of dams on river sediment load in some other parts of the world are shown in Table 7.1.

Table 7.1 *Estimates of reduction in sediment load (percent) in some major rivers as a result of sediment trapping by dams (modified after Walling, 2006).*

Colorado (United States)	100
Don (Russia)	64
Ebro (Spain)	92
Kizil Irmak (Turkey)	98
Nile (Egypt)	100
Orange (South Africa)	81
Rio Grande (United States)	96
Volta (Ghana)	92

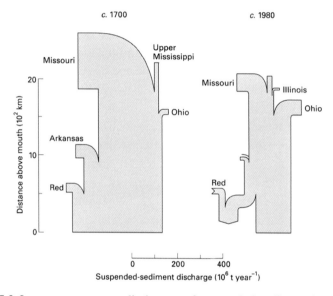

Figure 7.3 Long-term average discharges of suspended sediment in the lower Mississippi (after Meade and Parker, 1985, with modifications).

A dramatic exemplification of the effects of large dams are the data for the Colorado River in the United States. Prior to 1930 it carried an annual load of around 125–150 million tons of suspended sediment to its delta in the Gulf of California. Following the installation of a series of dams it now discharges neither sediment nor water to the sea (Schwarz et al., 1991). There have also been marked changes in the amount of sediment passing along the Missouri and Mississippi rivers (Figure 7.3). Before 1900, this system carried an estimated 400 million tons of sediment per year to coastal Louisiana. Between 1987 and 2006 this had fallen to an average of 145 million tons per year. About half of this decline was accounted for by sediment trapping behind dams, and the rest by such actions as bank revetments and soil erosion management (Meade and Moody, 2010). Also in the United States, the Columbia River has undergone a greater than 50 percent reduction in total sediment transport since the mid-nineteenth century, largely because of flow regulation (Naik and Jay, 2011).

In India, Rao et al. (2010) report that dams on the Krishna and Godavari have also caused major changes in discharges and sediment loads. Between 1951 and 1959, the former discharged 61.9 km^3 of water, but this had fallen to only 11.8 km^3 by 2000–8. Its suspended sediment load was 9 million tons during 1966–9 but was as low as 0.4 million tons by 2000–5. The Indus has also undergone a marked reduction in flow and sediment transport, especially since the completion of the Tarbela Dam in Pakistan in the early 1970s. Between 1931 and 1947, the Indus at Kotri had a flow of about 70–150 km^3 per year, and a sediment discharge of 190–330 million tons per year. For the period from 1962 to 1986 the respective figures were 10–100 km^3 per year and 10–130 million tons per year (Meadows and Meadows, 1999). The figures given by Renaud and Syvitski (2013) are as follows: the annual water and sediment discharges between 1931 and 1954 averaged 107 km^3 and 193 Gt, respectively. These discharge rates during the period 1993–2003 dropped by an order of magnitude to 10 km^3 and 13 Gt, respectively. One-fifth of the Indus delta plain has been eroded since the river was first dammed in 1932 (Giosan et al., 2014). Gupta et al. (2012) have compared current with historical sediment loads for a range of Asian rivers. For rivers in peninsular India, the overall decline was 67 percent, 74 percent, 87 percent, 80 percent, 41 percent, and 95 percent for the Mahanadi, Godavari, Krishna, Cauvery, Tapti, and the Narmada rivers, respectively. For rivers flowing into the Arabian Sea from India, the overall decline in annual sediment flux was 95 percent for the Narmada, 41 percent for the Tapi, 68 percent for the Mahi, and 96 percent for the Sabarmati. In addition, the Ganga had lost 31 percent, and the Indus, 93 percent. Gupta et al. (2012) also produced a graph of the trends in sediment flux reductions with rising number of dams across the rivers concerned (Figure 7.4).

In China, in 2003–5 the sediment load in the upper Yangtze was only 17 percent of that in the 1950s–60s (Xu et al., 2007). The Three Gorges Dam (Figure 7.5) has had a particularly marked impact since it began to impound sediment and water in 2003 (Xu and Milliman, 2009; Dai and Lu, 2014) and sediment starvation downstream has caused channel incision to occur (Dai and Liu, 2013). Figure 7.6 shows the change in sediment load in the Yangtze before 2000 AD and in the period since the construction of the Three Gorges Dam (2001–10).

There have also been massive changes in the sediment load of the Yellow River, associated with dam construction (Ran et al., 2013; Figure 7.7). As a result, the northern shore of its delta has retreated 300 m each year for the past 35 years

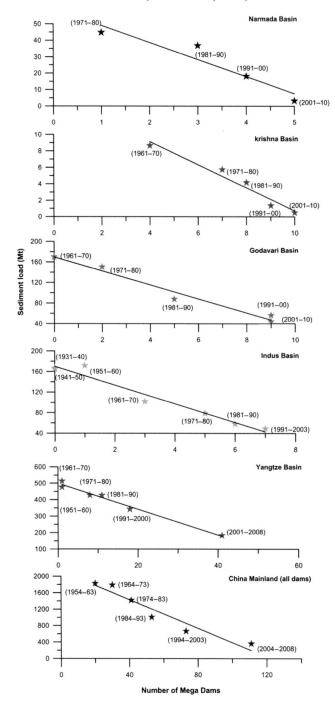

Figure 7.4 Annual sediment loads for some major rivers in relation to dam construction (from Gupta et al., 2012, figure 3).

Figure 7.5 The Three Gorges Dam on the Yangtze in China has transformed the flow and sediment budgets downstream. Scale bar 3.69 m. ©2014Digital Globe and ©2014 Mapabc.com.

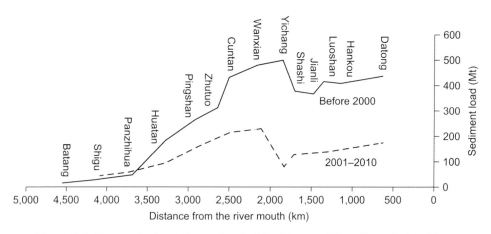

Figure 7.6 Changes in the sediment load of the Yangtze River (from Dai and Lu, 2014, figure 2).

(Giosan et al., 2014). Data on changing sediment loads in the Pearl River (Zhujiang) are provided by Wu et al. (2012), who found that from 1994 to 2009 sediment load decreased by 83 percent.

On a global basis, Syvitski et al. (2005) calculated that sediment retention behind dams has reduced the net flux of sediment reaching the world's coasts by

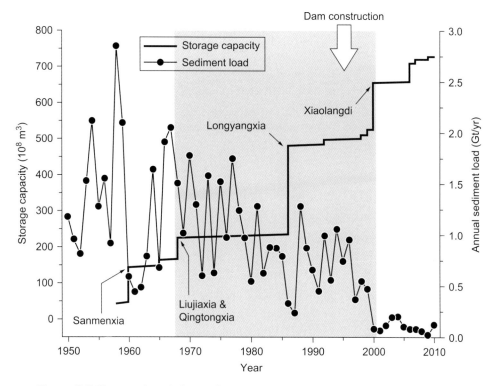

Figure 7.7 Temporal variations of reservoir storage capacity (left) and annual sediment load (right) for the Yellow River basin and sediment load at Huayuan-kou gauging station. The gray background denotes the period of large-scale dam construction. Also shown are the key reservoirs constructed on the Yellow River main stem channel (from Ran et al., figure 4).

c. 1.4 billion metric tons per year. Reservoirs behind dams now trap c. 26 percent of the global sediment delivery to the coastal ocean (Syvitski and Milliman, 2007). Conversely, Syvitski et al. (2005) have calculated that humans have simultaneously increased the sediment transport by global rivers through soil erosion by c. 2.3 billion metric tons per year. On balance, therefore, global river sediment loads have increased more as a result of soil erosion than they have been reduced by dam retention. Wilkinson and McElroy (2007) calculated that "natural" sediment fluxes to the world's rivers are about 21 Gt yr^{-1} and that "anthropogenic" losses may be around 75 Gt yr^{-1}.

However, differences exist between different rivers and areas. In central Japan, for instance, sediment loads declined in the last decades of the twentieth century because land that had formerly been intensively cultivated had become urbanized (Siakeu et al., 2004). In the case of some major Chinese rivers draining into the western Pacific (Chu et al. 2009), recent reductions in sediment load have been caused by a combination of factors: dam retention (56 percent), soil and water conservation (23 percent), water abstraction (15 percent) and in-channel sand mining (6 percent). This is also the case for the Sacramento River in California (Wright and Schoellhamer, 2004). Its sediment load fell by about one-half between 1957 and 2001 and this has been attributed to a number of factors: depletion in the amount of old hydraulic mining debris, trapping of sediment in reservoirs, riverbank protection, the construction of levees, and altered land use (e.g. urbanization replacing agriculture).

7.3 Channel Changes: Deliberate

Humans have modified the channels of many rivers, though it is not always easy to differentiate between human effects and those of climatic fluctuations and natural flood events (Liébault and Piégay, 2002; Radoane et al., 2013; Segura-Beltrán and Sanchis-Ibor, 2013). Besides having an inherent geomorphological fascination, changes in channel capacity are important because they have a substantial impact on flood hazard (Slater et al., 2015).

River engineering effects river flows and channel forms (see, e.g., Brookes, 1985; Korpak, 2007), as has been indicated for the Rhine and Mississippi by Hudson et al. (2008). Channel changes have been undertaken deliberately, as when levees are constructed or river courses are channelized and straitened (Figure 7.8; see also Chapter 3). Channelization has occurred along approximately 300,000 km of streams in the United States (Kroes and Hupp, 2010).

Both for improving navigation and flood control humans have deliberately straightened many channels. The elimination of meanders contributes to flood control in two ways. First, the resultant shortened course increases both the gradient and the flow velocity, and the floodwaters erode and deepen the channel, thereby increasing its flood capacity. Second, it prevents some overbank floods on the outside of curves, against which the swiftest current is thrown and where the water surface rises highest.

Deliberate channelization, as on some Tennessee rivers, by altering flows of water and sediment, can lead to unintended consequences, such as swamp formation (Shankman and Smith, 2004). Working in the United Kingdom, Brookes (1987), found that enlargement of channel cross-sections through erosion had occurred downstream from a variety of channelization works. He explained this in terms of increased flood flows causing higher stream velocities, which in turn caused erosion, thereby increasing channel width and/or depth. Channelization may also lead to changes in sediment deposition and sediment types (e.g. Nakamura et al., 1997), with sediment coarsening having been encountered in some studies (e.g. Talbot and Lapointe, 2002). Gradient over-steepening and

Figure 7.8 The ultimate in channelization. The Los Angeles River corseted in concrete and squeezed between railroad tracks in downtown Los Angeles.

channel narrowing, caused by channelization, lead to formation of a river system that has a steep, straight, narrow, and deep channel. Reduced floodplain water storage and self-acceleration of flow concentrated in the channel zone make flood waves progressively more flashy (Wyzga, 1996).

Many channels may be deliberately confined by the emplacement of riprap (rock armor). It is a particularly common form of revetment to protect against undesired lateral channel migration. Potential consequential morphological effects of riprap placement in a gravel-bed river include inhibition of local sediment supply to the channel and consequent bed scour and substrate coarsening (Reid and Church, 2015).

Urban and Rhoads (2003), working in Illinois, placed the impacts of channelization in a longer time context. They claimed (p. 794) that

[H]umans are now the dominant agent of stream-channel change in the upper Embarras River watershed, a typical agricultural watershed in east central Illinois, USA. Human modification of streams over the last several decades has exceeded the efficacy of geomorphological processes by one to two orders of magnitude. Channelization involves short-duration expenditure of massive equivalent amounts of energy to induce sudden change in the position or geometry of streams: that is, these events are highly efficacious. More important, channelization has a high degree of geomorphological effectiveness: it alters stream channels at rates that greatly exceed background rates of change over a certain time scale, radically restructures the morphology and dynamics of the entire fluvial system, and produces morphological effects that persist over the same time scale as the recurrence interval of the disturbance. Put simply, humans have become catastrophic agents of fluvial change.

7.4 Accidental Channel Changes

However, as with so many other instances of anthropogeomorphic change, much channel change is the inadvertent result of modification of the amounts of water and sediment that come down channels, or the modification of riparian vegetation (Figure 7.9).

Channel changes in recent decades may result both from climatic fluctuations, such as El Niño Southern Oscillation (ENSO), and human activities (Teo and Marren, 2015). The complexity and diversity of causes of stream-channel change is reviewed by Downs and Gregory (2004) and Church et al. (2009), while Wohl (2015) discusses the many human actions that control river flows and sediments and thus affect channel forms. Major changes in the configuration of channels can be either because of human-induced changes in stream discharge or sediment load: both parameters affect channel capacity. As Downs

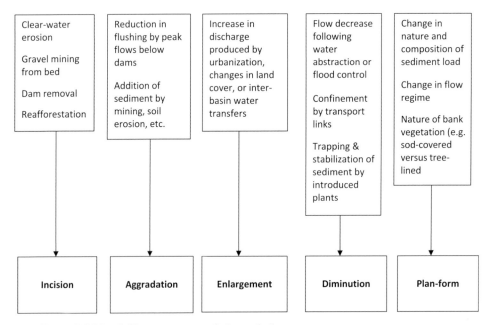

Figure 7.9 Nondeliberate causes of channel change.

et al. (2013) point out, many channel changes are the result of cumulative impact. Adjustments in alluvial channel morphology arise from the influence of numerous "drivers for change" operating at multiple spatial and temporal scales. Thus, causal understanding of channel morphodynamic response in the historical period requires knowledge of a suite of drivers for change rather than a focus on a single causal influence, whether natural or human in origin. In Britain, Macklin et al. (2013) have analyzed the complex history of channel incision and note that climatic fluctuations explain some of the observed phases during the Holocene, but that in the last 1,000 years in addition to climate change, factors that are likely to have contributed to increasing extreme event discharges within this period include large-scale channelization to improve navigation or the siting of mills, the rapid expansion of arable cropland during the medieval period, and locally to urban and industrial developments. Increasing runoff through the conversion of land from woodland to farmland at a time of wetter climate (e.g. 900–800 years ago) would, in combination, have resulted in more frequent floods capable of entraining coarser bed and bank material, promoting channel entrenchment.

A good review of channel changes in Italy is provided by Surian and Rinaldi (2003), who found that in the twentieth century, and particularly since the 1950s–1960s, most Italian rivers have experienced two types of channel adjustment: incision, which is commonly of the order of 3–4 m, but in some cases even more than 10 m; and narrowing of the active channel, in some cases up to 50 percent (or even more). In some rivers, these adjustments, which frequently occur together, have led to changes in channel pattern from braided to wandering. The causes of these adjustments included land-use changes, channelization, construction of dams, and sediment mining, all of which have been particularly severe since the 1950s. The main effect of these interventions on fluvial processes was a dramatic reduction in sediment supply. The same effect has occurred in Poland, where reafforestation in the Wisłoka basin has caused a reduction in sediment supply and consequent incision (Lach and Wyzga, 2002). However, the causes of channel change in Poland have been complex and spatially variable, and Zawiejska and Wyzga (2010), working in the Dunajec basin in the Polish Carpathians, found that in its lower course the river was considerably narrowed, shortened, and embanked between the 1880s and the 1920s. Here, bed degradation commenced in the late nineteenth century and resulted in 3.1 m of incision. In its middle course, however, relatively little change occurred where the river is confined by valley sides and where bedrock exposures prevent channel incision. In the reaches within intramontane basins, channelization works carried out in the 1950s–70s considerably narrowed the river and transformed its multithread channel into a single-thread, straight channel. Rapid bed degradation induced by the works resulted in up to 2 m of incision over the second half of the century. However, some of this was attributed to afforestation, for between 1931 and 2000 forest cover increased from 26 percent to 42 percent, while arable land was reduced from 45 percent to 17 percent.

In the United States, the Platte River of the Great Plains has shown a remarkable change in its characteristics in the last two centuries (Horn et al., 2012). The river evolved from an open, braided condition to a stream with multiple, stable ana-branches, and heavily vegetated banks and islands. The mean channel area of the central Platte River decreased by an average of 46 percent from 1938 to 2006, and the mean widths of the individual channels in 1858 were 539 percent greater than in 2006. Channel constriction, from the construction of dams and diversion canals for irrigation, and encroachment of riparian vegetation into formerly open channels, seem to have been primarily responsible for these changes. Damming and flow diversion through canals along the river led to the stabilization of bars and the shrinkage of channels.

Elsewhere in the United States, equally far-reaching changes in channel form are produced by land-use changes and the introduction of soil conservation measures. Trimble (1974; 2008a) showed how the river basins of Georgia were modified through human agency between 1700 (the time of European settlement) and the present. Initially, clearing of the land for cultivation caused massive slope erosion, which resulted in the transfer of large quantities of sediment into channels and floodplains, which thus aggraded because of the accumulation of what is widely termed "postsettlement alluvium." The phase of intense erosive land use persisted and was particularly strong during the nineteenth century and the first decades of the twentieth century, but thereafter conservation measures, reservoir construction, and a reduction in the intensity of agricultural land use caused streams to carry a less heavy sediment load, so that they became less turbid, and incision took place into their floodplain sediments. By means of this active streambed erosion, streams incised themselves into the modern alluvium, lowering their beds by as much as 3–4 m.

In southwest Wisconsin, a broadly comparable picture of channel change was documented by Knox (1977). There, as in the Upper Mississippi Valley (Knox, 1987), it is possible to identify stages of channel modification associated with various stages of land use, culminating in decreased overbank sedimentation as a result of better land management in the last half century. One localized effect on channels in the United States is that associated with the floating of logged timber downstream (see Ruffing et al., 2015).

In some parts of the United States channel losses have been recorded. Diversion, piping, and burial of stream channels are all common causes of stream loss during land development, as are impoundment of stream channels for reservoirs and the excavation of ponds (Julian et al., 2015).

In Australia, dramatic channel incision occurred both on valley sides and on valley floors, often within a few decades of European settlement (Rutherford, 2000). Most of this historical incision was triggered by disturbances to the material making up valley floors. Drains, cattle tracks, roads, and other forms of flow concentration made the valley floors more susceptible to higher runoff from cleared and trampled catchments and from storms. Other causes of channel change identified by Rutherford included channel enlargement by sand and gravel extraction, changes in flow as a result of dam construction, channel invasion by exotic vegetation, removal of riparian vegetation, and bank erosion by boats. The last of these processes was investigated by Bradbury et al. (1995) on the Gordon River in Tasmania. Here erosion by boat wash had caused severe bank erosion, but the mean measured rate of erosion of estuarine banks slowed from 210 to 19 mm yr^{-1} with the introduction of a 9 knot speed limit. In areas where cruise vessels continue to operate, alluvial banks were eroded at a mean rate of 11 mm yr^{-1}. Very similar alluvial banks no longer subject to commercial cruise boat traffic eroded at the slower mean rate of 3 mm yr^{-1}. In addition, the mean rate of bank retreat slowed from 112 to 13 mm yr^{-1} with the exclusion of cruise vessels from the leveed section of the river.

Water mills are powered by water, largely diverted from rivers. These mills have been used in a large number of industrial activities, ranging from the milling of cereal grains and the weaving of cloth, to their lesser-known uses for the boring of cannons, the grinding of the ingredients for gunpowder, and the powering of bellows for blast furnaces (Bishop and Muñoz-Salinas, 2013). Their construction has been identified as a potent cause of historical channel change in the eastern United States (Wohl and Merritts, 2007). Walter and Merritts (2008), using old maps and archives, showed that whereas before European settlement the streams of the region were small anabranching channels within extensive vegetated wetlands, after the construction of tens of thousands of seventeenth- to nineteenth-century mill dams, 1–5 m of slackwater sedimentation occurred and buried the presettlement wetlands with fine sediment. Where such mill dams have been removed, accelerated bank erosion has occurred (Pizzuto and O'Neal, 2009). Water mills and their associated leats are also widespread and of considerable antiquity in England and Scotland and have also had a profound effect on river characteristics (Downward and Skinner, 2005). At the time of the Domesday Book (1086) there were thousands of mills in England (Hodgen, 1939) and their numbers increased until c. 1300 (Langdon, 1991).

7.5 Effects of Urbanization

Navratil et al. (2013) and Hawley et al. (2012) discuss the role of urbanization in channel enlargement and incision, for urban development causes flashier, larger, more erosive discharges that occur with increased frequency and duration. In an analysis of the literature, Chin (2006) stressed how variable river channel response to urbanization could be, but found that enlargement was reported in ~75 percent of the studies recording morphological results with cross-sectional areas generally increasing 2–3 times and on occasion as much as fifteen times. Other urban-induced channel changes include reductions in sinuosity which commonly result from artificial straitening, a tendency (not invariable) for bed material to coarsen from scouring of fines and increased competence because of higher peak flows, and an overall increase in drainage densities. Trimble (2003) provided a good historical analysis of how the San Diego Creek in Orange County, California, has responded to flow and sediment yield changes related to the spread of both agriculture and urbanization. Also working in southern California, Taniguchi and Biggs (2015) found that channel response to urbanization differed according to bed composition: sand-bedded channels incised and enlarged more for a given percent impervious cover compared to gravel- and cobble-bedded channels, which widened but in general did not incise.

7.6 Effects of Transport Corridors

Valleys often carry roads and railways. Blanton and Marcus (2013) have considered the effects of such transport corridors on channel forms and have developed a number of hypotheses to look at the effects of what they call "transport discon-nections." They suggest, first of all, that these can impede the natural meandering and migration of channels across their floodplains. This in turn disrupts the erosion and cut-and-fill alluviation that creates high habitat and biological diversity across the active channel and floodplain. Within the channel, confining structures tend to concentrate energy, which leads to higher shear stress and stream power that can wash out riffles and degrade low-velocity habitats such as pools and alcoves. Truncated meanders and lower sinuosity can also be associated with confinement by transport links. Wetted channel areas and widths can be smaller in transportation-impacted systems, resulting in less channel complexity and the presence of fewer bars and islands. Ponds, sloughs, oxbows, and paleochannels with hydrophilic vegetation can lose their water supply as they are disconnected from the main channel and therefore can contract or disappear. The disconnected floodplain can contain a proportionally smaller area of riparian forest, a narrower riparian zone, and a lower proportion of channel banks with gallery forest.

7.7 Effects of Mining on Stream Channels

Mining impacts upon channels in a variety of ways. A distinction needs to be drawn between the impact of waste material resulting from mining being dumped into channels, and the impact of gravel and sand mining from the beds of rivers themselves.

With regard to the former process, in a classic study, Gilbert (1917) demonstrated that hydraulic gold mining in the Sierra Nevada Mountains of California led to the addition of vast quantities of sediments into the valleys draining the range. Much of this remains to this day (James, 1989), and some of it may be contaminated with the mercury that was used as part of the process of collecting the gold (Lecce et al., 2008). This legacy sediment in itself raised channel bed levels, changed their configurations, and caused the flooding of lands that had previously been immune. The build-up of channels in California was sometimes so great that it enabled rivers to cut across valley spurs (James, 2004). Of no lesser significance was the fact that the rivers transported vast quantities of debris into the estuarine bays of the San Francisco system and caused shoaling, which in turn diminished the tidal prism of the bay.

There have been numerous other studies of channel aggradation as a result of inputs of mining waste, some of which is heavily polluted with heavy metals and the like (Taylor and Little, 2013). For example, Thorndycraft et al. (2004) recognized that on Dartmoor, England, stream aggradation occurred as a result of tin mining in Roman and medieval times. In North Carolina, Lecce and Pavlowsky (2014) described the increases in sedimentation rates that occurred during the mining phase, while in Norway, Langedal (1997) assessed the effects of molybdenum mining on stream aggradation. Knighton (1989) found that in tin mining regions of Tasmania, aggradation had occurred together with a 300 percent expansion in channel width. Enlargement of channels has also been reported to have resulted from increased stream flows as a result of runoff from bare spoil heaps (Touysinhthiphonexay and Gardner, 1984), while Graf (1979) has described channel incision and arroyo formation resulting from vegetation removal in mining areas in Colorado.

River beds and floodplains provide a source of sand and gravels for use as building aggregates. Among the consequences of in-channel mining are incision, knickpoint migration upstream, channel instability and bank erosion, increased bank-full widths, and bed armoring caused by a sediment deficit and selective outwashing of fines (Brown et al., 1998; Rinaldi et al., 2005). Sometimes, however, the effects are rapidly removed by subsequent flood events (Rempel and Church, 2009). Useful case studies come from the Mekong Delta in Vietnam (Brunier et al., 2014); Alaska (Givear, 1995); the French Alps (Marston et al., 2003); the Rhone

(Petit et al., 1996); the Piave and Orco rivers in Italy (Comiti et al., 2011; Brestolani et al., 2015); and the Gállego River of Spain (Martín-Vide et al., 2010).

7.8 Changing Riparian Vegetation, Animal Activity, and Stream Channels

Riparian vegetation is an important control of river-bank stability and thus of channel form (Tickner et al., 2001; Teo and Marren, 2015; Zaimes and Schutz, 2015) while forms and flows in turn affect riparian vegetation (Hupp and Osterkamp, 1996). Humans modify bank vegetation in a range of ways. Land use immediately adjacent to the river (e.g. cultivation of crops) may increase sediment deposition and eutrophication, while logging, grazing, and trampling, water extraction, and recreation also affect riparian zones. Such disturbances often occur in concert with, or act as triggers for, the proliferation of alien plants. The diversity and abundance of alien plants have increased in riparian zones throughout the world. River ecosystems are highly prone to their invasion, largely because of their dynamic hydrology and because rivers act as conduits for the efficient dispersal of propagules (Richardson et al., 2007). One area of great interest is the role of invasive *Tamarix* and other Eurasian aliens in the western United States (Friedman et al., 2005). The spread of these woody plants along dryland channels has caused channel narrowing to occur (e.g. Birken and Cooper, 2006). Equally, their removal has led to subsequent channel widening (Pollen-Bankhead et al. 2009; Vincent et al., 2009; Jaeger and Wohl, 2011).

The establishment or reestablishment of riparian forest has been implicated with channel narrowing in southeastern France during the twentieth century (Liébault and Piégay, 2002). Much, however, depends on the nature of the invasive vegetation, and Greenwood and Kuhn (2014) have investigated the role of the invasive plant, *Impatiens glandulifera* (Himalayan balsam), which is now found in many river catchments in most European countries. Its intolerance to cold weather and rapid seasonal dieback, has implicated it in promoting erosion along the riparian zone. During seasonal dieback stems lose their turgidity, foliage shrivels, and whole stands simultaneously collapse within a very short period of time. When this happens, the protection afforded to the underlying soil by the previously dense canopy rapidly vanishes, leaving areas along the riparian zone devoid, or partially devoid, of protective vegetation cover. There is, indeed, a major general question about the ways in which different vegetation types affect channel form (Trimble, 2004). Are tree-lined banks more stable than those flowing through grassland? On the one hand, tree roots stabilize banks and their removal might be expected to cause channels to become wider and shallower (Brooks and Brierley, 1997). On the other hand, forests produce log-jams which can cause aggradation or concentrate flow onto channel banks, thereby leading to their erosion. These issues are discussed in Trimble (1997b) and Montgomery (1997).

The presence of animals has also influenced channels. As Butler (2006) has recounted, the extirpation of species can affect river forms and dynamics. In particular, he has drawn attention to the role of beavers and the effects of their near-extirpation during the period following European contact and expansion across North America. Although modern beaver ponds entrap hundreds of millions to a few billion cubic meters of sediment, these values, he argued, pale in significance compared to the values associated with beavers on the precontact landscape when beaver ponds entrapped hundreds of billions of cubic meters of sediment. Widespread removal of North American beavers by fur trappers led to increased stream incision and attendant changes from relatively clear-flowing to sediment-laden streams.

7.9 Stream Restoration

Stream restoration describes a set of activities that help improve the environmental health of a river or stream and which have an impact on channels. It has become an important approach to river management since the 1980s (Adams et al., 2004), and has been adopted in Australia (Brooks and Lake, 2007), Europe (Buijse et al., 2002), Japan (Nakamura et al. 2006), the United Kingdom (McDonald et al., 2004) and the United States (Palmer et al., 2007). The purpose of management has shifted from simple anthropocentric, utilitarian needs associated with river channel engineering for flood and erosion/sedimentation control toward incorporation of a range of goals traditionally optional, but now required (e.g. ecological concerns). Table 7.2 lists some of these. Restoration objectives vary broadly from small-scale projects targeting populations of individual species to entire watershed recovery projects that may be ongoing for several decades. The aim may be to restore the "natural" state and functioning of the river system (Dufour and Piégay, 2009) but this is not always easy to determine. The wish is to improve biodiversity, water quality, recreation, flood management, and landscape development, and, preferably, to do this in economically effective and aesthetically pleasing ways. Restoration activities may range from a simple removal of a disturbance which inhibits natural stream function (e.g. repairing or replacing a culvert, or removing barriers to fish such as weirs), to stabilization of stream banks, to riparian zone restoration and constructed wetlands, to the installation of engineered log jams (Erskine et al., 2012; Nichols and Ketcheson, 2013) and to the reestablishment of meandering systems. It is important to appreciate that as with so many modifications of one part of the environment, there may be a suite of consequential changes in another part. Stream adaptation to restoration may liberate large amounts of sediment which can impact upon downstream reaches (Sear et al., 2009).

Table 7.2 *River restoration scenarios based on five ecosystem amenities that commonly motivate restoration projects (from Wohl et al., 2005, table 1).*

Amenity of interest	Key conditions	Components to model	Potential restorative actions
Clean water	Water/sediment chemistry Pathogen density	Contaminant/pathogen loading Water/sediment transport Pathogen population dynamics	Clean-up point – sources of pollution Alter land use in catchment
Uncontaminated food	Body-loads of contaminants	Contaminant loading Water/sediment transport Food organism/contaminant contact Food organism metabolism of contaminant	Clean up contaminant sources Constrain contaminant contact with food organism
Aesthetic appeal	Water clarity Bank stability Channel shape Riparian/aquatic vegetation	Nutrient loading Water/sediment transport Suspended solids dynamics Flow (disturbance) dynamics Flow/vegetation interactions Native/exotic vegetation interactions	Alter land/water use in catchment Reinstate natural channel shape Reinstate natural flow regime Manipulate sediment composition Manipulate vegetation composition
Rare or valued biota	Water/sediment chemistry Habitat structure Flow regime Production dynamics Other nonhuman biota	Contaminant loading Water/sediment transport Organism/contaminant contact Habitat requirements/limitations Organism/flow interactions Trophic requirements/limitations Interactions with competitors, predators, parasites	Clean up contaminant sources Alter land/water use in catchment Reinstate natural habitat structure Reinstate natural flow regime Reinstate natural productivity Stock target biota Reduce biota with adverse effects
Productive fishery	Water/sediment chemistry Habitat structure Flow regime Production dynamics Other nonhuman biota Harvest regime	Contaminant loading Water/sediment transport Organism/contaminant contact Habitat requirements/limitations Organism/flow interactions Trophic requirements/limitations Interactions with competitors, predators, parasites Impacts of harvest	Clean up contaminant sources Alter land/water use in catchment Manipulate habitat structure Manipulate flow regime Manipulate system productivity Stock target biota Reduce biota with adverse effects Reduce harvest

7.10 Dam Removal Effects

Related to stream restoration and to the developing aversion to hard engineering is dam removal. In recent years – partly because some old dams have become unsafe, ecologically unacceptable, or redundant – they have been demolished. This has had a series of impacts on channels (Grant and Lewis, 2015; Poeppl et al., ; 2015; Table 7.3).

Most of the dams involved have been just a few meters high, though one example from Australia was 15 m high (Neave et al., 2009) and two examples from the United States were 32–64 m high (East et al., 2014). The consequences of dam removal include: incision into the sedimentary fill that had accumulated in the reservoir behind the dam; migration of a knickpoint upstream; deposition of liberated sediment; and the formation of bars and the like downstream. However, the precise consequences vary greatly depending on the nature and erodibility of the fill and the regime of the river. Some rivers appear to remove their fill rapidly and almost entirely, while others do not.

7.11 Holocene Floodplain Sedimentation Related to Accelerated Erosion

Macklin et al. (2014) have used the term "Anthropocene Alluvium" (AA) to describe floodplain sediments that have been generated by human activities in the Holocene. They have suggested the indicators of AA shown in Table 7.4. In British and Irish catchments, Foulds and Macklin (2006) suggest that geomorphic instability linked with Holocene land-use changes was especially intense in the Bronze and Iron Ages (Table 7.5). Macklin et al. (2014) found that the oldest AA units in the United Kingdom dated to the Early Bronze Age (c. 4400 cal BP) and that there was an apparent 1,500-year lag between the adoption of agriculture (c. 6000 cal BP) in the United Kingdom and any impact on floodplain sedimentation. The earliest environmental human impacts on river channel and floodplain systems in the United Kingdom may, they argue, have been hydrological rather than sedimentological. The medieval period is confirmed as an important one for the accelerated sedimentation of fine-grained materials, notably in the smallest catchments (see also, Macklin et al. 2010, and Figure 7.10), and this can be related to the agricultural revolution of the Middle Ages. There have numerous other studies of the history of erosion and alluviation in the Holocene in Europe and Britain (see, e.g., Notebaert and Vertsraeten, 2010; Dusar et al., 2011). In Belgium, Broothaerts et al. (2014) found that typically, peat formation – in a marshy environment during the Early and Middle Holocene – was replaced by clastic overbank deposition as accelerated anthropogenic erosion and sedimentation occurred. Their results indicated that the end of peat growth and the transition

Table 7.3 *Selected studies of the consequences geomorphological effects of dam removal in the United States.*

Source	Location	Height (m)
Pearson et al., 2011	Merrimack, New Hampshire	4
Rapid incision and clearance of fill		
Kibler et al., 2011	Brownsville, Oregon	1.8–3.4
Release of gravel fill causing bars and pools and riffles downstream		
Burroughs et al., 2009	Pine River, Michigan	5
Big headcut, but only 12% of sediment fill eroded after ten years		
Major et al., 2008; 2012; Cui et al., 2014	Marmot Dam, Oregon	14
Knickpoint migrates 1.5 km upstream, and creation of multiple thread channel over 2 km downstream		
Doyle et al., 2003	Koshonong, Wisconsin	3.3
Little downstream aggradation as limited erosion from dam area		
Doyle et al., 2003	Baraboo, Wisconsin	2
Channel incision upstream and temporary aggradation of point bar downstream		
Cheng and Granata, 2007	Sandusky, Ohio	2.2
Less than 1% of sediment stored in reservoir was transported downstream		
Wildman and MacBroom, 2005	Anaconda, Connecticut	3.35
Only c one-third of sediment removed after five years. Channel widening upstream of dam but incision limited by original channel armoring		
Rumschlag and Peck, 2007	Muroe, Ohio	3.66
Vertical incision to preimpoundment substrate within one month. 1 m of aggradation downstream		
East et al., 2014	Elwha River, Washington	32–64
Downstream dispersion of a sediment wave caused widespread bed aggradation of ~1 m (greater where pools filled), changed the river from pool-riffle to braided morphology, and decreased the slope of the lowermost river. The newly deposited sediment, which was finer than most of the predam-removal bed, formed new bars, prompting aggradational channel avulsion.		
Harris and Evans, 2014	Secor Dam, Ohio	2.5
Downstream migration of a sediment wave at rates up to 0.5 m/hr. The overall effect was erosion of the former reservoir to a distance of 150 m upstream of the former dam		
Gelfenbaum et al., 2015	Elwha and Glines Canyon Dam, Washington	32–64
The removal of the Elwha and Glines Canyon dams, initiated in September 2011, induced massive increases in river sediment supply. Measurements of beach topography and near shore bathymetry show that ~2.5 million m^3 of sediment was deposited during the first two years of dam removal, which is ~100 times greater than deposition rates measured prior to dam removal. The coastal geomorphology of the active delta changed rapidly from a wave-dominated shape to a river-dominated shape as sediment supply to the coast increased after dam removal began.		

Table 7.4 *Criteria for identifying UK anthropogenic alluvium (modified from Macklin et al., 2014, table 1). Full-size table.*

Type of evidence	Definition
Color change	Change in sediment color resulting from a change in composition or provenance
Stratification change	Change from massive to layered alluvium or vice versa depending on sedimentary context
Artifacts	Includes objects made or modified by human agency (e.g. fence stakes, pottery) and waste materials (animal bones, charcoal)
Textural change	Abrupt change in grain size and/or organic content; change from peat to mineral sediment; rapid sedimentation
Biological evidence	Evidence from pollen, mollusca, and coleoptera for anthropogenic modification of the landscape (woodland clearance and cultivation); supported by environmental magnetism and charcoal
Contaminants	Elevated concentrations of pollutants from metal mining (e.g. Pb, Sn) or industry (e.g. coal/coke fragments)

Table 7.5 *Accelerated sedimentation in Britain in prehistoric and historic times.*

Location	Source	Evidence and date
Howgill Fells	Harvey et al. (1981)	Debris cone production following tenth century AD introduction of sheep farming
Upper Thames Basin	Robinson and Lambrick (1984)	River alluviation in Late Bronze Age and early Iron Age
Lake District	Pennington (1981)	Accelerated lake sedimentation at 5000 BP as a result of Neolithic agriculture
Mid-Wales	Macklin and Lewin (1986)	Floodplain sedimentation as a result of early Iron Age sedentary agriculture
Brecon Beacons	Jones et al. (1985)	Lake sedimentation increase after 5000 BP at Llangorse due to forest clearance
Weald	Burrin (1985)	Valley alluviation from Neolithic onward until early Iron Age
Bowland Fells	Harvey and Renwick (1987)	Valley terraces at 5000–2000 BP (Bronze or Iron Age settlement) and after 1000 BP (Viking settlement)
Southern England	Bell (1982)	Fills in dry valleys: Bronze and Iron Age
Callaly Moor, Northumberland	Macklin et al. (1991)	Valley fill sediments of Late Neolithic to Bronze Age
Semer Water and Raydale, North Yorkshire	Chiverrell et al. (2007)	Bronze Age

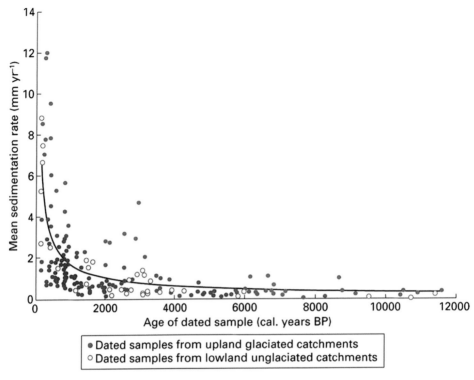

Figure 7.10 Holocene floodplain sedimentation rates in Great Britain plotted for upland (glaciated) and lowland (unglaciated) catchments (modified after Macklin et al. 2010, figure 18).

toward clastic overbank deposition was diachronous at the catchment scale, ranging between 6,500 and 500 cal yr BP.

Work in Bavaria, Germany, by Heine et al. (2005), showed how agricultural intensification led to both slope colluviation and floodplain sedimentation. Also working in Germany, Lang (2003) suggested that soil erosion led to phases of accelerated erosion, colluviation, and alluvial sedimentation in the Bronze Age, the Iron Age/Roman period, and at c AD 1000. Similarly, Dreibrodt et al. (2010) found that erosion in Germany was at a maximum in the Late Bronze Age and pre-Roman Iron Age (c. 1600 BC–1000 AD), high and late medieval times (c. 1000–1350 AD) and late modern times (from c. 1500 until today). At a lake in Denmark, Rasmussen and Bradshaw (2005) found higher rates of minerogenic accumulation during the latter part of the Late Neolithic, and indicated that this implied increased pressure on the soils resulting in increased erosion rates. The accumulation rate of minerogenic material also suggested increasing erosion throughout the Early and Late Bronze Age, with a distinct peak at the end of the latter period.

A survey of the Lake Jasień area (northern Poland) by Majewski (2014) demonstrated that economic activity in the early Iron Age and the pre-Roman period, caused deforestation that may have led to activation of hillslope processes. In the French Alps, studies of lake sedimentation showed that during the Roman period, the frequency of erosive events was the highest of the last 10,000 years, and that the most intense grazing pressure on landscape and the erosion occurred during the Roman occupation (Giguet-Covex et al., 2014). In another study in the Alps, Arnaud et al. (2012) identified a peak of erosion in the Bronze Age associated with the onset of millet cultivation. In Anatolia, Turkey, high rates of anthropogenic erosion occurred particularly in the Early Bronze Age (Marsh and Kealhofer, 2014). Bertran (2004) calculated erosion rates in two small catchments of the Quercy region (southwestern France) from colluvial deposits trapped in karstic depressions. These showed a progressive increase during the Holocene. Rates lower than 20 t km^{-2} yr^{-1} were found for the period that covers the beginning of the Holocene to the early Iron Age. After the early Iron Age, erosion increased significantly and the mean rate reached 80 t km^{-2} yr^{-1} before medieval times. The associated coarse-grained, crudely stratified colluvium is thought to reflect plowing of the surrounding fields. Erosion rates increased to c. 130 t km^{-2} yr^{-1} during medieval and modem times.

Hoffmann et al. (2009), working on sedimentation rates, suggested that human-induced accelerated soil erosion had been a feature of the last 3,000 years and they showed that on the Rhine, Mississippi, and Yellow Rivers there has been a massive rise over that sort of time period (Figure 7.11).

The combination of slope clearance for agriculture with times of intense storms would be an especially powerful stimulus to soil erosion, and catastrophic soil erosion in Central Europe was identified for the first half of the fourteen century and in the mid-eighteenth to the early nineteenth century by Dotterweich (2008). Zolitschka et al. (2003), however, while recognizing the possible impact that climatic fluctuations such as the Little Ice Age may have had in Germany, argue that the general lack of synchroneity in the sedimentation record over the last 5,000 years points to the importance of human activities.

An example of using sediment accumulation rates (SARs) to infer long-term rates of erosion is provided by Hughes et al.'s (1991) study of Kuk Swamp in Papua New Guinea. They identified low rates of erosion until 9000 BP, when, with the onset of the first phase of forest clearance, erosion rates increased from 0.15 cm per thousand years to about 1.2 cm per thousand years. Rates remained relatively stable until the last few decades when, following European contact, the extension of anthropogenic grasslands, subsistence gardens and coffee plantations has produced a rate that is very markedly higher: 34 cm per thousand years.

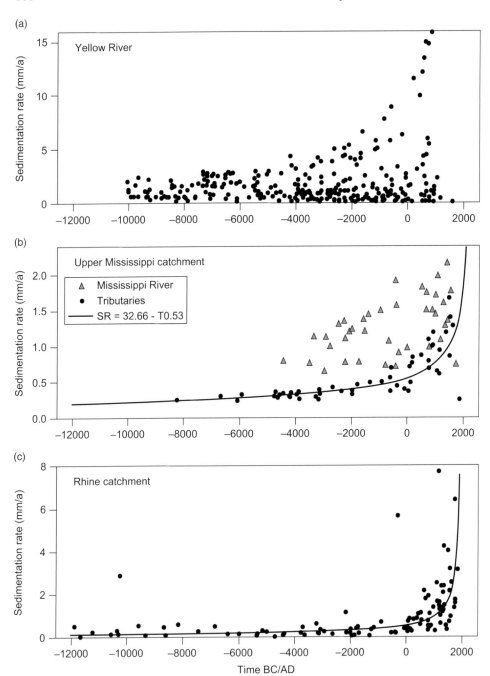

Figure 7.11 Rates of catchment sedimentation for some major rivers (from Hoffmann et al., 2009, figure 7).

In New Zealand, anthropogenic land-use changes have been restricted to the last ~800 years, with the settlement occurring in two phases, first by Polynesian immigrants, and then by European colonists within the last 200 years. A good long-term study of the response rates of erosion to land-use changes associated with these demographic changes is provided by a study undertaken on North Island by Page and Trustrum (1997). During the last 2,000 years of human settlement their catchment underwent a change from indigenous forest to fern/scrub following Polynesian settlement (c. 560 BP) and then a change to pasture following European settlement (AD 1878). SARs under European pastoral land use were between five and six times the rates that occurred under fern/scrub and between eight and seventeen times the rate under indigenous forest. In a broadly comparable study in another part of New Zealand, Sheffield et al. (1995) looked at rates of infilling of an estuary. In pre-Polynesian times SARs were 0.1 mm yr^{-1}, during Polynesian times the rate climbed to 0.3 mm yr^{-1}, while since European land clearance in the 1880s the rate has shot up to 11 mm yr^{-1} (see also Nichol et al., 2000). Richardson et al. (2014) working in Kaeo, northern New Zealand, found that fluvial sedimentation commenced around ~7,600 years ago, followed by floodplain aggradation at a rate of 0.3 to 0.7 mm yr^{-1} Floodplain sedimentation accelerated following Polynesian settlement, with aggradation rates of 3.3–10.1 mm yr^{-1}. An SAR of at least 13.5 mm yr^{-1} has characterized the last century or so of European farming. Other examples of such trends in New Zealand are provided by Glade (2003; Table 7.6).

An interesting study of erosion in Kenya over the last 300 years is provided by Fleitmann et al., (2007). They analyzed Ba/Ca records from *Porites* corals from the Malindi coral reef, which document the flux of suspended sediment from the

Table 7.6 *Rates of sediment accumulation (mm yr^{-1}) in New Zealand following European settlement and forest clearance.*

Site	European pasture rate	Polynesian or pre-Polynesian rate	Factor of increase
Whangape (estuary)	1.7–4.6	0.1–0.5	9.2–17
Rapongaere (swamp)	3.6	0.3	12
Poverty Bay (continental shelf)	3.7	0.3	12.3
Lake Tutira (freshwater lake)	14.0	2.1	6.7
Wellington Harbor (near coast)	38.2	2.1	18.2
Abel Tasman (coastal wetland)	1.6–2.7	0.5–1.7	1.6–3.2

Sabaki River with a subweekly resolution. They found that while the sediment flux was almost constant between 1700 and 1900, a continuous rise was observed after 1900, first due to British settlements and afterwards due to steadily increasing demographic pressure on land use. The peak in suspended sediment load and hence soil erosion occurred between 1974 and 1980 when there was a five to tenfold increase relative to natural background levels. They attributed this to the combined effects of dramatically increasing population, unregulated land use, deforestation, and severe droughts in the early 1970s.

In Central America, there have been various studies of long-term erosion and sedimentation rates. Beach et al. (2006) have suggested that rates were high in the Maya Lowlands in the pre-Classic Period (c. 1000 BC–AD 50) and in Late Classic times (c. AD 550–900). Similarly, O'Hara et al. (1993), working on cores from Lake Patzcuaro in Mexico, identified high rates in the pre-Classic/Early Classic period (2500–1300 BP) and in later postclassic times (850–350 BP). However, Fisher et al. (2003), who also worked in Lake Patzcuaro, argued that some high rates of erosion were associated not with intense land use, but with some phases of land abandonment, as in post-Hispanic times. Regional degradation was initiated from the disruption of a human-modified environment dependent on human labor lost to European-introduced disease. It was not the consequence of the introduction of European crops, animals, technologies, or land tenure.

In extreme cases, and in small catchments, increased SARS consequent upon land-cover changes can lead to almost total infilling of channels, and Sousa et al. (2015) reported on the evolution of coastal brooks in the southwest of Spain from the early seventeenth century to the end of the twentieth century. They found that during the second half of the twentieth century, the average annual rate of thalweg regression almost quadrupled to 432.2 m yr^{-1}, mainly due to anthropogenic impacts associated with logging.

7.12 Case Studies

7.12.1 Channel Incision and Aggradation: Arroyos and Other Gullies in the United States

In the southwestern United States many broad valleys became incised by as much as 20 m (Figure 7.12) between 1865 and 1915, with the peak occurring in the 1880s (Cooke and Reeves, 1976; Summa-Nelson and Rittenour, 2012). This incision, aided by headward erosion (DeLong et al., 2014), created valley-bottom gullies or trenches (*arroyos*) which had a rapid and detrimental effect on the flat, fertile, and easily irrigated valley floors, which are the most desirable sites for settlement and economic activity in a harsh environment. Some arroyos can be over 50 m wide and tens or even hundreds of kilometers long.

Figure 7.12 An arroyo, Laguna Creek, to the west of Kayenta, Arizona, United States.

There has been much debate as to the causes of this incision (Bull, 1997; Elliott et al., 1999; DeLong et al. 2011; Harvey and Pederson, 2011). First, it is possible, as Schumm et al. (1984) pointed out, that it could result from some intrinsic natural geomorphological threshold (such as stream gradient) being crossed, rather than as a result of climatic change or human influence. Under this autogenic argument, conditions of valley-floor stability decrease slowly over time until some triggering event initiates incision of the previously "stable" reach.

Second, humans, through timber-felling, overgrazing, harvesting hay, causing compaction along trails, channeling of runoff from trails and railways, disrupting valley-bottom sods by animals' feet, and facilitating the invasion of grasslands by scrub, could also have caused the entrenchment. The apparent coincidence of settlement and arroyo development in the late nineteenth century tended to support this view.

Third, studies of the long-term history of the fills shows that there have been repeated phases of aggradation and incision and that some of these took place before the influence of humans could have been significant (Huckleberry and Duff, 2008; French, 2009). For instance, Elliott et al. (1999), recognized various

Holocene phases of channel incision at 700–1200 BP, 1700–2300 BP, and 6500–7400 BP. There has been an increasing appreciation of the scale and frequency of climatic changes in the Holocene (McFadden and McAuliffe, 1997). Waters and Haynes (2001) argued that arroyos first appeared in the American southwest after c. 8,000 years ago, and that a dramatic increase in cutting and filling episodes occurred after c. 4,000 years ago. They believe that this intensification could be related to a change in the frequency and strength of El Niño events.

There are various climate-related hypotheses. Huntington (1914) argued that valley filling could be a consequence of a shift to more arid conditions. These, he reasoned, would cause a reduction in vegetation cover which in turn would promote rapid removal of soil from devegetated hillslopes during storms, and would overload streams with sediment. With a return to more humid conditions, vegetation would be reestablished, sediment yields would fall, and entrenchment of valley fills occur. Bryan (1928) put forward an alternative and contradictory explanation, arguing that a slight tendency toward drier conditions, by depleting vegetation cover and reducing soil infiltration capacity, would produce significant increases in erosive storm runoff, which would cause incision to occur. Leopold (1951) advanced another climatic interpretation, involving a change in rainfall intensity rather than quantity. He indicated that a reduced frequency of low-intensity rains would weaken the vegetation cover, while an increased frequency of heavy rains at the same time would increase the incidence of erosion. Support for this came from Balling and Wells (1990), who attributed early twentieth-century arroyo trenching in New Mexico to a run of years with intense and erosive rainfall that succeeded a phase of drought conditions in which the protective ability of the vegetation had declined. In a similar vein, Graf et al. (1991) argued that periods of prolonged summer drought would reduce forest cover, increase slope erosion, and thus cause sediment to accumulate in the valleys, promoting aggradation. Mann and Meltzer (2007) recognized that in New Mexico incision occurred in the medieval Warm Period and aggradation in the Little Ice Age. They argued that when the North American Monsoon system is strong, more frequent summer thunderstorms cause increased flooding. Wetter summers over periods of decades to centuries cause the vegetation cover to increase, which reduces sediment input from hillslopes at the same time that floods are eroding the valley fills. Incision therefore occurs. Hereford (1986) put forward the idea that large floods have been important, arguing that erosion and entrenchment result from larger flood regimes, with streams having a large sediment transport capacity. With lower flood regimes, however, a reduction in channel width and sediment storage occurs. He suggested that if there are no floods then no alluviation of floodplains is possible. DeLong et al. (2011), working on the complex stratigraphy of the Cuyama River in west-central California, believed channel aggradation occurred during periods of relative aridity and low peak discharge events, while wet periods, possibly floods after drought, led to fluvial incision. They suggested that the widespread arroyo cutting that was initiated near the turn of the twentieth century in the southwestern United States may indicate that regional climate variation "primed" the fluvial systems in such a way that other, local triggers readily initiated channel bed incision. In small catchments, fires may have caused accelerated slope erosion and subsequent valley alluviation (Jones et al., 2010). On Santa Cruz Island, California, Perroy et al (2012) found that shortly after the introduction of sheep in 1853, localized sedimentation rates on the Pozo floodplain increased by two orders of magnitude from 0.4 mm yr^{-1} to ~25 mm yr^{-1}. Accelerated sedimentation was followed by arroyo formation at ca. 1878 and rapid expansion of the incipient gully network. They argued that accelerated sedimentation due to overgrazing, and an unusually large 1878 rainstorm event, set the stage for arroyo formation in the Pozo watershed between 1875 and 1886. They hypothesized that even in the absence of modern human disturbance, down cutting would have occurred due to intrinsic hillslope stability thresholds.

It is possible that arroyo incision and alluviation resulted from a range of causes (Gonzalez, 2001) – autogenic, climatic, and anthropogenic – and that the timing of incision or aggradation will have varied from area to area and that individual arroyos will have had unique histories. It is also likely that different reaches of river valleys may have responded differently to environmental changes. For instance, upstream incision may have created sediment pulses that led to downstream aggradation (DeLong et al., 2011; Harvey et al., 2011; Gellis et al., 2012).

7.12.2 Mediterranean Alluviation and Incision

In the Mediterranean lands there have also been controversies surrounding the age and causes of alternating phases of aggradation and erosion in valley bottoms. In a classic and influential study, undertaken before many modern techniques of dating and environmental reconstruction were available, Vita-Finzi (1969) suggested that at some stage during historical times many of the streams in the Mediterranean area, which had hitherto been engaged primarily in down cutting, began to aggrade. Renewed down cutting, still seemingly in operation today, has since incised the channels into the alluvial fill. He proposed that the reversal of the down cutting trend in the Middle Ages was both ubiquitous and confined in time, and that some universal and time-specific agency was required to explain it. He believed that vegetation removal by humans was not a medieval innovation and that some other mechanism was required. He proposed that the mechanism was precipitation change during the Little Ice Age (AD 1550–1850). However, Butzer

(1974) disputed this. He reported plenty of post-Classical and pre-1500 alluviation (which could not therefore be ascribed to the Little Ice Age), and he doubted whether Vita-Finzi's dating was precise enough to warrant a 1550–1850 date. Instead, he suggested that humans were responsible for multiple phases of accelerated erosion from slopes and accelerated sedimentation in valley bottoms from as early as the middle of the first millennium BC. This interpretation was favored by van Andel et al. (1990) who detected an intermittent and complex record of cut-and-fill episodes during the late Holocene in Greece. They believed that this evidence is compatible with a model of the control of timing and intensity of landscape destabilization by local economic and political conditions. There does, however, still remain the possibility that the Holocene younger alluvium has a climatic rather than an anthropogenic origin (see Macklin and Woodward, 2009, p. 335). In Jordan, Cordova et al., (2008) suggested that both climatic and anthropogenic processes were involved in the complex history of Holocene alluviation and that in some cases their influence may have been complementary. Plainly, major uncertainties remain as to the relative importance of climatic and human factors.

On the other hand, there is evidence that flood events rather than climatic changes *per se* may have played a role in the sequence of alluviation and incision (see Macklin and Woodward, 2009, p 338). It is also clear that this sequence is more complex than was formerly thought. Barker and Hunt (1995), for example, found an early Roman phase of alluviation in Central Italy and related this to land clearance. Carmona and Ruiz (2011) also found a Roman alluviation phase in eastern Spain. In southern Italy, Boenzi et al. (2008) argued that Holocene filling and gullying in the Basento river valley occurred as a succession of cut-and-fill episodes. A first phase of accumulation occurred in the Late Neolithic, which was followed by down cutting (4,500–3,700 years ago). A second depositional phase took place in the Greek–Roman period (2,800–1,620 years ago). Another down-cutting phase took place 1,620–1,500 years ago, followed by a deposition phase between 1,440 and 1,000 years ago. After that, yet another phase of deep incision took place. They suggested that these cut-and-fill episodes were predominantly climate-driven. In Macedonia, Greece, Lespez (2003), believed that distinct phases of stream aggradation, soil erosion, and landscape stability in the Drama basin over the past 7,000 years could be tied directly to long-term land-use changes. During the Middle and Late Holocene, 1–4 m of alluvial fill accumulated. Low levels of alluvial aggradation were recorded during the Late Neolithic and the Early Bronze Age (7400–4000 cal BP), but in the center of the basin, the rate of aggradation has doubled since the Late Bronze Age (0.5–0.6 mm yr^{-1}) compared to the Holocene as a whole (0.25 mm yr^{-1}). Moderate rates of alluvial fill were experienced during the Late Bronze Age (3600–3000 cal BP). Two further historical phases of alluvial aggradation and soil erosion were identified: in the Antique and Early Byzantine Era (third century BC to seventh century AD) and, more significantly, in the Ottoman period (beginning of the fifteenth to the twentieth century AD).

7.12.3 *Madagascan Lavakas*

Spectacular gullies, locally called lavaka, occur in the rolling saprolite-mantled terrains of Madagascar (Figure 7.13). Here too there have been debates about cultural versus natural causation (Wells and Andriamihaja, 1993). Proponents of cultural causes have argued that since humans arrived on the island in the last 2,000 years, there has been excessive cattle grazing, removal of forest for charcoal and for slash-and-burn cultivation, devastating winter (dry season) burning of grasslands, and erosion along tracks and trails. However, the situation is more complex than that, and the lavaka are polygenetic. Tales of ecological mayhem and catastrophic erosion may have been exaggerated (Kull, 2000). Tectonism and natural climatic factors may be at least as important, and given the climatic and soil types of the island many lavaka are a natural part of the landscape's evolution (Zavada et al., 2009). Some of them also clearly predate primary (i.e. uncut) rain forest. Recently, Cox et al. (2009) have pointed out that there are no sediment generation rate data available for lavakas and no data quantifying the relationship between lavaka activity and human influence. Cited estimates, in the tens of

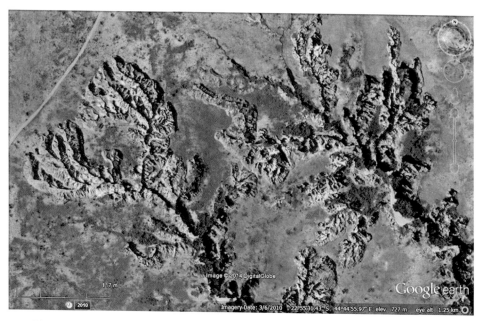

Figure 7.13 Lavakas in Madagascar. Scale bar 117 m. ©2014 DigitalGlobe.

thousands of tons per square kilometer per year, either represent small-scale plot studies in areas preidentified as zones of rapid erosion, which typically greatly overestimate regional erosion rates, or are reported without supporting information on data collection methods. Moreover, their ^{10}Be data further suggest that lavakas were major sediment contributors before humans significantly altered the landscape. Mietton et al. (2014) used optical dating to show that many lavaka were indeed more ancient than human settlement. Cox et al. (2010) have argued that if lavakas are primarily the result of human activities, then, all else being equal, the most heavily used areas should be most affected; but comparison of areas with the same geology, topography, and climate shows differently. They indicate that the most densely populated, heavily grazed, and devegetated regions surrounding major cities have only low to moderate lavaka densities, whereas some of the emptiest lands, almost devoid of cattle, grassland burning, or habitation have among the highest lavaka concentrations.

7.13 The Future: River Flows and Channels under Climate Change

Climate changes are but one of many possible controls of changing stream runoff. Nonetheless, global warming will affect rivers in many ways. Increasing temperatures will melt snow and ice and promote greater losses of moisture through evapotranspiration. There will be changes in the amount, intensity, duration, and timing of precipitation, which will also affect the size of snowpacks and river flows. Vegetation cover will respond to temperature and precipitation changes, as will land use. Vegetation assemblages will also be affected by an increasing frequency of fires (Flannigan et al., 2013; Wang et al., 2015) which will affect runoff. Westerling et al. (2011), working in the Yellowstone area of the United States, have suggested that "Continued warming could completely transform ... fire regimes by the mid-21st century, with profound consequences for many species and for ecosystem services including aesthetics, hydrology, and carbon storage. The conditions associated with extreme fire seasons are expected to become much more frequent, with fire occurrence and area burned exceeding that observed in the historical record or reconstructed from paleo-proxy records for the past 10,000 y." Likewise, Pechony and Shindell (2010) have remarked "Our future projections indicate an impending shift to a temperature-driven global fire regime in the 21st century, creating an unprecedentedly fire-prone environment. These results suggest a possibility that in the future climate will play a considerably stronger role in driving global fire trends, outweighing direct human influence on fire (both ignition and suppression), a reversal from the situation during the last two centuries." Fire frequency could also be affected by an increased incidence of lightning strikes, and for the United States, Romps et al.

(2014) estimated they may increase by 12 ± 5 percent per °C of global warming and about 50 percent over this century.

Higher atmospheric CO_2 levels may stimulate plant growth and lead to changes in plant water use efficiency and thus to transpiration and runoff (Morgan et al., 2004; Gedney et al., 2006; Keenan et al., 2013). Global warming may also affect soil properties (such as organic matter content) which could alter runoff generation processes. Climate change will cause human interventions in the hydrological system with, for example, greater use of irrigation in areas subject to increased drought risk, and the continued spread of engineering controls on flooding and erosion.

On a global basis it is possible runoff will increase because of a global increase in precipitation (Douville et al., 2002). Historical discharge records indicate that global runoff increases by c. 4 percent for each 1°C rise in temperature (Labat et al., 2004).

Arnell (2002) attempted to map future runoff trends on a global basis and found that there was a large range in responses. Some areas will become markedly prone to greatly reduced annual runoff, while others will see an enhancement of flows. However, the patterning at a global scale indicates that by the end decades of this century, high latitudes in the Northern Hemisphere, together with parts of Central Africa and Central Asia, will have higher annual runoff levels, whereas Australia, southern Africa, northwest India, the Middle East, and the Mediterranean basin will show reduced runoff levels. There is some tendency, to which the Taklamakan of Central Asia appears to be a major exception, for major deserts (e.g. Namib, Kalahari, Australian, Thar, Arabian, Patagonian, and North Sahara) to become even drier. More recently, Sperna-Weiland et al. (2011) have analyzed available General Circulation Models (GCMs) and suggest that there will be consistent decreases in discharge for southern Europe, southern Australia, parts of Africa, and southwestern South America, while discharge will increase slightly in some monsoonal areas and in the Arctic. Arnell and Gosling (2013) reported that more than two-thirds of climate models project a significant increase in average annual runoff across almost a quarter of the land surface, and a significant decrease over 14 percent, with considerably higher degrees of consistency in some regions. They also reported that most climate models project increases in runoff in Canada and high-latitude eastern Europe and Siberia, and decreases in runoff in Central Europe, around the Mediterranean, the Levant, Central America, and Brazil. Another global assessment by van Vleit et al. (2013) came up with a broadly similar picture of change, with an increase in mean annual river discharge in the high northern latitudes and large parts of the tropical (monsoon) region, and a decrease for the mid northern latitude region (United States, Central America, Southern and Central Europe, Southeast Asia) and the southern latitudes (southern parts of South America, Africa, and Australia). In the United States, models tend to suggest substantial reductions in runoff in the southwest and central United States, but increases in the northeast (Karl et al., 2009). Liu et al. (2011) have suggested that in the Yellow River Basin of China the effects of increased evapotranspiration will be more than offset by increases in precipitation so that by 2080 that river's annual discharge could increase by 35–43 percent.

Hurricanes are major drivers of flows and extreme floods in tropical rivers (Figure 7.14). That said, some caution is necessary with regard to the extent to which warming will stimulate tropical cyclone activity (see IPCC, 2013, p. 1248), and Knutson et al. (2008) argued that there might not be any great increase in hurricane frequency driven by increases in atmospheric greenhouse gas concentrations. Haig et al. (2014), working in Western Australia, found a sharp decrease in cyclone activity after 1960 in Western Australia. This was in contrast to the increasing frequency and destructiveness of Northern Hemisphere tropical cyclones since 1970 in the Atlantic Ocean and the western North Pacific. They noted that other studies project a decrease in the frequency of tropical cyclones toward the end of the twenty-first century in the southwest Pacific, southern Indian, and Australian regions.

Figure 7.14 The consequences for river flow of a large tropical cyclone is illustrated by the effects of Cyclone Domoina in Swaziland in 1984. It demolished a bridge over the Usutu River. Such events could become more prevalent in a warmer world as the intensity, frequency, and latitudinal spread of such storms occurs.

However, the relationship between sea surface temperature increase and increasing global hurricane activity has been confirmed by Hoyos et al. (2006) and Saunders and Lea (2008), while Santer et al. (2006) have demonstrated that human factors have caused the increase in sea surface temperatures and cyclone development in both the Atlantic and Pacific regions. Webster et al. (2005) and Elsner et al. (2008) have shown that the strongest tropical cyclones have become more intense in recent decades, and Marciano et al. (2015) suggest that one cause for this is that precipitation is projected to increase with warming owing to increased atmospheric water vapor content. This presents the possibility for enhancement of cyclone intensity through increased lower-tropospheric diabatic potential vorticity generation, with enhanced latent heat release becoming responsible for increases in future cyclone intensity. Tsuboki et al. (2014) believe that the most intense future super typhoons or cyclones could attain wind speeds of 85–90 m s^{-1} and minimum central pressures of 860 hPa.

Nijssen et al. (2001) suggested that the largest changes in the hydrological cycle will occur in snow-dominated basins of mid to higher latitudes (Figure 7.15), and in particular there are likely to be marked changes in the amplitude and phase of

Figure 7.15 In high altitude and high to mid latitude areas, the presence of snowpack is an important control of runoff, as in this river in central Sweden.

the annual water cycle (Arora and Boer 2001). Pederson et al., (2011) discussed the history of snowpack in North America and believed that the rapid reduction in snowpack since the 1980s is due to unparalleled springtime warming. There has been a fundamental shift from precipitation to temperature as the dominant influence on snowpack in the North American cordillera. That said, changes in precipitation totals may also be significant and in some areas may work in the same or opposite direction as warming (Adam et al., 2009; Stewart, 2009; Costa-Cabral et al., 2013).

Changes in land cover and land use, by altering evapotranspiration, will also have an impact on future stream flows and will probably lead to an overall increase at the global scale (Sterling et al., 2013).

The other major control on future river flow in cold regions will be the melting of glaciers. Glaciers in areas like the Karakorams (Figure 7.16) and Himalayas are important sources of flow to rivers like the Indus and the Ganga. If melting occurs, discharges may initially increase, but as the glacier mass declines through time, so will stream flows (Bradley et al., 2006; Rees and Collins, 2006; Immer Zeel et al., 2010). Some areas will show an initial increase in runoff, and there will also probably be a concomitant increase in flood frequency (Hirabayashi, et al., 2013). Bliss et al. (2014) undertook a global survey of streamflow response to glacier

Figure 7.16 The Hunza River in the Karakoram Mountains of northern Pakistan, a major tributary of the Indus, receives much of its flow from melting glaciers.

melting from now to the end of the century. They suggested that the magnitude and sign of trends in annual runoff totals will differ considerably depending on the balance between enhanced melt and the reduction of the glacier reservoir by glacier retreat and shrinkage. In their analysis, most regions exhibited a fairly steady decline in runoff, demonstrating that they have passed their peak runoff. For example, runoff from glaciers in Western Canada and the United States declined by 72 percent between 2003–2022 and 2080–2099. Low latitudes exhibited the fastest runoff decline (96 percent) due to the rapid and near complete volume loss of ice. In some regions, runoff will be stable for a few decades and then will decline (by 29 percent in Alaska and by 25 percent in the southern portion of Arctic Canada). Iceland, Svalbard, and South Asia will experience increasing runoff until the middle of the century, peaking 22 percent, 54 percent, and 27 percent higher than the initial period and declining thereafter, ending 30 percent, 10 percent, and 11 percent below their initial values, respectively. However, in the northern portion of Arctic Canada and the Russian Arctic, the runoff will steadily increase throughout most of the twenty-first century, ending 36 percent and 85 percent higher than it was at its start.

One possible response of streams to changes in runoff is that the drainage density (channel length per unit area) will alter. After considering the relationship between mean annual rainfall and drainage density in Africa, de Wit and Stankiewicz (2006) proposed that in areas with 500 mm per year, a 10 percent decrease in precipitation could reduce drainage density by as much as 50 percent. Changes in river flows will affect channel dimensions and will have implications for the positions of international boundaries (Grainger and Conway, 2014).

If future climate changes modify river flood regimes, then the consequences for river channels and biota will be substantial (Death et al., 2015). Changes in geomorphology arising from extreme floods could determine the distribution, size, and variability of habitats such as pools and riffles, substratum composition, and the extent of deposited fine sediment. Death et al. (2015) see this as a major research priority.

7.14 Lake Level Changes

One of the results of human modification of river regimes is that lake levels have changed, though once again it is not always possible to distinguish between the part played by humans and that played by natural climatic changes.

A case study of the combined effects of climate change and anthropogenic pressures is Bosten Lake (Guo et al., 2015), the largest inland freshwater lake in China. This has shown three dramatic fluctuations of lake level over the past fifty years, with the causes leading to the dramatic fluctuations in lake level being different in each period. In the first period (1958–87) the lake level showed a continuously declining trend due to an increase in lake evaporation. In the second period (1988–2002), the lake level rose rapidly by 4.6 m because outflow was controlled by humans and the water inflow to the lake increased at the same time. In the third period from 2003–10, the emergency project of transferring water to the Tarim River led to increased lake outflow, while inflow decreased at the same time due to a reduction in precipitation, and these factors resulted in the lake level decreasing sharply by 3.8 m.

A lake basin for which there are particularly long records is the Valencia Basin in Venezuela. It was the declining level of the waters in its lake which so struck von Humboldt in 1800 (see Chapter 1). He recorded its level as being about 422 m above sea level. The 1968 level was 17 m lower. Humboldt believed that the cause of the declining level was deforestation, and this has been supported by Böckh (1973), who points also to the abstraction of water for irrigation. This remarkable fall in level meant that the lake ceased to have an overflow into the River Orinoco. It has as a consequence become subject to a build-up in salinity, and is now eight times more saline than it was 250 years ago.

Even the world's largest lake, the Caspian, has been modified by human activities. The most important change was the fall of 3 m in its level between 1929 and the late 1970s. This was undoubtedly partly the product of climatic change, for winter precipitation in the northern Volga Basin, the chief flow-generating area of the Caspian, was generally below normal for that period. Nonetheless, human actions contributed to this fall, particularly since the 1950s, because of reservoir formation, irrigation, municipal and industrial withdrawals, and agricultural practices. An amelioration of climate since the late 1970s has caused some two meters of recovery in the level of the lake (Sapozhnikov et al., 2010).

Interbasin water transfers cause changes in lakes, as illustrated by the desiccation that is taking place in the Aral Sea (Saiko and Zonn, 2000; Shi et al., 2014; Figure 7.17). Between 1960 and 1990, largely because of diversions of river flow, it lost more than 40 percent of its area and about 60 percent of its volume. Its level fell by more than 14 m. By 2002 its level had fallen another 6 m, and by 2008 its area was just 15.7 percent of that in 1961 (Kravstova and Tarasenko, 2010). This has lowered the artesian water table over a band 80–170 km in width, has exposed 24,000 km^2 of former lake bed to desiccation, and has created dry surfaces from which salts are deflated to be transported in dust storms (see Chapter 8).

Water abstraction from the Jordan River has caused a decline in the level of the Dead Sea, producing recessional shorelines (Figure 7.18). During the last four to five decades increasing amounts of water have been diverted from surface and groundwater sources in its catchment. It was at its highest level in 1896, reaching

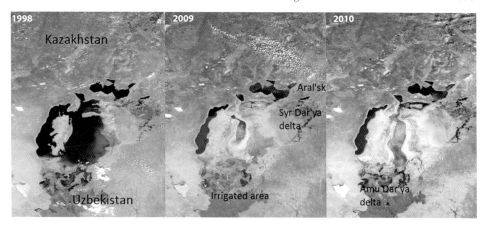

Figure 7.17 Changing extent of the Aral Sea, 1998 to 2010.

Figure 7.18 Recessional shorelines on the shore of the Dead Sea, Jordan.

an elevation of ~388.4 m below mean sea level (m.b.m.s.l.) and ~390 m in the early 1920s. Since then it has almost constantly been dropping, reaching a level of 426 m.b.m.s.l. in 2013. Since the late 1990s its level has been decreasing by approximately 1 m per year (Filin et al., 2014). This has accelerated large-scale environmental deterioration, including soil erosion, rapid headcut migration, gully incision, and widespread development of sinkholes. (Yechieli et al., 2006; Closson et al., 2007) (see Chapter 4).

Another lake in the Middle East that has shown recent catastrophic desiccation is Urmia in Iran. According to Tourian et al. (2015) it was the largest inland body of salt water in the Middle East and the largest permanent hypersaline lake in the world with an area varying from 5,200 to 6,000 km^2 in the twentieth century. The lake has lost about 70 percent of its surface area over the last fourteen years, partly because of drought but also because of an increase in groundwater exploitation that has caused the water table to fall.

The levels of lakes in East Africa have also been responding to a range of pressures. For example, abstraction of water to supply the cut-flower industry has led to a fall in the level of Lake Naivasha in Kenya (Mekonnen et al., 2012), while in Ethiopia two lakes, Haromaya and Adele, have dried up completely because of changes in land use in their catchments, water abstraction for irrigation and municipal supply, and siltation (Alemayehu et al., 2007). In the Ethiopian Rift Valley south of Addis Ababa, the level of Lake Abiyata has fallen some meters because of water abstraction connected with the potash industry (Ayenew and Legesse, 2007; Seyoum et al., 2015).

In the American southwest, water abstraction for municipal supply and irrigation has led to catastrophic falls in the levels of Mono and Owens Lakes (Gill, 1996), which in turn has caused an increase in dust storm activity from their desiccated floors. In the case of Mono, since the first diversions of its tributaries to quench the thirst of growing Los Angeles in 1941, its water level has dropped by 11 m. Water abstraction for irrigation has led to declines in the levels of lakes in the Great Basin, where, beginning in the mid-nineteenth century, the introduction of agriculture upstream of Walker Lake resulted in the water from inflowing rivers being diverted. As a result, its level dropped approximately 40 m between 1882 and 1994. In Mexico, the country's largest lake, Chapala, began to shrink in the 1970s, corresponding with increased agricultural development in the Río Lerma watershed. Since then, it has lost more than 80 percent of its water. Between 1986 and 2001, Chapala shrank in size from 1,048 km^2 to 812 km^2. Also in Mexico, Lake Texcoco has been drained to reduce flooding risk and to provide land for agricultural expansion.

However, not all lakes have shrunk in response to human activities. The Salton Sea in southern California became a lake because of the diversion of water into it from the Colorado River for irrigation. While it varies in dimensions and area with fluctuations in agricultural runoff and rainfall, the Salton Sea today has an average extent of 24 km by 56 km. With an estimated surface area of c. 900 km^2 it is now the largest lake in California.

7.14.1 The Future: Lakes

Future climatic changes will directly affect lake levels, but will also have indirect effects through the increasing levels of water abstraction for irrigation that may be stimulated by a drier climate (Jeppesen et al., 2015). Climatic changes will be particularly important for the closed depressions which are widespread in arid lands and in tectonic rifts. Their water levels and salinity characteristics respond rapidly and profoundly to climatic changes (Grimm et al., 1997). In the early 1960s, prior to the development of the Sahel Drought, Lake Chad had an area of 23,500 km^2, but by the 1980s, as its level fell, it had split into two separate basins and had an area of only 1,500 km^2 (Figure 7.19). The Caspian Sea was at 29.10 m below sea level in 1977, but in 1995 had risen to 26.65 m below sea level, an increase of 2.45 m in just seventeen years. Similarly impressive changes have occurred in recent decades in the level of the Great Salt Lake in Utah, with a particularly rapid rise taking place between 1964 and 1985 of nearly 6 m.

From the early 1950s to the middle 1980s the total area of lakes in China with an individual area of over 1 km^2 declined from 2,800 to 2,300 km^2 and the whole area of China's lakes has been reduced from 80,600 km^2 to 70,988 km^2. An increasingly warm and dry climate was the principal cause of the reduced lake area on the Qingzang Plateau, Northwest China, the Inner Mongolian Plateau, and the North China Plain. In the south of China, however, human impacts, such as water abstraction, played a greater role (Ma et al., 2010).

Figure 7.19 The changing level of Lake Chad since 1972. Based on data in http://commons.wikimedia.org/wiki/File:Lake_Chad.jpg. (Accessed November 20, 2014).

Lakes will respond to the temperature and precipitation changes that may result from future global warming. Falls of up to 9 m by the end of the present century have been predicted for the Caspian (Elguindi and Giorgi, 2006). Rather more modest falls of fractions of a meter have been projected for the Great Lakes of North America (Angel and Kunkel, 2010; Hayhoe et al., 2010).

Lakes will also change in permafrost areas and it is likely that as glaciers downwaste and retreat there will be more marginal lakes and associated outburst floods of the types that have been recorded in recent decades (see Chapter 10).

7.14.2 Other Possible Future Human Impacts on Lakes

In the future, in addition to the effects of climatic change, major geoengineering schemes may take place. There have been proposals for augmentation of lake volumes, either by means of river diversions or by transferring seawater to them through tunnels or canals (Cathcart, 1983). Among such plans have been those to flood the salt lakes of the Kalahari by transferring water from the rivers of central Africa; and the schemes to transfer Mediterranean water to the Dead Sea and to the Quattara Depression in Egypt. There have also been proposals that the level of the Aral Sea could be restored by transferring water from the Black Sea into the Caspian and then into the Aral (Badescu and Cathcart, 2011).

More generally, however, continuing population increases, urbanization, and the need to increase food production by means of irrigation, will ensure that lakes will show changes in level, extent, and sedimentation in coming decades.

7.15 Conclusions

Humans have greatly modified river flows, sediment loads, and channel forms, sometimes deliberately and sometimes unintentionally. It has often proved difficult to predict the fluvial response to human disturbance because of the complexity that is inherent in fluvial systems, the natural variability of flow and sediment transport, the numerous feedback loops that exist, and because of the complex interplay between society and geomorphic processes in managed fluvial systems. Humans have also modified lake levels. In the future, climate changes will further modify runoff and flood characteristics, but the effects of these also need to be seen in the context of changes in future land cover, land use, irrigation, water withdrawals, the extent of impervious surfaces in cities and on transport links, and groundwater exploitation (García-Ruiz et. al., 2011; Caldwell et al., 2012). Lakes will also respond to future climatic and land-use changes.

8

Aeolian Processes and Forms in the Anthropocene

8.1 Introduction

Considerable interest in the role of humans in altering aeolian landscapes developed in the mid to late nineteenth century. Notable here was Marsh's *Man and Nature* (1864; see also Chapter 1). He devoted fifty-four pages to dunes, called taming of dunes, "a geographical revolution," talked of "the warfare man wages with the sand hills," and asked, "In what degree the naked condition of most dunes is to be ascribed to the improvidence and indiscretion of man." Later, Sokolov (1884; 1894) drew the attention of his Russian and German audiences to pressures that were exerted on dunes and noted the hazards posed by dune reactivation and migration. The Dust Bowl of the 1930s in the United States provided a later major stimulus to studies of wind erosion, dust-storm generation, and control methods. There was a burgeoning concern with desertification in other drylands, but it has also become evident that wind erosion is a potent force on agricultural soils in large tracts of Europe (Borrelli et al., 2014). Modeling the effects of land cover and land-use changes on the aeolian environment has become a major research priority (Li et al., 2014).

The study of aeolian dust in ocean and ice core sediments, the analysis of long-term meteorological data (Goudie, 1983), the recording of trajectories of dust storms on satellite images, the development of new forms of dust trap, the recognition of dust activity on Mars, and the use of field wind tunnels, caused a huge expansion of interest in the 1970s and 1980s (see, e.g., Morales, 1979; Goudie, 1978; and Péwé, 1981). It became realized that dust had a whole suite of impacts on the Earth's environment at local, regional, and global scales (Goudie and Middleton, 2006; Knippertz and Stuut, 2014), and also on human health (Goudie, 2014).

This chapter discusses two issues relating to aeolian geomorphology within the Anthropocene: how humans have affected dust generation, and how they have

179

modified sand dunes. The former topic is returned to in Chapter 11, when we discuss the impact of dust loadings on the Earth System. However, further thought also deserves to be given to the effects of land-use change on closed dryland depressions of probable aeolian origin – pans or playas – as, for example, in the High Plains of the United States, where depression volumes and hydroperiods have been greatly reduced by sediment runoff from land converted from natural grassland to rangeland (Luo et al., 1997; Tsai et al., 2007; Smith et al., 2011). Similar degradation has also affected some of the closed depressions of northeast Spain (Castaneda and Herrero, 2008).

8.2 Dust Storms and Wind Erosion

Dust storms result from the entrainment of fine particles from dry, deflated surfaces, and are one indicator of soil erosion by wind (Goudie and Middleton, 2006; Zobeck et al., 2013; Goudie, 2013a). Human pressures that influence their incidence include disturbance of desert surfaces, which often results in the lowering of the threshold shear velocity, making the surface more susceptible to wind attack (Gillies, 2013; Brungard et al., 2015). Disturbance factors include vehicular traffic and track formation, (Goossens et al., 2012), increasing wildfires (Gabet, 2014), military activity (Oliver, 1946; van Donk et al, 2003; Gillies et al., 2007), removal of vegetation cover for wood supply, fragmentation or removal of biological crusts (Belnap and Gillette, 1997), and grazing and crop production. Another important anthropogenic effect is the desiccation of lakes and soil surfaces by interbasin water transfers and ground water depletion (see, e.g., Ravi et al., 2011). Desiccated lake beds, such as those of the Aral Sea (Indoitu et al., 2015; Figure 8.1) and Owens Lake in California, are now major sources of dust (Gill, 1996). Humans may also produce new surfaces from which dust can be generated, including feedlots (Rogge et al., 2006), construction sites (Pianalto and Yool, 2013), and mine tailings (Stovern et al., 2014).

Anthropogenic Causes of Increased Emissions of Dust

Construction sites
Dirt roads
Feedbacks creating drought accentuation
Feedlots
Fires
Lake desiccation resulting from extraction of water and interbasin water transfers

Military activity (including artillery blasts, helicopter dust generation, mine laying, etc.)

Mine tailings

Reactivation of sand dunes

Surface disturbance of soil crusts and stone pavements by vehicles, plowing, warfare, etc.

Vegetation removal by overgrazing, deforestation, etc.

Some studies have estimated that up to 50 percent of the current atmospheric dust load originates from anthropogenically disturbed surfaces (see, e.g., Tegen and Fung, 1995). However, a more recent study (Tegen et al., 2004), has suggested that this may be an over-estimate and that dust from agricultural areas contributes <10 percent to the global dust load. Likewise, studies of dust over North Africa

Figure 8.1 MODIS image of the Aral Sea, April 2003, showing dust plume blowing off its desiccated floor (courtesy of NASA).

using remote sensing (Brooks and Le Grand, 2000) suggest that there is little or no evidence that dust production is associated with widespread land degradation. On the other hand, a more recent global review of the importance of different human pressures is provided by Ginoux et al. (2012). They estimated that natural dust sources globally accounted for 75 percent of emissions and anthropogenic for 25 percent. North Africa accounted for 55 percent of global dust emissions but only 8 percent were anthropogenic. Elsewhere, anthropogenic dust emissions could be much higher (e.g. 75 percent, in Australia).

In many parts of the world, dust-storm frequencies are changing in response to land use and climatic changes, though it has often proved difficult to disentangle the importance of these two main factors. Nonetheless, human activities may have had an important effect on dust storms for a considerable time. Von Suchodoletz et al. (2010) even speculated that humans intensified dust-storm activity in the northwest Sahara as early as 7,000–8,000 years (i.e. very early in the Anthropocene). The history of dust-storm activity before the era of meteorological records can be obtained from various types of core evidence and environmental reconstruction. This has shown that in some areas the incidence of dust storms has increased. Neff et al. (2008), for instance, used analyses of lake cores in the San Juan Mountains of southwestern Colorado, United States, to show that dust levels increased by 500 percent above the late Holocene average following the increased western settlement and livestock grazing during the nineteenth and early twentieth centuries. Increasing amounts of dust deposition in peat bogs in the United States also seem to have coincided with the spread of European agriculture. A core from the Antarctic Peninsula (McConnell et al., 2007) showed a doubling in dust deposition in the twentieth century, and this is explained by increasing temperatures, decreasing relative humidity, and widespread desertification in the source region – Patagonia and northern Argentina. Marx et al. (2014) used a core from Australia to show that since the 1880s rates of wind erosion were ten times higher than background Holocene levels. Finally, analysis of a 3,200-year marine core off West Africa showed a marked increase in dust activity at the beginning of the nineteenth century, which was a time that saw the advent of commercial agriculture (including the clearing of ground for groundnut production) in the Sahel (Mulitza et al., 2010).

The US Dustbowl of the 1930s demonstrated the serious nature of some past aeolian episodes within the Anthropocene. It was caused by a combination of (1) a major drought caused by a series of hot, dry years which depleted the vegetation cover and made the soils dry enough to be susceptible to wind erosion, and (2) by years of overgrazing and unsatisfactory farming techniques (Lee and Gill, 2015). The Dust Bowl may have had a feedback effect on the drought itself (Cook et al., 2009). The most severe dust storms ("black blizzards") occurred between 1933 and 1938. At Amarillo, Texas, at the height of the period, one month had 23 days with at least ten hours of airborne dust. For comparison, the long-term average for this part of Texas is just six dust storms a year. Blame for this dramatic disaster has largely been laid at the feet of the pioneering farmers – "sod busters" – who plowed up the Great Plains. For although dust storms are frequent in the area during dry years, and the 1930s was a period of drought, the scale and extent of the 1930s events were unprecedented. The waves of settlers that arrived in the area from 1914 to 1930, in conjunction with the increasing use of mechanized agriculture, catalyzed by high wheat prices, led to exceptionally large-scale wind erosion when drought hit the plains in 1931. In 1937, the US Soil Conservation Service estimated that 43 percent of a 6.5 million ha area in the heart of the Dust Bowl had been seriously damaged by wind erosion (Goudie and Middleton, 1992).

Even in the relatively moist environment of Britain, susceptible organic soils and Pleistocene coversands may be subject to "soil blows." One of the factors identified as causing this in the 1960s was the cultivation of sugar beet, a crop that requires a fine tilth and which, compared to other crops, tends to leave the soil relatively bare in early summer, when there can be long spells of dry, windy weather (Pollard and Miller, 1968; Robinson, 1968).

In the mid-twentieth century, a major aeolian episode, evident in meteorological records (Sazhin, 1988) and ice cores (Olivier et al., 2006), occurred in the former Soviet Union. After Khrushchev's "Virgin Lands" program of agricultural expansion in the 1950s, in the Omsk region, where steppe was replaced by fields, dust-storm frequency went up on average by two and a half times. In China, forced migration and agricultural expansion in the arid west during Chairman Mao Zedong's so-called "Great Leap Forward" of the late 1950s and early 1960s also led to an increase in dust-storm frequencies (Ta et al., 2006).

It remains to be established, however, if meteorological data for the last six decades or so, indicate whether or not increasing dust-storm frequencies are the norm across the globe (Goudie, 2014). Some areas have indeed shown increasing trends (e.g. the eastern Mediterranean, the Gobi of Mongolia, the western United States, and Korea). However, others have shown declining trends in the late twentieth century (e.g. China, the Canary Islands, Turkmenistan, Kazakhstan, Central Asia, Pakistan, and parts of the US High Plains and Utah). In Australia the declining trend was followed by a spike of activity in the early years of the present century. Other areas (e.g. the Kalahari, southwestern Iran, and Seistan) have shown marked fluctuations upwards and downwards in response to such factors as lake flooding and desiccation or climatic fluctuations such as sunspot cycles. Using a variety of data sources, Mahowald et al. (2010) estimated the global picture of changes in dust-storm activity for the twentieth century. They suggest a doubling of atmospheric desert dust loadings took place and they ascribe much of this to human impacts.

In the West African Sahel, where drought had persisted since the mid-1960s, analysis of wind, precipitation, and visibility data by Ozer (2003) showed that there have been remarkable changes in dust emissions since the late 1940s. He indicated that during the predrought conditions, which existed from the late 1940s to the late 1960s, yearly dust production was 126×10^6 tons. It rose to 317×10^6 tons during the 1970s and has been $1{,}275 \times 10^6$ tons since 1980, a tenfold increase. Variability in Sahel dust emissions has also been linked not only to droughts, but also to stronger winds, to changes in the North Atlantic Oscillation (Engelstaedter et al., 2006), to North Atlantic sea surface temperatures (Wong et al., 2008) and to the Atlantic Multidecadal Oscillation (Jilbert et al., 2010). The role of human population increase was, however, championed by Moulin and Chiapello (2006).

Dust emissions from West Africa and their transmission across the Atlantic have recently shown some decline (Ridley et al., 2014). Chin et al. (2014) attributed this to (1) an increase of the sea surface temperature in the North Atlantic, which may drive the decrease of the wind velocity over North Africa, which reduces the dust emissions, and (2) the increase of precipitation over the tropical North Atlantic, which enhances dust removal during transport. When the North Atlantic Ocean was cold from the late 1960s to the early 1990s, the Sahel received less rainfall and the tropical North Atlantic experienced a high concentration of dust. The opposite was true when it was warm before the late 1960s and after the early 1990s (Wang et al., 2012). Another possible cause is that weaker winds have resulted from increased roughness and reduced turbulence, associated with the observed increase in vegetation cover in the Sahel (Cowie et al., 2013).

Humans have developed techniques for wind erosion and dust-storm control (Middleton, 1990; Riksen et al., 2003a; Sterk, 2003; Nordstrom and Hotta, 2004; Ravi et al., 2011). Four main categories of methods have been used.

- Agronomic measures: crops residues, mulches, etc.
- Soil management: restricted tillage operations
- Mechanical methods: fences, windbreaks, shelterbelts, etc.
- Lake bed stabilization: by irrigation

Of these methods, the planting of shelterbelts has a long history, and in the United States, for example, the government established a tree nursery in the Nebraska Sand Hills in 1902 to pioneer the mass production of conifer seedlings for farm protection on the Great Plains and the Forest Service conducted experiments on the most effective form of planting (Gardner, 2009). Poplars and willows are used in many parts of the world (Isebrands and Richardson, 2014). The Chinese have also instituted a major shelterbelt program, though its effectiveness has been the subject of debate (Wang et al., 2010). Nonetheless, Tan and Li (2015) have shown that

China's "Green Great Wall" (also known as the "Three-North Shelterbelt Program"), launched in 1978, caused a reduction in dust-storm intensity. In any event, over much of the Taklimakan Desert, dust-storm frequencies have declined since the 1960s, though this may partly be for climatic reasons (Yang et al., 2015).

8.3 Sand Dunes

Humans have impacted on sand dunes in a variety of ways. For example, rivers can be sources and sinks of dune sediments, so that if their channels and flow regimes are altered, dunes may be affected (Draut, 2012). More generally, sand dunes have been reactivated on many desert margins by human activities (Figures 8.2 and 8.3).

Relict and stable dunes dating back to dry phases of the Pleistocene and Holocene have become active, as in parts of China (Yang et al., 2007; Liu and Wang, 2014), the Kalahari (Thomas and Twyman, 2004), the Sahel of West Africa (Mortimore, 1989), the High Plains of the United States (Goudie and Middleton,

Figure 8.2 Dunes in the Molopo Valley between Botswana and South Africa. Located on the naturally active boundary for dune movement, these dunes are susceptible both to droughts and to overgrazing, as here in the vicinity of a well.

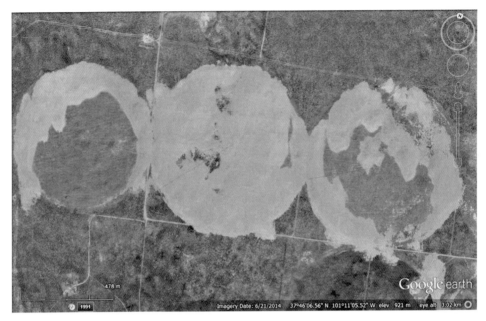

Figure 8.3 Dunes from the Late Pleistocene or Holocene dry phase in Kansas have been reactivated by exposure of fields irrigated by center-pivots to wind action after crops have been harvested. Scale bar 478 m. ©Google Earth, 2014.

1992), and India (Kumar et al., 1993). On the edge of the Rub 'Al Khali in Arabia, very rapid dune accretion may have been caused by vegetation removal for smelting of copper and by domestic stock during the Abassid period (750–1250 AD; Stokes et al., 2003).

Along coastlines, dunes have been modified by human activities, not least around the Mediterranean (Feola et al., 2011; Ciccarelli, 2014). Examples are given in Table 8.1. Some of these activities have a long history, and there are numerous examples of coastal and inland dune systems being transformed quite early in the Holocene.

As nonflooding sites located at the junction of land and sea, coastal dunes have always been favored sites for human occupation, as indicated by the widespread occurrence of prehistoric middens (Knight and Burningham, 2011). In Germany, dunes may have been activated as early as the Mesolithic (Tolksdorf et al., 2013; Nicolay et al., 2014) and showed reactivation to land use changes later in the Holocene (Völkel et al, 2011; Küster et al., 2014). The inland dunes of Hungary (Kiss et al., 2012) were also modified by various phases of human activity during the Subatlantic and some changes in the European sand belt may date back to the Neolithic (Tolksdorf and Kaiser, 2012). However, it has not always been easy to discriminate between phases of increased dune activity caused by changes in

Table 8.1 *Examples of the causes of coastal dune destabilization.*

Cause	Example
Access tracks	Purvis et al., 2015
Afforestation	Isermann et al., 2007
Aggregate extraction	Catto, 2002
Agriculture	Breckle et al., 2008
Beach raking	Nordstrom et al., 2012
Deforestation	Tsoar and Blumberg, 2002; Clemmensen et al., 2007
Fire	Strong et al., 2010
Fluvial flow and sediment supply interruption	Draut, 2012
Golf courses	Meulen and Salman, 1996
Grazing	Cosyns et al., 2001
Groundwater disturbance	Laity, 2003
Hunting	Catto, 2002
Inundation and eutrophication by waste water	Meulen and Jungerius, 1989
Invasive plants	Hilton, 2006; del Vecchio et al., 2013; Zarnetske et al, 2012
Littoral drift interruption by jetties, etc.	Gomez-Pina et al., 2002; Flor-Blanco et al., 2015
Military activity and warfare	Ruz et al., 2005
Mining and sand extraction	Lubke, 2013, Navarro-Pons et al., 2007
Nutrient enrichment	Provoost et al., 2011; Brunbjerg et al., 2014
Overgrazing by domestic stock	Seifan, 2009; Blanco et al., 2008
Pipeline landfalls	Ritchie and Gimingham, 1989
Plantations	Corona et al., 1988
Rabbits, introduction of	Bigelow et al., 2005; Provoost et al., 2011
Recreation	Meulen and Salman, 1996
Sewage outfall construction	Carmo et al., 2010
Shoreline stabilization	Warren, 2013
Timber exploitation and wood gathering	Corona et al., 1988
Tourism	Catto, 2002
Training wall construction and dredge spoil dumping	Pye and Neal, 1994
Trampling by humans	Hesp et al., 2010
Urbanization	Stancheva et al., 2011; Hernández-Calvento et al., 2014
Vehicular disturbance	Thompson and Shlacher, 2008; Jewell et al., 2014

climate (including windiness) and those caused by human activities (Gilbertson et al., 1999; Clarke and Rendell, 2009; Roskin et al., 2013). Indeed, phases of dune instability may occur when both appropriate climatic conditions and human pressures coincide (Beerten et al., 2014). Polynesian migrants appear to have modified dune systems in New Zealand (Horrocks et al., 2007; Hesp, 2001) and aboriginal peoples seem to have induced dune activity in the Great Plains of Canada (Wolfe et al., 2007) and on the Channel Islands of California (Erlandson et al., 2005). On the eastern seaboard of the United States a whole suite of human activities, including the depredations of voracious hogs, caused large-scale dune mobilization in the nineteenth century, as is graphically described by Senter (2003).

In Europe the tempo of coastal dune change caused by humans probably increased in medieval times (Provoost et al., 2011) and into the nineteenth century (Buynevich et al., 2007), but in some areas has been further accelerated in the recent decades of the Anthropocene as a result of vehicular disturbance, tourism, the spread of accidentally introduced plants such as marram (*Ammophila*), animal introductions, eutrophication by nitrates in waste water or polluted air, groundwater pumping, and urbanization. In addition, there have been increasing attempts to try and cause deliberate modification of dune systems (Ranwell, and Boar, 1986) by deliberate planting of trees and grasses, or by beach nourishment. Planting of dune grasses and trees has been undertaken on European coastal dunes since the Middle Ages, and sand fences have been utilized since at least the sixteenth century (Pye and Tsoar, 1990). Some of these techniques aim to stabilize dune surfaces to reduce dune migration while others seek to restore dune activity (Arens et al., 2013; Rhind and Jones, 2009; Martínez et al., 2013a, b). Methods have been developed for dune and sand movement management (Watson, 1990; Viles and Spencer, 1995) including:

- Dune removal
- Reshaping, trenching, and sod cutting
- Covering with gravel, spraying with oil, chemical stabilizers, etc.
- Sand fences, checkerboards, etc. (Figure 8.4)
- Planting of grasses (e.g. *Ammophila*) or trees and shrubs (e.g. *Hippophaë rhamnoides* and *Artemisia halodendron*)
- Shelterbelts and windbreaks (e.g. of poplars and willows)
- Beach protection or nourishment
- Fertilizer application to promote plant growth
- Control of grazing and trampling

Studies of the comparative effectiveness of these techniques have been undertaken (Breckle et al., 2008; Warren, 2013). For example, Zhang et al. (2004) found that the best means of stabilizing moving dunes in Inner Mongolia, China, were wheat straw checkerboards and the planting of *Artemisia halodendron*. This finding was confirmed by a study in the Kerqin Sandy land (Li et al., 2009). In northeast China, Miyasaka et al. (2014) found that tree planting was more effective in stabilizing

Figure 8.4 Sand fences on barchans dunes at Walvis Bay, Namibia.

dunes than shrub planting or exclusion of grazing. Along a major highway in the Taklamakan desert, checkerboards, reed fences, and nylon nets were found to be effective (Dong et al., 2004). On the desiccated bed of the Aral Sea, some salt tolerant plants, including *Salsola richteri* and *Calligonum caput-medusae*, have proved to be effective in sand stabilization (Shomurodov et al., 2013). In northwest Nigeria, Raji et al. (2004) found that shelterbelts were the most effective technique, and were superior to mechanical fencing. Success has also been claimed for chemical stabilizers (Han et al., 2007) and for geotextiles (Escalente and Pimentel, 2008), but some devices are prohibitively expensive (e.g. chemical fixers) while others (e.g. checkerboards) are not.

However, attempts to fix and control dunes may be ecologically or aesthetically undesirable and there is an increasing realization that dunes, especially on coasts, should in some cases be rejuvenated or returned to their natural state by such techniques as control of groundwater, beach replenishment, reductions in nutrient enrichment, and removal of invasive plants and plantations (Rhind and Jones, 2009; Martínez et al., 2013a; Clarke and Rendell, 2015). In Wales, for example, many coastal dune areas have become relatively more stable in the last six decades in response to climate change, reductions in grazing pressure, and nutrient enrichment. This has had detrimental ecological effects on dune organisms (Pye et al., 2014), including the loss of rare plants and invertebrates.

8.4 Future Anthropocene Climate Changes and the Aeolian Environment

Global climates will change substantially in coming decades and will impact aeolian environments (Goudie, 2013b). The Intergovernmental Panel on Climate Change (IPCC, 2007) suggested that in drylands temperatures could increase by between 1 and 7 °C by 2017–2100 compared to 1961–1990, and that precipitation levels could decrease by as much as 10–20 percent in the case of the Sahara but increase by as much as 10–15 percent in the Chinese deserts. In southwest Australia precipitation could decrease by as much as 40 percent (Delworth and Zeng, 2014). Most areas that are currently dry, such as the West African Sahel (Sylla et al., 2010) may see enhanced aridity because of reductions in precipitation. Zeng and Yoon (2009) suggested that as conditions become drier and vegetation cover is reduced, there may be vegetation-albedo feedbacks which will serve to enhance any aridity trend. By 2099 their model suggests that globally the warm desert area may expand by 8.5 million km^2 or 34 percent.

Future climatic change may be important for dust-storm activity. If soil moisture declines in response to changes in precipitation and/or temperature, dust-storm activity could increase (Wheaton, 1990). A comparison between the Dust Bowl years of the 1930s and model predictions of precipitation and temperature in the Great Plains of the United States indicates that mean conditions could be similar to or worse than those of the 1930s (Rosenzweig and Hillel, 1993). If dust-storm activity were to increase this could have a feedback effect on precipitation that would lead to further decreases in soil moisture (Miller and Tegen, 1998). Munson et al. (2011) argued that with increased drought there will be a reduction in perennial vegetation cover in the Colorado Plateau and thus an increase in aeolian activity. In the Bodélé depression of the Central Sahara, in spite of the possibility of higher rainfall amounts predicted by some models, higher wind velocities may increase dust activity in coming decades (Washington et al., 2009).

By contrast in northern China, dust-storm activity has decreased in recent warming decades, partially in response to changes in the atmospheric circulation and associated wind conditions (Jiang et al., 2009), and therefore might decrease still further in a warming world (Zhu et al., 2008). Wind velocities have fallen as warming has occurred (Wang et al., 2007), though some recovery has taken place in recent years (Lin et al., 2013). This "atmospheric stilling" has been a feature of recent warming decades in Australia and elsewhere (Vautard et al., 2010), though the pattern is not always clear, with regional and temporal differences being evident in the Iberian Peninsula (Azorin-Molina et al., 2013) and in China (Lin et al., 2013). Modeling studies have suggested that more generally in low latitudes, extreme wind events will become less frequent with global warming, and this has been confirmed for the United States (Breslow and Sailor, 2002). One explanation for this is that according to General Circulation Models (GCMs) (1) high latitudes will warm more than low; (2) there will, therefore, be a smaller equator-to-pole temperature difference; and (3) this means weaker monsoonal wind speeds will prevail. Another factor of potential importance in China is the earlier greening of vegetation resulting from higher spring temperatures. This may reduce springtime dust activity (Fan et al., 2014). Thus, future climate changes may stimulate more dust-storm activity in some areas (e.g. the Great Plains of the United States) and dampen it down in others (e.g. northwest China).

Sand dunes, because of the crucial relationships between vegetation cover and sand movement, have in the past proved susceptible to climate change. Some areas, such as the Kalahari margins, or the High Plains of the United States have been especially prone to changes in precipitation and/or wind velocity because of their location in climatic zones that are close to a threshold between dune stability and activity. Indeed, the occurrence of severe droughts is very important in determining the degree of dune activity in the western United States (Hanson et al., 2009). Dunes have repeatedly switched between activity and stability in response to Holocene droughts and these may become more prevalent over wide areas (Dai, 2011).

Detailed scenarios for dune remobilization have been developed for the mega--Kalahari by Thomas et al. (2005; Figure 8.5). Much of this vast region is currently vegetated and stable, but GCMs suggest that by the end of this century all dune fields, from South Africa in the south to Zambia and Angola in the north, will be reactivated. This could disrupt pastoral and agricultural systems and could also expose fine material to wind attack and dust-storm generation (Bhattachan et al., 2013; 2014). However, the methods used to estimate future dune field mobility are still full of problems and more research is needed before we can have full confidence in them (Knight et al., 2004). Contra Thomas et al. (2005), Ashkenazy et al. (2011) argued that the Kalahari is unlikely to be subjected to sufficiently dry or windy conditions for its dunes to become greatly mobilized by the end of this century.

As well as responding to local disturbances to surface conditions, coastal dunes will be impacted upon by future sea-level rise, which will cause beach retreat and overtopping to occur. Saye and Pye (2007) have modeled how the dunes of the Welsh coast may react to these changes. As sea levels climb, coastal protection structures may prevent dune movement inland. Moreover, groundwater levels may be altered, either in response to climate change or to rising sea levels (Clarke and Ayutthaya, 2010). Either way, this will impact upon dune slack habitats (Curreli et al., 2013). Dune vegetation may be directly affected by changes in temperature and precipitation (Mendoza-González et al., 2013). Coastal dunes might also be eroded by increases in storm attack, though not all models indicate that storminess will increase (Winter et al., 2012).

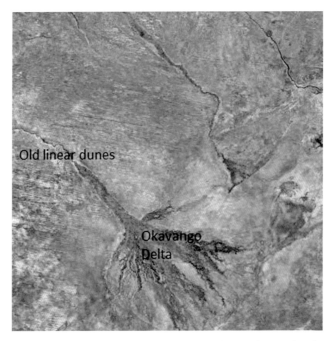

Figure 8.5 Ancient and predominantly stable linear dunes in the northern Kalahari. Some models suggest that these may become reactivated if conditions become drier (Landsat, courtesy of NASA).

8.5 Conclusions

As we saw in Chapter 1, Steffen et al. (2007) identified three stages in the Anthropocene: Stage 1, c. 1800–1945, which they called "The Industrial Era"; Stage 2, 1945 to c. 2015, which they called "The Great Acceleration"; and Stage 3, which may now be starting, when people have become aware of the importance of the human impact and may thus start stewardship of the Earth System. Although it has often proved difficult to discriminate between those changes caused by natural environmental changes and those that have resulted from the human impact, this chapter has shown that there may have been some anthropogenic intensification of both dust-storm generation and dune activity before Stage 1, that both phenomena have been altered as a result of human activities in recent centuries (Stage 1) and decades (Stage 2), and that Stage 3, the period of stewardship, has already started.

9

Coastal Processes and Forms in the Anthropocene

9.1 Coastal Change

Because settlements, industries, transport facilities, and recreational developments are so concentrated on coasts, the pressures placed on coastal landforms are often acute (Nordstrom, 1994; Nordstrom, 2000; Evans, 2008), and the consequences of excessive erosion serious (Figure 9.1). While most areas are subject to some degree of natural erosion and accretion, the balance can be upset by human activity in a whole range of different ways. For example, land use changes and hydraulic mining (Schoellhamer et al., 2013) can lead to pulses of accelerated siltation of coastal inlets (Poirier et al., 2011) and estuaries (Raharimahafa and Kusky, 2010). Changes to coastal dunes are discussed in Chapter 8.

Long-term erosion of beaches (and other soft coasts) is already a widespread phenomenon at the regional and global scales and sea-level rise over the last 100 years has been linked to these changes (Hinkel et al., 2013). However, humans have seldom attempted to *accelerate* coastal erosion deliberately. More usually, this phenomenon is an unexpected and unwelcome result of various engineering projects. Indeed, sometimes coast erosion has been accelerated as a result of human efforts to reduce it, as for example when the positioning of a groyne to stop erosion in one place causes erosion down drift (Figure 9.2). Sea walls have effects on adjacent beaches. As reflective structures they may encourage the generation of standing waves which can enhance mobilization of sediment to create scour. They also represent a barrier which interrupts exchange and supply of sediment between the natural hinterland and the beach system (Hanley et al., 2014).

9.1.1 Beach Erosion

Many soft coastlines are actively eroding. For example, 68 percent of the coastline of New England and the mid-Atlantic region of the United States has undergone erosion in recent decades (Hapke et al., 2013). The reasons for this trend are many

Figure 9.1 The beach at Malibu, California, has become very narrow as a result of ongoing erosion. Such a beach will be very vulnerable in the face of sea-level rise.

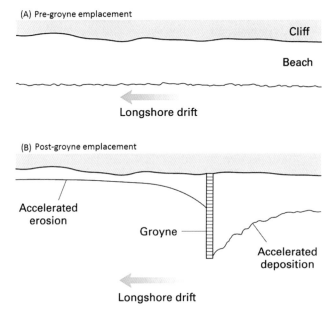

Figure 9.2 Interruption of longshore drift by the construction of a groyne causes both accelerated deposition and accelerated erosion.

Table 9.1 *Causes of beach erosion (possible human interventions in brackets).*

Beach sediment wastage through long-term attrition
Change in wave incidence and energy
Degradation of backing dunes (e.g. by urbanization, deforestation, and trampling)
Interception of sediment delivery by longshore drift (e.g. by pier and breakwater)
Lack of protective vegetation (e.g. by deforestation)
Loss of beach sediment by drift along shore
Loss of permafrost and of sea ice protection against wave attack (e.g. by warming)
Reduction in sediment supply from eroding cliffs (e.g. by sea wall construction)
Reduction in sediment supply from inland dunes (e.g. by dune stabilization)
Reduction in sediment supply from the sea floor (e.g. by dredging)
Reduction of fluvial sediment supply to coast (e.g. by dam and reservoir construction)
Removal of beach sediment (e.g. by mining)
Removal of beach sediment by runoff
Rise in the beach water-table (e.g. by irrigation)
Submergence and increased wave attack (e.g. as result of rising sea levels or subsidence)
Wave reflection (e.g. from sea walls)

(Bird and Lewis, 2015), and some of the possible causes of beach erosion, together with some human interventions that create them, are shown in Table 9.1.

A good beach offers coastal protection. If material is mined from it, accelerated cliff retreat may take place. The classic example of this was the mining of 660,000 tons of shingle from the beach at Hallsands in Devon, England, in 1887 to provide material for the construction of dockyards at Plymouth. The shingle proved to be undergoing little or no natural replenishment and in consequence the shore level was reduced by about 4 m. The loss of the protective shingle soon resulted in cliff erosion, and the village was very largely destroyed. More recently, Jonah et al. (2015) have investigated the effects of sand mining along the Cape Coast of Ghana, and have found that these activities are directly related to the rate of local coastline erosion. They predicted that at the current level of sand mining, the Cape Coast area may eventually experience a loss of most recreational beaches which may eventually lead to a loss of the tourism industry as well as make coastal communities vulnerable to flooding. Offshore dredging can also deprive beaches of their sediment and lead to accelerating retreat, as along some of the Baltic coastline of Germany (Kortekaas et al., 2010).

Another common cause of beach and cliff erosion at one point is coast protection at another (Brown et al., 2011). As already noted, a wide beach protects the cliffs behind, and beach formation is often deliberately encouraged by the construction of groynes and other "hard engineering" structures (Figure 9.3). Nordstrom (2014) categorized such structures into five types: groynes; bulkheads, seawalls, and revetments; breakwaters and sills; artificial headlands; and sand

Figure 9.3 Groynes along the coast of the English Channel near Hastings, showing their effect on sediment movement from the west. Scale bar 174 m. ©Google Earth 2014.

fences. However, these structures sometimes merely displace the erosion (possibly in an even more marked form) further along the coast. In Dorset, southern England, the construction of the Cobb breakwater at Lyme Regis has caused accumulation updrift, and beach narrowing down drift (Figure 9.4). Likewise, the construction of some sea walls, erected to reduce coastal erosion and flooding, has had the opposite effect to the one intended. Given the extent to which artificial structures have spread along the world's coastlines, this is a serious matter (Walker, 1988). As Higgitt and Lee (2001) have pointed out, coastal protection works reduce sediment inputs to the coastline by reducing rates of cliff recession, while groynes, harbor breakwaters, and other shoreline structures disrupt longshore sediment transport.

Problems of coastal erosion are exacerbated because much of the reservoir of sand and shingle that creates beaches is a relict asset. It was deposited on continental shelves during the last glacial maximum (around 20,000 year BP), when sea level was about 120 m below its present level. It was then transported shoreward and incorporated into present-day beaches during the phase of rapidly rising postglacial sea levels until about 6000 BP. Since that time, except for minor oscillations of a few meters, world sea levels have been stable and much less material is, as a consequence, being added to beaches and shingle complexes via this mechanism.

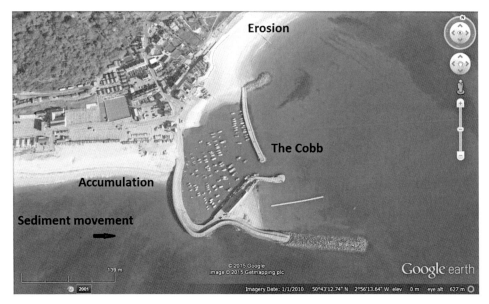

Figure 9.4 The effect of the building of the Cobb at Lyme Regis, Dorset, southern England. Scale bar 139 m. ©Google Image 2015, ©Getmapping plc. 2015.

In some areas, sediment-laden rivers carry material into the coastal zone which becomes incorporated into beaches by means of longshore drift. Thus any change in the sediment load of rivers, created, for example, by damming, may result in a change in the sediment budget of neighboring beaches. In Texas, where over the last century four times as much coastal land has been lost as has been gained, one of the main reasons for this change is believed to be the reduction in the suspended loads of some of the rivers discharging into the Gulf of Mexico. In 1961–70, these rivers carried on average only about one-fifth of what they carried in 1931–40. On the eastern seaboard of the United States, no less dramatic changes in sediment delivery to the coastline have occurred (Figure 9.5). On a global basis, large dams may retain 25–30 percent of the global flux of river sediment (Vörösmarty et al., 2003; see also Chapter 7).

A case study of the potential effects of dams on coastal sediment budgets is provided for California by Willis and Griggs (2003). Given that rivers provide the great bulk of beach material (75–90 percent) in the state, the reduction in sediment discharge by dammed rivers can have highly adverse consequences. Almost a quarter of the beaches are down coast from rivers that have had sediment supplies diminished by one-third or more. Most of those threatened beaches are in southern California where much of the state's tourism and recreation activities are concentrated (see Figure 9.1).

Anthropogenic vegetation modification may create increased coastal erosion potential. This was illustrated for the hurricane-afflicted coast of Belize by Stoddart

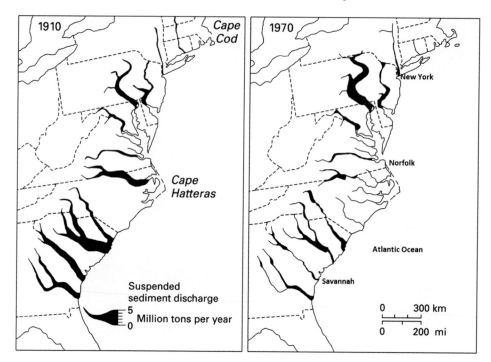

Figure 9.5. The decline in suspended sediment discharge to the eastern seaboard of the United States between 1910 and 1970 (modified after Meade and Trimble, 1974).

(1971). He showed that natural, dense vegetation thickets on low, sand islands (*cays*) acted as a baffle against waves and served as a sediment trap for coral blocks, shingle, and sand transported during extreme storms. However, on many islands the natural vegetation had been replaced by coconut plantations. These had an open structure easily penetrated by sea water, tended to have little or no ground vegetation (thus exposing the cay surface to stripping and channeling), and had a dense but shallow root net easily undermined by marginal sapping. Stoddart found that where the natural vegetation had been replaced before a storm (Hurricane Hattie), erosion and beach retreat led to decreases in height after the storm, whereas where natural vegetation remained, banking of storm sediment against the vegetation hedge led to an increase in height.

Other examples of markedly accelerated coastal erosion and flooding result from anthropogenic degradation of dune ridges (see also Chapter 8). Frontal dunes are a natural defense against erosion. Many of those areas in eastern England which most effectively resisted the great North Sea storm and surge of 1952 were those where humans had not intervened to weaken the coastal dune belt.

Many beach protection schemes now involve beach nourishment (by the artificial addition of appropriate sediments to build up the beach), or employ miscellaneous sand by-passing techniques (including pumping and dredging) whereby sediments are transferred from the accumulation side of an artificial barrier to the erosional side (Bird, 1996). So-called soft means of coastal protection, rather than hard engineering structures such as sea walls or groynes, have become more commonly used (Temmerman et al., 2013). For example, the encouragement of dune formation and promotion of salt marsh accretion are becoming recognized as being aesthetically pleasing, effective, and economically advantageous. Softer engineering approaches to the maintenance of sand coastlines often include some form of sediment nourishment (Hanley et al., 2014). A dramatic example is taking place in the Netherlands, where the "sand engine" (*"De Zandmotor"*) involves the emplacement of sand into the sublittoral zone which is then reworked by waves and currents, providing a long-term input into the local sediment budget to maintain sediment supply to sand bars, beaches, and dune systems many kilometers distant (Stive et al., 2013). The extent of coastal change in Europe generally, and the methods used to manage beaches are discussed by Pranzini et al. (2015).

9.1.2 Delta Erosion

Many deltas have been formed or have grown considerably in the wake of human interventions that liberated large amounts of sediments in their catchments, especially in Europe, with many iconic examples in the Mediterranean such as the Ebro, Ombrone, Po, Rhône, and Tiber. Southern European deltas appear to have grown in Roman times as a result of increased river sediment loads (Maselli and Trincardi, 2013). However, more recently, delta erosion has become a pervasive and developing problem (Renaud and Syvitski, 2013), not least in South and Southeast Asia. This is partly because deltas sink under their own weight of sediment, suffer from compaction of organic sediments, have been modified by the construction of embankments and channel diversions, and may be undergoing accelerated subsidence because of the removal of fluids (e.g. groundwater and gas). They are also subjected to ongoing sea-level rise (SLR). They also suffer from a reduced amount of sediment nourishment following dam construction upstream and from a reduction in the number of distributary channels as a consequence of a need to support navigation in a limited number of larger channels (Syvitski and Saito, 2007; Svyitski et al. 2009). Indeed, such actions have already been implicated in the changes that have taken place in the morphology of the Nile Delta and its lagoons over the last half century (El Banna and Frihy, 2009), and in the morphology and rates of recession and progradation, as has been shown, for example, in Jabaloy-Sanchez et al.'s (2010) study of the Adra River delta in southeast Spain and by the study of the Patía Delta in northwest South America

Figure 9.6 The fragile birdsfoot delta of the Mississippi (courtesy of NASA).

(Restrepo, 2012). Anthony (2015) has argued that the present-day massive swing toward significant reductions in fluvial sediment supply to deltas may signify the ultimate demise of many of them in the coming decades through a process of delta shoreline straightening by waves, in addition to accelerated sinking.

In their analysis of erosion loss in the Mississippi Delta, Walker et al. (1987) suggested it was the result of a variety of complex interactions among a number of physical, chemical, biological, and cultural processes: channelization, worldwide sea-level changes, subsidence resulting from sediment loading by the delta of the underlying crust, changes in the sites of deltaic sedimentation as the delta evolves, catastrophic storm surges, subsidence resulting from subsurface fluid withdrawal, and changes in the amount of sediment carried by the Mississippi in response to land use changes and river engineering (Tweel and Turner, 2012). Plainly, however, as seen from space the birdsfoot delta is a very fragile structure (Figure 9.6).

In China the construction of around 50,000 dams in the Yangtze basin has caused sediment starvation and severe erosion of its delta (Yang et al., 2011; Dai and Lu, 2014). The trends of sediment loads in the Yellow, Yangtze, Pearl, Red, and Mekong are shown in Figure 9.7. In contrast to the relatively slow historical increase in sediment flux during the period 2000–1000 year BP, the recent sediment flux has been decreased at an accelerating rate over centennial scales (Wang et al., 2011).

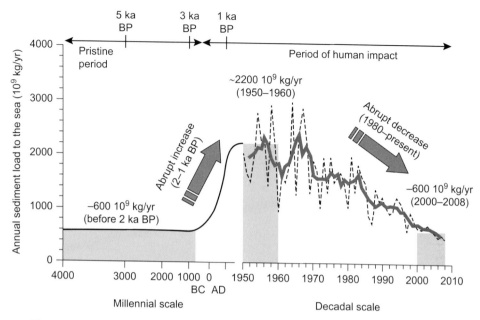

Figure 9.7 Change in total sediment flux from five major rivers (Yellow, Yangtze, Pearl, Red, and Mekong) to the western Pacific Ocean at the millennial scale (6000 year BP to 1950) and decadal scale (1950–2008), illustrating historical increases and recent decreases due to human interventions in the river basins (From Wang et al., 2011, figure 14).

Over a century ago the Nile Delta began to retreat. The Rosetta mouth lost about 1.6 km of its length from 1898 to 1954. The imbalance between sedimentation and erosion appears to have started with the Delta Barrages (1861) and culminated with the Aswan High Dam itself a century later. In addition, large amounts of sediment are retained in an extremely dense network of irrigation channels and drains that has been developed in the Nile Delta itself (Stanley, 1996). Thus much of the Egyptian coast is now "undernourished" with sediment. However, extensive coastal protection works have been installed at the Rosetta mouth, and these have had some success in moderating erosion rates (Ghoneim et al., 2015), at least on the short term.

9.2 Coral Reefs

Coral reefs, among the most diverse, productive, and beautiful communities in the world, are currently undergoing profound change (see Sheppard et al., 2009; Birkeland, 2015) in response to a number of stressors (Figure 9.8).

Accelerated sedimentation and increased land erosion resulting from poor land management (Fabricius, 2005), together with dredging, is probably

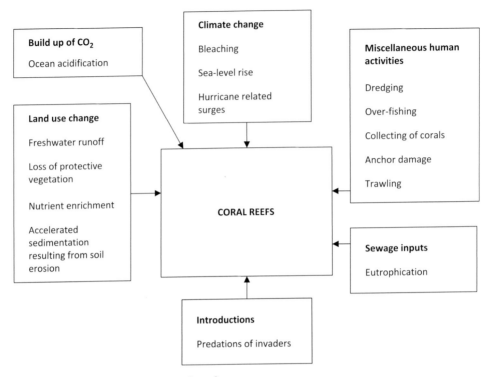

Figure 9.8 Stresses upon coral reefs.

responsible for more damage to reef communities than all the other forms of human insult combined. Quantifying the impact of anthropogenic sediment delivery from agricultural land use change against the high variability of natural sediment loads in tropical rivers is a challenging research area (Bartley et al., 2014). Nonetheless, sedimentation effects are likely to be diverse, for although some coral species can tolerate very high sedimentation rates and turbidity and have shown recovery from short-term or low levels of sedimentation, most coral reef organisms are negatively affected by smothering (sedimentation) and reduced light availability for photosynthesis due to turbidity in the water column. The exact responses to sedimentation depend on the coral species, duration and amount of sedimentation, and sediment composition.

An interesting example of the role of accelerated sedimentation is provided by geochemical studies of long-lived corals from Australia's Great Barrier Reef (McCulloch et al., 2003). These showed that since 1870 and the start of European settlement in its catchment, the Burdekin River has carried five to ten times more sediment to the reef than it did previously.

Not all accelerated sedimentation is the result of deforestation and agriculture. Analysis of sediment cores from St. Lucia in the Caribbean (Bégin et al., 2014) indicated that accumulation rates of terrigenous sediment, originating from the upstream watersheds, and calcareous sediment, likely arising from dead corals, increased two- to threefold over the last three to four decades. The great majority (83–95 percent) of sediment yield was attributable to unpaved and degraded roads.

Runoff from the land, which may contain nitrates and other nutrients derived from fertilizers can also cause nutrient enrichment. Equally, accelerated runoff of freshwater into lagoons can reduce salinity levels below the level of tolerance of reef communities.

Sewage, some derived from tourist complexes, is the second worst form of stress to which coral reefs are exposed, for oxygen-consuming substances in sewage result in reduced levels of oxygen in the water of lagoons (Risk, 2014; Edinger and Risk, 2013). The detrimental effects of oxygen starvation are compounded by the fact that sewage may cause nutrient enrichment to stimulate algal growth and feed various predators, which in turn can overwhelm coral. Reef surfaces may be transformed into "weedy algal lawns" (DeGeorges et al., 2010).

Reefs have also been impacted upon by accelerated coral bleaching associated with global warming and by increasing rates of ocean acidification caused by increasing atmospheric and oceanic CO_2 levels (Pandolfi et al., 2011). These two may interact synergistically, and thus may negatively influence survival, growth, reproduction, larval development, settlement, and postsettlement development of corals. Interactions with local stress factors such as pollution, sedimentation, and overfishing are compounding the effects of climate change (Atweberhan et al., 2013). Coral bleaching follows anomalously high seawater temperatures, and such episodes have increased steadily over the last three decades in both frequency and intensity. There is, however, great variation in reef recovery in the aftermath of bleaching events. Where there is sufficient survival of existing colony tissue, recovery can occur within a few years, but in other cases little recovery of coral cover has been observed even after five to ten years and, most commonly, recovery of coral cover requires about a decade. Coral species that are more susceptible to severe bleaching, such as the branching *Acropora* that forms much of the habitat complexity of Indo-Pacific reefs, will probably be reduced in abundance relative to species that exhibit less sensitivity, such as slower-growing genera with massive or encrusting growth forms (Pandolfi et al., 2011).

One of the reasons why all these stresses may be especially serious for reefs is that they have long life-spans, so that it can take an extended time for them to recover from damage. Whatever may be the causes, in recent decades severe declines in coral cover on reefs have been identified, though there are debates as to just how serious the degradation has been, not least in the context of the Great Barrier Reef (GBR) of Australia. Hughes et al. (2011) suggested that there has been nearly 40 percent loss of coral cover on the inner reefs of the GBR, whereas

Sweatman et al. (2011) believed that losses of GBR corals over the last forty years have probably been exaggerated. In the Indo-Pacific region as a whole, coral cover declined from 42.5 percent in the early 1980s to only 22.1 percent by 2003, an average annual rate of 1 percent. In the Caribbean between 1977 and 2001 the annual rate was even higher – 1.5 percent (Bruno and Selig, 2007). Coral cover in the Caribbean has fallen slowly over the past twenty-five to thirty years from an average of roughly 50 percent to only 10 percent (Hughes et al., 2013). Some heavily degraded reefs are now little more than a mix of rubble, seaweed, and slime (Pandolfi et al., 2005).

9.3 Estuaries

Estuaries have been subjected to numerous human influences (Irabien et al., 2015), including the construction of jetties at their mouths, as along the Cantabrian coast of Spain (Flor-Blanco et al., 2015). In France the mudflats of the Seine estuary have been reduced in extent because of the erection of dykes and other construction works connected with navigation. The Northern mudflat of the estuary is the most affected by these human activities, and their surface area has been reduced by 62 percent during the last twenty-seven years (Cuvilliez et al., 2009). Pye and Blott (2014) report that most UK estuaries have been affected to some extent by embanking and land-claim, resulting in a significant reduction in the tidally influenced area and total estuary volume. In estuaries such as the Severn and Thames, embanking began at least as early as Roman times and was particularly important between the thirteenth and nineteenth centuries. The active tidally flooded area in many estuaries in eastern and southern Britain is now less than 50 percent of the former (preembanking) tidally active area. In extreme cases, including the Blyth estuary in Suffolk, up to 95 percent of the original tidally active area had been enclosed and claimed for agricultural purposes by the early nineteenth century. The effect of constructing training walls to improve navigation has also been demonstrated in many British estuaries, including the Lune, Wyre, Mersey, Ribble, and Dee (see, e.g., van der Wal et al., 2002; Blott et al., 2006). A major effect is to reduce the natural migration of the low water channels, thereby allowing sediment accretion and eventual saltmarsh development on one or more sides of the estuary (provided a sufficient supply of sediment is available). Increased intertidal and subtidal sedimentation outside the trained (and sometimes dredged) channel, in turn, leads to reduced tidal capacity, a reduction in tidal current velocities and further sedimentation through a process of positive feedback. The dredging of sediment from navigation channels and docks, with disposal at offshore locations, has also contributed to a net loss of sediment from some estuaries, notably those adjoining Southampton Water and the Solent, the Thames, and the Medway.

Serious sedimentation of bays and estuaries caused by human activity occurred on the eastern coast of America following European settlement and agriculture. As Gottschalk (1945, p. 219) wrote:

Both historical and geological evidence indicates that the preagricultural rate of silting of eastern tidal estuaries was low. The history of sedimentation of ports in the Chesapeake Bay area is an epic of the effects of uncontrolled erosion since the beginning of the wholesale land clearing and cultivation more than three centuries ago.

He calculated that at the head of the Chesapeake Bay 65 million m^3 of sediment were deposited between 1846 and 1938. The average depth of water over an area of 83 km^{-2} was reduced by 0.76 m. New land comprising 318 ha was added to the state of Maryland. The dramatic effects of deforestation and agriculture on sedimentation rates in Chesapeake Bay, United States, in the nineteenth century, and the subsequent reduction caused by dam construction, are shown in Figure 9.9 (Pasternack et al., 2001).

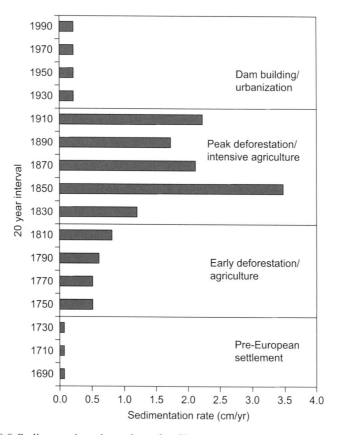

Figure 9.9 Sedimentation chronology for Chesapeake Bay, United States (modified after Pasternack et al., 2001, figure 10).

Jalowska et al. (2015) provide a long-term history of change from the Roanoke bayhead delta in North Carolina, United States. After the mid-1600s AD, when the first European settlers began to clear forest and to farm the drainage basin, the delta rapidly accreted and the interdistributary bay filled with sediment from increased agricultural runoff. Regression was also facilitated by the low rates of SLR at that time (-0.01 to 0.047 cm yr^{-1}). An episode of bayhead delta retreat was then initiated during the nineteenth century and continues today. This is because improved agricultural practices and dam construction have decreased the amount of sediment delivered to the bayhead delta. Additionally, the rate of SLR has increased to 0.21 cm yr^{-1}.

Jaffe et al. (2007) give a useful survey of changing rates of sedimentation in San Pablo Bay, California, United States, over the last century and a half. Their analysis of historical bathymetric surveys revealed large changes in morphology and sedimentation from 1856 to 1983. In 1856, the morphology of this bay was complex, with a broad main channel, a major side channel connecting to the Petaluma River, and an ebb–tidal delta crossing shallow parts of the bay. In 1983, its morphology was simpler because all channels except the main one had filled with sediment and erosion had planed the shallows, creating a uniform gently sloping surface. They found that the timing and patterns of geomorphological change and deposition and erosion of sediment were influenced by human activities that altered sediment delivery from rivers. From 1856 to 1887, high sediment delivery (14.1×10^6 m^3 yr^{-1}) to San Francisco Bay during the hydraulic gold-mining period in the Sierra Nevada resulted in net deposition of $259 \pm 14 \times 10^6$ m^3 in San Pablo Bay (see also Chapter 7). This rapid deposition filled channels and increased intertidal mudflat area by 60 percent (37.4 ± 3.4–60.6 ± 6.2 km^2). However, from 1951 to 1983, $23 \pm 3 \times 10^6$ m^3 of sediment was eroded from San Pablo Bay as sediment delivery from the Sacramento and San Joaquin Rivers decreased to 2.8×10^6 m^3 yr^{-1} because of damming of rivers, riverbank protection, and altered land use. Intertidal mudflat area in 1983 was 31.8 ± 3.9 km^2, similar to that in 1856.

More generally, there are many examples of estuarine siltation resulting from deforestation of the river catchments that feed them, and these are reviewed by Poirier et al. (2011), who found that in many of the works they cited two- to six-fold increases in sedimentation rates were reported, as was the formation of mud drapes. Soil erosion has increased the sediment loadings of the Betsiboka, Madagascar's largest river. It is a major conduit for transporting lateritic soils and sediments derived from the highlands to the sea. These entrained lateritic sediments color the river a blood-red hue (Figure 9.10). As a result, Bombetoka Bay has significantly changed during the past thirty years, with a dramatic increase in the amount of sediment moved by the river, and deposited in the estuary and in offshore delta lobes (Raharimahafa and Kusky, 2010).

Figure 9.10 Estuarine infilling in northwest Madagascar. Scale bar 27 km.
© 2014 Google Earth.

In other cases, such as the Ord Estuary of northwestern Australia, accelerated siltation has resulted from reduced flood scour following dam construction upstream (Wolanski et al., 2001).

In many estuaries and bays, oyster reefs are an important component (Kirby, 2004; Rodriguez et al., 2014; Zu Ermgassen et al., 2012). As Beck et al. (2011) have explained, native oyster reefs once dominated many estuaries, ecologically and economically, but centuries of resource extraction, exacerbated by coastal degradation, have pushed them to the brink of functional extinction worldwide. They examined the condition of oyster reefs across 144 bays and 44 ecoregions and their comparisons of past with present abundances indicated that in many bays, more than 99 percent of oyster reefs have been lost and are functionally extinct. Overall, they estimated that 85 percent have been lost globally.

9.4 Salt Marshes, Mangrove Swamps, and Seagrasses

Salt marshes are important habitats for wildlife and also play a very substantial role in coastal protection. Many of them have been much modified by human actions, including draining and ditching (Figure 9.11). In some parts of the world, salt marsh growth is currently being deliberately encouraged by removing coastal

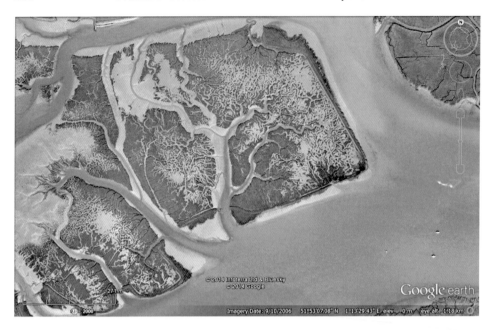

Figure 9.11 Salt marshes between Harwich and Walton-on-the-Naze, Essex, eastern England. The top figure shows a relatively natural marsh with an intricate drainage system. The bottom figure shows a marsh with extensive ditching. Scale bar 271 m. ©2014 Infoterra Ltd and Bluesky, ©Google 2014.

embankments and defense structures and allowing natural sediment accretion to construct a new marsh where previously there was agricultural land. The "managed realignment" or "deembankment" of coastal defenses (Garbutt et al., 2006; Esteves, 2014) by breaching the former sea defense line to create a new intertidal surface for saltmarsh development has been increasingly implemented in Northwest Europe and North America with 150 realignment trials identified since the early 1990s (Friess et al., 2014). Rising sea levels and subsiding land, coupled with the high cost of maintaining hard coastal defenses, have led coastal managers to look for these more cost effective and sustainable methods of coastal protection.

In Britain, the nature of some salt marshes, and the rate at which they accrete, has been transformed by the introduction of *Spartina alterniflora* from the east coast of North America, possibly in shipping ballast. The crossing of this plant species with the native *Spartina maritima*, produced an invasive cord-grass of which there were two forms: *Spartina townsendii* and *Spartina anglica*, the latter of which is now the main species. It appeared first on Southampton Water in 1870 and then spread rapidly to other salt marshes in Britain: partly because of natural spread and partly because of deliberate planting (Doody, 1984). The plant has often been effective at excluding other species and also at trapping sediment. Rates of accretion can therefore be as high as 8–10 cm per year (Ranwell, 1964). There is evidence that this has caused progressive silting of estuaries such as those of the Dee (Marker, 1967).

Spartina invasion has also occurred in other parts of the world including the Netherlands, Australia, New Zealand, California, and China (Gedan et al., 2009; Strong and Ayres, 2009). However, for reasons that are still debated, many *Spartina* marshes have suffered dieback, which has sometimes led to marsh recession (Hughes and Paramor, 2004). Among the hypotheses that have been put forward to explain this phenomenon are the role of rising sea level, pathogenic fungi, increased wave attack, competition from and bioturbation by the polychaete, *Nereis diversicolor*, grazing by the crab *Sesarma reticulatum* (Bertness and Silliman, 2008), and the onset of waterlogging and anaerobic conditions on mature marsh (Hübner et al., 2010).

However, *Spartina* invasion is but one of many possible causes of changes in marsh accretion rates (Gedan et al., 2011). Other factors include runoff and sediment erosion from the land as a result of land cover changes (Mattheus et al., 2010); the construction of dykes and embankments; the excavation of canals; the throwing up of spoil banks; land reclamation schemes; the excavation of grid ditches to help control mosquitoes; tidal restriction by bridges and berms; and subsidence resulting from fluid abstraction. These are reviewed in the context of the United States by Kennish (2001), in the context of the Arabian Gulf by Burt (2014), and globally by Gedan et al. (2009).

Some North American salt marshes expanded rapidly during the eighteenth and nineteenth centuries due to increased rates of sediment delivery following deforestation associated with European settlement (Kirwan et al., 2011). However, since then other human impacts have occurred. The history of marshes in the Cape Cod area has been described by Coverdale et al. (2013). Prior to 1939, one of the most conspicuous disturbances in New England marshes was mosquito ditching. In the period from 1939 to 1976, the permanent human population of Cape Cod nearly tripled, increasing from 37,000 in 1940 to more than 100,000 in 1976. This boom triggered increased destruction of salt marsh habitat to facilitate coastal development and recreation. Significant loss of marsh due to the native herbivorous purple marsh crab *Sesarma reticulatum* occurred after 1976. On some parts of Cape Cod, *Sesarma reticulatum* had denuded 50 percent of marsh creek banks by 2008. The increase in its abundance was triggered by recreational fishing of its predators. Most recently, marshes on Cape Cod have begun to recover from this effect and this recovery has coincided with the invasion of another exotic, the European green crab, *Carcinus maenas*, into heavily burrowed creek banks in die-off marshes. It may reduce herbivore populations and so facilitate the recovery of this heavily degraded system (Bertness and Coverdale, 2013).

One particular type of coastal ecosystem coming under increasing pressure from human activities is the mangrove forest characteristic of tropical intertidal zones (Polidoro et al., 2010). They are being degraded and destroyed on a large scale in many parts of the world, either through exploitation of their wood resources or because of their conversion to agriculture, aquaculture, salt-evaporation ponds, or housing developments. Data from FAO suggest that the world's mangrove forests, which covered 19.8 million ha in 1980, have now been reduced to only 14.7 million ha, with the annual loss running at about 1 percent per year (compared to 1.7 percent a year from 1980 to 1990; www.fao.org/DOCREP/005/Y7581E/y7581e04.htm). Further data are presented in Table 9.2. This shows that in southeast Asia, with the exception of Bangladesh, most countries showed a decline in mangrove cover between 1980 and 2005.

Attempts are being made in many countries to undertake mangrove restoration (Balke and Friess, 2015), but these are not always successful.

In some parts of the world, meadows of seagrass can, over time, trap sediment, build upwards, and create various forms of reef. This is the case with the *Posidonia oceanica* meadows of the Mediterranean (Bonhomme et al., 2015). These are highly vulnerable to human activities, (e.g. coastal development, establishment of port facilities, small pleasure boats which unwittingly plow furrows at low tide, to cross the reef, deliberate dredging of channels to allow boats to moor within the lagoon, and artificial beaches). On a global basis, seagrass loss rates are comparable to those reported for mangroves, coral reefs,

Table 9.2 *Country-specific temporal variation of mangrove cover (km²) (Modified from DasGupta and Shaw, 2013).*

Country	1980	2005	2005 area as percent of 1980 area
Pakistan	3,450	1,570	45.5
India	5,067	4,480	88.4
Bangladesh	4,280	4,760	111.2
Myanmar (Burma)	5,550	5,070	91.4
Indonesia	42,000	29,000	69.1
Malaysia	6,740	5,650	83.8
Vietnam	2,692	1,570	58.3
Philippines	2,950	2,400	83.1
Thailand	2,800	2,400	85.7

Table 9.3 *Main future climate drivers for coastal systems.*

Driver	Effects
CO_2 concentration increase	Ocean acidification
Sea surface temperature increase	Change in ocean circulation, reduced sea ice in high latitudes, coral bleaching, stimulation of coral growth in cooler seas, poleward species migration, more algal blooms, and red tides
Sea level	Inundation, flood and storm surge damage, beach erosion, saltwater intrusion up estuaries etc., rising water tables, impeded drainage, wetland loss
Storm intensity	Increased extreme water levels and wave heights
Storm frequency	Altered surges and storm waves
Wave climate	Changes in wave conditions, leading to altered patterns of erosion and accretion; reorientation of beach plan form
Runoff	Altered flood risk in coastal lowlands, water quality/salinity, fluvial sediment supply, and circulation and nutrient supply

and tropical rainforests and place seagrass meadows among the most threatened ecosystems on Earth (Waycott et al., 2009).

9.5 Future Sea-Level Rise

In the future, coastlines will be subjected to a variety of driving forces (Table 9.3). One of these is SLR, for if temperatures climb in coming decades, so will global sea levels. Even if carbon emissions are controlled, the effects of global warming will continue well after 2100 because of residual effects (Levermann et al., 2013).

Globally, sea levels are currently rising because of the thermal expansion of seawater (the steric effect), the melting of the cryosphere (glaciers, ice sheets, and permafrost) and miscellaneous anthropogenic impacts on the hydrological cycle, such as irrigation, the creation of reservoirs, and the pumping of groundwater (which modifies how much water is stored on land; Gornitz et al., 1997; Gregory et al., 2013; Clark et al., 2015). In addition, regionally, sea level is affected by wind- and buoyancy-driven ocean currents associated with the redistribution of heat and salt in the ocean. For example, the northward flowing Gulf Stream is balanced by higher sea level to its east and lower sea level along the east coast of the United States. If the transport of the Gulf Stream diminished due to changes in the buoyancy-driven and/or the wind-driven ocean circulation, then sea level along the eastern seaboard of the United States would increase (Hu and Deser, 2013).

Between 1910 and 1960, annual rates of global SLR ranged from 0.25 to 0.75 mm while those between 1960 and 1990 ranged from 0.60 and 1.09 mm. From 1993 to 2010, the rate accelerated to c. 3.2 mm per year. The steric effect over the period 1910–1990 accounts for about one-third of the observed eustatic change over that period (Cazenave and Llovel, 2010). The shrinkage of glaciers as warming occurs releases meltwater that flows into the oceans, also causing SLR (Radić and Hock, 2011).

Estimates about how much SLR is likely to occur by the end of this century have tended to be revised downward through time (French et al., 1995; Pirazzoli, 1996). They have now settled at best estimates of 50 to 100 cm by 2100. This implies rates of SLR of 5–10 mm per year, which compares with a rate of about 1.5–2.0 mm during the twentieth century (Miller and Douglas, 2004).

However, signs that the polar ice sheets are currently melting at an accelerating rate suggest that their contribution to twenty-first-century SLR may be greater than the IPCC report of 2007 had proposed (Rignot, et al., 2011; Bamber and Aspinall, 2013; Nick et al., 2013). By the end of the present century, sea levels may be 0.5 to 1.4 m above the 1990 level (Rahmstorf, 2007), or even 0.75–1.90 m (Vermeer and Rahmstorf, 2009). Nicholls et al. (2011) believe that with a 4 °C or more rise in temperature, a credible upper bound for twenty-first-century SLR is 2 m. These suggested values are, however, higher than those employed by the IPCC in its 2013 report (Church et al., 2013). It gives values of 0.26–0.98 m for the amount of SLR in 2081–2100 compared to 1986–2005.

The consequences of increasing global sea levels will be magnified in areas which are undergoing natural subsidence and those where, because of groundwater and hydrocarbon removal, the land is sinking (see Chapter 4). This is the case in the Ganges–Brahmaputra–Meghna delta where the rate of subsidence (8.8 mm per year) comfortably exceeds the current rate of global SLR (Brown and Nicholls, 2015).

9.5.1 Consequences of Future Sea-Level Rise.

Inundation. Rising sea levels will impact upon low-lying shorelines, including islands (Bellard et al., 2013), wetlands (Webb et al., 2013; Van der Noort, 2013), mangrove swamps (Krauss et al., 2013), muddy coasts (Anderson et al., 2014), coral reefs, lagoons, deltas, barrier islands, and beaches. In addition, phenomena like storm surges may increase if hurricane activity becomes more severe (Grinsted et al., 2013; Woodruff et al., 2013; Yang et al., 2014). Many world cities are also under threat (Hallegatte et al., 2013). An early review is provided by Bird (1993).

Florida, United States, provides an excellent illustration of the threats posed to low-lying areas. Approximately 10 percent of the state's land area is less than 1 m above present sea level. A 0.6 m rise in sea level would inundate about 70 percent of the total land surface of the Florida Keys, 71 percent of the population and 68 percent of property, while a 1.5 m rise would inundate 91 percent of the land surface, 71 percent of the population, and 68 percent of the property (Noss, 2011). Severe consequences have also been proposed as likely for California (Revell et al., 2011; Heberger et al., 2011) and the Gulf Coast (Anderson et al., 2014).

Coastal marshes and wetlands. Coastal marshes may be susceptible to the effects of increased CO_2, SLR, storm frequency and intensity, changes in ambient temperature, and ocean physical changes, including elevated sea temperature, and acidification (Morzaria-Luna et al., 2014).

Coastal wetlands, including mangrove swamps, are potentially highly vulnerable to SLR (McFadden et al., 2007; Gedan et al., 2011). For example, Park et al. (1986) undertook a survey of coastal wetlands in the United States and suggested that sea-level change could, by 2100, lead to a loss of between 22 and 56 percent of the 1975 wetland area, according to the degree of SLR that takes place. At a global scale, Nicholls et al. (1999) suggested that by the 2080s SLR could cause the loss of up to 22 percent of the world's coastal wetlands. When combined with other losses due to direct human action, up to 70 percent of them could be lost by that date. Similar predictions by Blankespoor et al. (2014) confirm this analysis.

Tidal wetland systems are expected to shift landward depending on the coupling between ecology and geomorphology, the rate of SLR, and sediment transport dynamics. Indeed, changes in marshes are already taking place in response to SLR. Working on the Gulf Coast of Florida in the United States, Raabe and Stumpf (2015) found that tidal wetland systems had shifted landward over the last 120 years. Loss of tidal marsh at the shoreline was -43 km^2, representing a 9 percent loss to open water. They documented 1.2 m yr^{-1} of marsh shoreline erosion, and 2.3 m yr^{-1} inland marsh expansion. A total of 82 km^2 of forest converted to marsh and 66 km^2 of forest converted to forest-to-marsh transitional habitat. The result was a net regional gain of 105 km^2 of intertidal area, an increase of 23 percent, constituting a marine transgression of coastal lowland. Loss of salt marsh due to a combination of SLR and the expansion of mangrove has also been identified in Texas (Armitage et al., 2015).

Marshes will be especially vulnerable where sea defenses and other barriers prevent the landward migration of marshes as SLRs takes place. In such locations what is termed "coastal squeeze" occurs (Doody, 2013).

Although coastal wetlands have long been considered vulnerable to SLR, recent work has identified feedbacks between plant growth and geomorphology that may allow some wetlands actively to resist its deleterious effects (Kirwan and Megonigal, 2013). The future susceptibility of marshes will be variable, as indicated in the following list.

Salt marsh sensitivity

Less sensitive
  Areas with high sediment input
  Areas with high-tidal ranges and thus high sediment transport potential
  Areas with effective organic accumulation
  Areas with plants that are effective sediment trappers
  Areas without excessive grazing and trampling
  Areas undergoing uplift

More sensitive
  Areas that are subsiding
  Areas with a low sediment input (e.g. cyclically abandoned delta areas)
  Areas with a low sediment input due to dam and reservoir construction
  Mangroves, which have a longer life cycle, therefore a slower response
  Areas that are constrained by seawalls, etc. (coastal squeeze)
  Microtidal areas, where a rise in sea level represents a larger proportion of total
    tidal range
  Reef settings, which have a lack of allogenic sediment

Organic and inorganic sediment supply is a crucial control of how marshes will respond to rising sea levels (Reed, 2002). On coasts with a limited supply, SLR will impede the normal process of marsh progradation, and increasing wave attack will start or accelerate erosion along their seaward margins. Tidal creeks will tend to become wider, deeper, and more extended headwards as the marsh is submerged. However, as Reed (1995) pointed out, salt marshes are highly dynamic features and in some situations may well be able to cope, even with quite rapid

SLR. Reed (1990) suggested that those marshes in riverine settings may receive sufficient inputs of sediment so that they are able to accrete rapidly enough to keep pace with projected SLR. Areas of high-tidal range, such as the marshes of the Severn Estuary in England/Wales, or the Tagus Estuary of Portugal, are also areas of high sediment-transport potential and may thus be less vulnerable to SLR (Simas et al., 2001). Likewise, some vegetation associations, for example, *Spartina* swards, may be relatively more effective than others at encouraging accretion, and organic matter accumulation may itself be significant in promoting vertical build-up of some marsh surfaces. Tall vegetation can enhance sediment deposition by reducing current flow and wave action, but herbivores shorten vegetation height and this could potentially reduce sediment deposition. A study by Elschot et al. (2013) looked at the effects of herbivore exclusion of both small (i.e. hare and goose) and large grazers (i.e. cattle) for marshes of different ages. Their results showed that both small and large herbivores can have a major impact on vegetation height, and that trampling by large grazers reduced marsh accretion rates by compacting the soil. This view was similar to that of Nolte et al. (2014), working on the Wadden Sea.

SLR will increase nearshore water depths and thereby modify wave refraction patterns. This means that wave energy amounts will also change at different points along a particular shoreline. Pethick (1993) maintained that this could be significant for the classic Scolt Head Island salt marshes of eastern England.

Mangrove swamps may be affected by anthropogenically accelerated SLR (Quisthoudt et al., 2013; Nitto et al., 2014; Figure 9.12), but may respond rather differently from other marshes in that they are composed of relatively long-lived trees and shrubs, which means that the speed of zonation change and adaptation will be less (Woodroffe, 1990). Swamps with low rates of sediment accretion, such as those on oceanic islands with a limited sediment supply, may be especially vulnerable (Ellison, 2015).

Low temperature is widely regarded as the primary control on the latitudinal limits of mangroves globally. Mangrove vegetation is essentially tropical and its distribution is constrained by sensitivity to freezing. The distributional limits of mangroves generally coincide with the 20 °C winter isotherm of seawater. Northernmost populations occur at ~32 °N latitude (Bermuda, Japan) and southernmost populations occur at ~37–8 °S latitude (Australia and New Zealand; Krauss et al., 2008). With increasing temperatures and fewer coastal freezes, mangroves may expand their latitudinal range, as for, instance, in the Gulf of Mexico (Comeaux et al., 2012; Armitage et al., 2015) and Florida (Cavanaugh et al., 2014). Indeed, there is now evidence that mangrove species have proliferated at or near their poleward limits on at least five continents over the past half century, at the expense of ordinary salt marsh (Saintilan et al., 2014).

Figure 9.12 A mangrove swamp, composed of *Avicennia marina*, from Ras Mohammed National Park, Sinai, Egypt.

Salt marsh, mangrove swamp, and sabkha regression caused by climatic change and SLR will compound the problems of wetland loss and degradation caused by other human activities (Kennish, 2001), including reclamation, ditching, diking, dredging, pollution, and sediment starvation. Nearly two-thirds of the sabkhas of the Arabian Gulf have been lost (Burt, 2014), including that of the classic Abu Dhabi example. In the 1960s, approximately 150 km of pristine coastal sabkha existed along the eastern and central portions of the Abu Dhabi coastline. Today, only 54 km remains (Lokier, 2013). Equally, more than half of the original salt marsh habitat in the United States has already been lost, and Shriner and Street (1998, p. 298) suggested that a 50 cm SLR would inundate approximately 50 percent of North American coastal wetlands in the twenty-first century.

Deltas. Deltas are currently being impacted upon by such processes as sediment starvation, but which may, like other low-lying coasts, be affected by SLR and storm surges. However, as Ibáñez et al. (2014) have pointed out, deltas may have certain characteristics that will enable them to respond positively to SLR. First, an increase in the aggradation and frequency of river channel avulsion with accelerated SLR facilitates the formation of new delta lobes in shallow areas and the

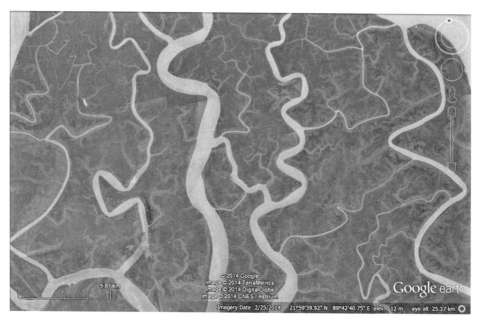

Figure 9.13 The Sundarbans of Bangladesh, an area that may be especially prone to sea-level rise. Scale bar 5.87 km. ©2014 Google, ©2014 Terrametrics, DigitalGlobe, CNES/Astrium.

subsequent increase in the efficiency of sediment deposition and retention. Second, an increase in the frequency and magnitude of flood events in delta plains as a consequence of an increased overflowing or crevassing through river natural levees may result in enhanced sediment inputs and trapping efficiency of the delta. An increase in the frequency and magnitude of overwash events in the delta fringe allows sandy beaches to quickly adapt to SLR.

On the other hand, some deltas appear to be subjected to especially high rates of SLR (Pethick and Orford, 2013). In the Sundarbans of Bangladesh (Figure 9.13), for example, the combined impact of land subsidence, eustatic SLR, tidal range amplification, and a decrease in fresh water input results in an average rate of increase in effective SLR of 14.1–17.2 mm yr^{-1}. The principal mechanism for this is the increased tidal range that occurs in estuary channels recently constricted by embankments. Of more general significance, however, is that some deltas, such as the Mekong (Van Mahn et al., 2015), now suffer from virtually no aggradation (because of upstream sediment trapping by dams and reservoirs), are suffering from accelerated compaction, and are sinking many times faster than the rate of sea-level increase (Syvitski and Kettner, 2011).

Cliffed coasts. As sea level rises, nearshore waters deepen, shore platforms become submerged, and deeper water allows larger waves to reach the bases of cliffs. Cliffs may thus suffer from accelerated rates of retreat, especially if they are made of susceptible materials (Brooks and Spencer, 2012; Ashton et al., 2011; Revell et al., 2011). They may also suffer from increased frequencies of coastal landslides. Cliffs on high latitude coasts (e.g. around the Arctic Ocean) might be especially seriously affected by global change, because of three factors that would work in the same direction. First, coasts formed of weak sediments that are currently cemented by permafrost would lose strength if warming caused the permafrost to melt and so would be more susceptible to erosion. Second, melting of sea ice would expose them to greater wave affects from open water. Third, warmer temperatures might generate larger storms.

Local erosional losses of up to 40 m per annum have been observed in some locations in both Siberia and Canada in recent years, while erosive losses of up to 600 m over the past few decades have occurred in Alaska (Parson et al., 2001). The number and total area of thaw slumps along the Yukon coastline increased by as much as 160 percent between 1952 and 2000 (Lantuit and Pollard, 2008). Overeem et al. (2011) measured an average retreat of 14.4 m yr^{-1} over a 3 km stretch of Beaufort coast in the two-year interval 2007–9, with rates as high as 30 m yr^{-1} locally. They indicated that these rates had accelerated over the past 50 years synchronously with Arctic-wide declines in sea ice extent, suggesting a causal relationship between the two. Documentation by Jones et al. (2009) of erosion rates along a 75 km stretch of the Alaskan Beaufort Sea revealed that the mean annual erosion rates doubled from ~7 m yr^{-1} for 1955–79 to ~14 m yr^{-1} for 2002–7. They found that the number of thermomechanical niche-forming episodes per annum, and thus the erosion rate, was a function of ice-free season duration, the number and type of storms impacting the coastline, sea level, and summertime sea-surface temperature. Mars and Houseknecht (2007), analyzed topographic maps and Landsat Thematic Mapper (TM) data of the low-lying Arctic coastal plain north of Teshekpuk Lake in the National Petroleum Reserve in Alaska, and found that coastal erosion had more than doubled along a segment of that coast from an average rate of 0.48 km^2 yr^{-1} during 1955–85 to 1.08 km^2 yr^{-1} during 1985–2005. Some areas underwent as much as 0.9 km of coastal erosion in the past fifty years. Coastal erosion had also breached numerous thermokarst lakes, causing initial draining of the lakes followed by progressive flooding by marine water. By this process, freshwater lakes had evolved into marine bays over time. Vermaire et al. (2013) found that along the coastline of the Mackenzie Delta in Canada, warming has been increasing the frequency and intensity of storms. After ~1980, surge activity has increased sharply in response to the rapid rise in mean annual temperature and the dramatic reduction in Arctic sea ice.

Beaches and barriers. Many beaches, especially those in closed bays, may suffer from accelerated erosion as sea level rises (Brunel and Sabatier, 2009). Bruun (1962) developed a widely cited and elegantly simple model of the response of a sandy beach to SLR in a situation where the beach was initially in equilibrium, neither gaining nor losing sediment (Bruun Rule). He implied that for an SLR of c. 1m, there might be a 100 m recession of a sandy coast. It is important to recognize, however, that there are some constraints on its applicability (Wells, 1995; Healy, 1996; List et al., 1997) and there are those who suggest that use of the Bruun Rule should now be abandoned (Cooper and Pilkey, 2004; Aagaard and Sorensen, 2012).

Barrier islands, such as those that line the southern North Sea and the eastern and Gulf Coast seaboard of the United States, are dynamic landforms. SLR effects on barriers may include an increased rate of landward migration of the barrier, decreased barrier width and elevation of barrier and sand dunes, increased frequency of storm overwash, increased frequency of barrier breaching and inlet formation, and widening and segmentation of the barrier (Eitner, 1996; Ashton et al., 2008; Williams, 2014).

Indeed, as sea level rises, barriers and other low-level coasts will be exposed to higher storm surges and greater flooding. Extreme storm surges may become more threatening around the southern North Sea (Woth et al., 2006). Locations in the northeastern United States, such as Nantucket Island, Martha's Vineyard, Cape Cod, and parts of Long Island and New Jersey are also regarded as very vulnerable (Ashton et al., 2008). In the New York area, Gornitz et al. (2002) have calculated that by the 2080s, the return period of the 100-year storm flood could be reduced to between 4 and 60 years (depending on location). Lin et al. (2012) have made similar predictions while in the case of Long Island, Shepard et al. (2012) have argued that even a modest and probable SLR of 0.5 m by 2080, would vastly increase the numbers of people (47 percent increase) and property loss (73 percent) impacted by storm surges. Equally, were hurricanes to become more frequent and widespread along the eastern and Gulf coasts of the United States, wave heights would tend to rise, as has happened during recent warming decades (Komar and Allan, 2008; Reed et al., 2015). Flooding would become a greater hazard (Frazier et al., 2010). For Corpus Christi, Texas, hurricane flooding is projected to increase by between 20 percent and 70 percent by the 2030s (Irish et al., 2010; Mousavi et al., 2011) and the effects of increased surge levels will be amplified along the Atlantic coast of the United States as this is an area where SLR appears to have been faster than the global average (Sallenger et al., 2012). SLR, combined with greater typhoon intensity, also means that storm surges pose an increasing threat to Tokyo, Japan, parts of which already lie below sea level (Hoshino et al., 2015).

9.5.2 Multiple Threats to Coral Reefs

Reefs face a suite of threats from climate change, ocean acidification, and SLR (Hoegh-Guldberg, 2011; Ateweberhan et al., 2013). Some of the predictions are dire. For instance, Sale wrote (2013, p. 325), "Coral reefs, as we knew them in the 1970s, are likely to have disappeared entirely from the planet by 2050, if current trends in human environmental impacts continue ... What will be left is eroding limestone benches, dominated by macroalgae, and with small isolated coral colonies." Equally, Spalding and Brown (2015, p. 771) have warned:

Without the stabilization of greenhouse gas concentrations, it seems inevitable that many of the world's coral reefs will become nonaccreting habitats – they will, based on most common definitions, cease to be coral reefs. This will happen more or less rapidly in different locations and will have concomitant and profound impacts on both biodiversity and people. As reefs decline, many of the millions of people who live near reefs will lose critical sources of food, as well as coastal protection and tourism revenues. Ironically, however, the main drivers of current reef decline – pollution, overfishing, sedimentation, and direct destruction – may be just as influential in the near term as climate drivers in the long term.

Hurricanes. One potential change is that of hurricane frequency, intensity, and distribution. This is discussed in Chapter 7. More strong hurricanes might build some coral islands up, erase others, and through high levels of runoff and sediment delivery they could change the turbidity and salinity of the water in which corals grow.

Increased sea-surface temperatures could have deleterious consequences for corals which are near their thermal maximum (Hoegh-Guldberg, 2001). Most coral species cannot tolerate temperatures greater than about 30 °C and even a rise in seawater temperature of 1–2 °C could adversely affect many shallow water coral species. As Hoegh-Guldberg (1999), Souter and Linden (2000), Sheppard (2003), and Baker et al. (2008) have suggested, continued warming trends superimposed on interannual and decadal patterns of variability are likely to increase the incidence of bleaching and coral mortality unless significant adaptation to increased temperatures occurs. Donner (2009) argued that the majority of the world's reefs would experience harmfully frequent thermal stress events before 2050, and Teneva et al. (2012) suggested that the most threatened reefs may be in the Central and Western Equatorial Pacific. On the other hand, some corals appear to have some ability to acclimatize to elevated temperatures (Palumbi et al., 2014) allowing them to inhabit reef areas with water temperatures far above their expected tolerances.

In cooler ocean regions, however, warmer sea-surface temperatures may lead to a rapid expansion in the range of tropical reef corals toward the poles, as appears to have taken place since the 1930s (Yamano et al., 2011).

Ocean acidification. Reefs may suffer from increasing ocean acidification (Royal Society, 2005; Chan and Connolly, 2013). This is because a proportion of the extra carbon dioxide being released into the atmosphere by the burning of fossil fuels and biomass is absorbed by sea water. As it combines with water it produces carbonic acid. The sea water becomes more acidic (i.e. it will have a lower pH than now). This absorption of carbon dioxide has already caused the pH of modern ocean surface waters to be about 0.1 lower than in preindustrial times. Ocean pH may fall an additional 0.3 by 2100 (Caldeira and Wickett, 2003), by which time the oceans may be more acidic than they have been for 25 million years. Several centuries from now, if these trends continue, ocean pH will be lower than at any time in the past 300 million years (Doney, 2006). Enhanced acidification will be particularly harmful to those organisms (hard corals, mollusks, vermetids, and plankton) that depend on the presence of carbonate ions to build their shells (or other hard parts) out of calcium bicarbonate (Orr et al. 2005a, b; Pelejero et al., 2010). Leclercq et al. (2000) have calculated that the calcification rate of scleractinian-dominated communities could decrease by 21 percent between the preindustrial period (1880) and the year (2065) at which atmospheric carbon dioxide concentrations will double. Thus, were calcification to decline, then reef-building capacity would also decline. It is also possible that ocean acidification will accelerate coral bleaching (Anthony et al., 2008). Rates of bioerosion and $CaCO_3$ dissolution could also increase (Andersson and Gledhill, 2013; DeCarlo et al., 2014). By reducing the growth potential and survivorship of corals, ocean warming and acidification are likely to change the competitive hierarchy of corals and macroalgae – at least indirectly by reducing the ability of corals to maintain or rapidly colonize available space following disturbances. At the local scale, over-fishing of herbivores can reduce the top-down control of macroalgae while nutrient enrichment may stimulate macroalgal growth rates – factors that work in combination to promote shifts toward algal dominance (Anthony et al., 2011). Consequently, concerns have been raised that ocean acidification, local stresses, and climate change could cause coral reefs to move from a condition of net $CaCO_3$ accretion to one of net erosion, which would have drastic consequences for their role and function as ecosystems within the Earth System.

SLR. During the early Late Pleistocene and Early Holocene, rates of SLR were sometimes 30–50 mm yr^{-1}, which is markedly higher than those generally predicted for coming decades of c. 5–8 mm yr^{-1}. In spite of this coral reefs grew upwards. Current data on reef accretion suggest a modal rate of vertical growth of 6–7 mm yr^{-1}. The implication of this is that it is unlikely that existing coral reefs will be extensively "drowned out" by any future SLR (Kench et al., 2009). However, because of the effects of coral bleaching, etc. there is some evidence in the Caribbean that rates of coral accretion have declined in recent years (Perry et al., 2013). This makes keeping up with SLR more difficult. In addition, reefs have a range of different relief environments, and low-lying reef islands on the rims of atolls may be especially vulnerable to the effects of SLR (Woodroffe, 2008). Moreover, in addition to the potential effects of submergence, there is the possibility that higher sea levels could promote accelerated erosion of reefs (Dickinson, 1999). SLR could also cause enhanced sedimentation and turbidity as tides ebb and flow over shallow fringing reefs (Field et al., 2011).

Wave overtopping. The future wave environments of coral reefs may be transformed (Quataert et al., 2015). Coral bleaching events, which are likely to become more numerous, kill off corals and result in a decrease in hydrodynamic roughness of coral reefs. Similarly, ocean acidification may reduce coral cover and thus hydrodynamic roughness. Together or independently, these processes will reduce bottom friction, which would increase wave heights, wave run-up, and wave-driven flooding. In addition, climate change may drive changes in extreme wave heights and their directions, which would further alter the resulting nearshore wave heights, water levels, and wave run-up.

Increasing sedimentation. Finally, reefs are very likely to have to withstand the effects of ever more intensive land cover change, including tropical forest removal, which will generate yet higher rates of damaging sedimentation (Maina et al., 2013).

9.6 Conclusions

During historical times, humans have become the dominant control on coastal change, primarily by altering river discharge and sediment supply to the coast, increasing rates of subsidence through fluid withdrawal, and through global climate change and associated accelerated SLR (Anderson et al., 2014). Many coastal landforms are essentially fragile (e.g. beaches, barriers, marshes, etc.) and many humans live in close proximity to the seas. Thus they are particularly likely to show major changes in the face of a range of human pressures.

10

Cryospheric Processes and Forms
in the Anthropocene

10.1 Thermokarst

In high latitudes and at high altitudes some areas are underlain by permanently frozen subsoil – permafrost. It is an especially important control of a wide range of geomorphological processes and phenomena, including slope stability, rates of erosion, and surface runoff. It can be disturbed and degraded by human activities, and in the tundra ground subsidence is associated with *thermokarst* development, this being irregular, hummocky terrain produced by the melting of the permafrost. It is characterized by lakes and deranged drainage. Its development is due primarily to the disruption of the thermal equilibrium of the permafrost and an increase in the depth of the active layer (French, 1976). When, for instance, surface vegetation is cleared for agricultural or constructional purposes the depth of thaw will tend to increase. The movement of tracked vehicles has been particularly harmful to surface vegetation, and deep rutting soon results from permafrost degradation. Similar effects may be produced by the siting of heated buildings on permafrost, by the laying of oil, sewer, and water pipes in or on the active layer (Lawson, 1986), and by the installation of seismic lines (Williams et al., 2013).

However, regions underlain by permafrost may be especially susceptible to current warming (Bowden, 2010). First, in higher latitudes, the amount of temperature increase predicted is greater than the global mean. Second, the nature (e.g. rain rather than snow) and amount of precipitation may also change substantially. Third, the northern limits of some very important vegetation zones, including boreal forest, shrub-tundra, and tundra may shift latitudinally by some hundreds of kilometers. Changes in snow cover and vegetation type may have a considerable impact on the state of permafrost because of their role in insulating the ground surface (Ling and Zhang, 2003).

Permafrost has indeed been degraded by recent warming, as in the Qinghai-Tibet Plateau of China (Jin et al., 2000). In Canada, between 1962 and 1988 the mean annual temperature along the Mackenzie Highway rose by 1 °C and over the same period the southern fringe of the discontinuous permafrost zone moved northwards by about 120 km (Kwong and Tau, 1994). In Alaska, Osterkamp and Romanovsky (1999) found that in the late 1980s to mid-1990s some areas experienced warming of the permafrost table of 0.5–1.5 °C and that associated thawing rates were about 0.1 m yr^{-1}. Jorgenson et al. (2006) identified that an increase in ice-wedge degradation and subsidence had occurred in northern Alaska since the 1980s in response to a 2–5 °C increase in mean annual temperatures. Increasing thaw settlement was also reported by Osterkamp et al. (2009) and Rowland et al. (2010), while Olthof et al. (2015) described lake expansion in the Tuktoyaktuk Coastal Plain of northwest Canada. Jones et al. (2011) noted an expansion of lakes in the continuous permafrost zone of the Seward Peninsula (Figure 10.1). Their assessment of lakes and ponds >0.1 ha using remotely sensed imagery from 1950–1, 1978, and 2006–7 revealed that the majority of thermokarst lakes were actively expanding as a result of surface permafrost degradation. However, as lakes expand the opportunity for drainage increases due to the encroachment toward a drainage gradient. Thus, total surface area of lakes in their study region declined by 15 percent due to the lateral drainage of several large lakes. In the discontinuous permafrost zone in Alaska, complete thawing under a pond or lake may lead to drainage integration that causes the pond to be drained, thereby reducing the number of ponds (Yoshikawa and Hinzman, 2003).

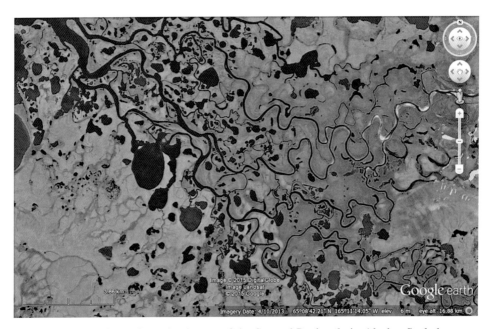

Figure 10.1 Thermokarst landscape of the Seward Peninsula in Alaska. Scale bar 3.86 km. ©2015 Digital Globe and Google Earth.

Sannel and Kuhry (2011) undertook studies in northern Canada, European Russia, and northern Sweden, and in their study the most extensive lateral expansion along lake margins from the mid-1970s to the mid-2000s occurred in large lakes (>20,000 m^2) in the Hudson Bay Lowlands and Rogovaya (Russia). They suggested that larger lakes are more likely to experience erosion as they have a longer fetch and thereby higher potential wave energy. They found that the maximum recorded lateral expansion rate was 7.3 m per decade in the Hudson Bay lowlands of Canada. Thermokarst degradation has also occurred in Siberia (Khvorostyanov, 2008), where satellite monitoring (1973–2008) by Kirpotin et al. (2009) revealed that in the zone of continuous permafrost thermokarst lakes expanded their areas by about 10–12 percent, but that in the zone of discontinuous permafrost lake drainage prevailed, confirming observations made in Alaska. The thawing of permafrost can also lead to rapid river bank erosion rates, as recorded by Kanevskiy et al. (2016) in Alaska.

Attempts have been made to assess the future distribution of permafrost. On a Northern Hemisphere basis, Nelson and Anisimov (1993) calculated the areas of continuous, discontinuous, and sporadic permafrost for the year 2050. They indicated an overall reduction of 16 percent by that date. In Canada, Woo et al., (1992, p. 297) suggested that if temperatures rise by 4–5 °C, "Permafrost in over half of what is now the discontinuous zone could be eliminated." Barry (1985) estimated that an average northward displacement of the southern permafrost boundary by 150 ± 50 km would be expected for each 1 °C warming so that a total *maximum* displacement of between 1,000 and 2,000 km would be possible. Jin et al. (2000) attempted to model the response of permafrost to different degrees of temperature rise in the Tibetan Plateau region and suggested that by the end of the twenty-first century, with a rise of 2.9 °C, the permafrost area will decrease by 58 percent. Similar estimates have been provided by Li et al. (2008). Locally, changing climate may promote fire occurrence, which can also contribute to permafrost degradation (Brown et al., 2015).

10.2 Glaciers

The second component of the cryosphere which we will consider is glacial ice. Humans have as yet had little direct influence on the state of glaciers, ice shelves, and ice sheets. One exception to this is the influence that mining debris can have on glacier state. In the case of the Kumtor mine in Kyrgyzstan, the dumping of mine spoil on receding and thinning glacier snouts initiated glacier speed-up events or surges. In addition to this, between 1999 and 2006 the Davidov Glacier had been artificially narrowed by spoil dumping, further accelerating its flow rate (Evans et al., 2015).

However, the situation will change in the future as a consequence of global warming. There are three main consequences of warming that may be discerned for ice sheets (Drewry, 1991): ice temperature rise and attendant ice flow changes; enhanced basal melting beneath ice shelves and related dynamical response; and changes in mass balance. Temperatures of the ice sheets will rise due to the transfer of heat from the atmosphere above, while ice shelves would have enhanced basal melt rates if sea-surface temperatures were to rise. This could lead to their thinning and weakening (Warner and Budd, 1990). Combined with reduced underpinning from grounding points as sea-level rises, this would result in a reduction of back-pressure on ice flowing from inland. Ice discharge through ice streams, the arteries of the ice sheets (Bennett, 2003), might therefore increase (Martin et al., 2015). Some studies (e.g. De Angelis and Skvarca, 2003; Rott, et al., 2010) have indeed found evidence of ice streams increasing their velocities when ice shelves have collapsed, but this is not invariably the case (Vaughan and Doake, 1996). None-theless, Cook and Vaughan (2010) have demonstrated the loss or substantial retreat of ice shelves around the Antarctic Peninsula in recent decades and the profound acceleration and thinning of ice streams that has occurred. Equally, Khazendar et al. (2015) have shown the accelerating change in glaciers following the collapse of the Larsen B ice shelf and they predict that the flow acceleration, ice-front retreat, and enhanced fracture of the remnant Larsen B ice shelf presage its approaching demise.

The West Antarctic Ice Sheet (WAIS), because it is a marine ice sheet that rests on a bed well below sea level, is likely to be much more unstable than the East Antarctic Ice Sheet. If the entire WAIS were discharged into the ocean, sea level would rise by 5 or 6 m. Fears have been expressed that it could be subject to rapid collapse, with disastrous consequences for coastal regions all over the world. Whether the WAIS will collapse catastrophically has been hotly debated (Oppenheimer, 1998; Joughlin and Alley, 2011), but there is some consensus that the WAIS will most probably not collapse in the next few centuries (Vaughan and Spouge, 2002).

The Greenland Ice Sheet (GIS) is rather different. Its state will depend in part on the rate of surface ablation. However, flow of the ice is also enhanced by rapid migration of surface meltwater to the ice bedrock interface. This coupling between surface melting and ice sheet flow provides a mechanism for rapid response of ice sheets to warming (Zwally et al., 2002; Parizek and Alley, 2004). Confirmation of this is provided by laser altimeter surveys, which show rapid ice thinning below 1,500 m (Krabill et al., 1999, Rignot and Thomas, 2002; Rignot and Kanagar-atnam, 2006). It appears that the outlet glaciers are thinning as a result of increased creep rates brought about by decreased basal friction consequent upon water penetrating to the bed of the glacier and the overall loss of mass of the GIS has accelerated in recent years (Thomas et al., 2006; van der Broeke et al., 2009). Straneo and Heimbach (2013) reported that mass loss from the GIS quadrupled from 1992–2001 (51 ± 65 Gt yr^{-1}) to 2002–2011 (211 ± 37 Gt yr^{-1}) (see Figure 10.2).

Many valley glaciers have retreated as a consequence of the climatic changes that have occurred since the ending of the Little Ice Age (LIA). Rates of annual retreat have often been around 20–70 m over extended periods of some decades in the case of the more active examples. Thus over the last century or so alpine glaciers in many areas have managed to retreat by some kilometers. The rate has not been constant, or the process uninterrupted, and it depends on such factors as topography, slope, size, altitude, accumulation rate, and ablation rate. Amount of debris cover is also important as it can offer some degree of insulation (Pellicciotti et al., 2014) and so leads to lower rates of retreat and volume loss compared to cleaner glaciers (Tennant and Menounos, 2013).

Data from the World Glacier Monitoring Service (WGMS) show that in the mountains of New Guinea nearly all the ice caps have disappeared since the late nineteenth century. In East Africa, the ice bodies of Mount Kilimanjaro shrank from c. 20 km^2 in 1880 to c. 2.5 km^2 in 2003. The reasons for the retreat of the East African glaciers include not only an increase in temperature, but also a relatively dry phase since the end of the nineteenth century (which led to less accumulation of snow) and a reduction in cloud cover (Mölg et al., 2003).

In South America, except for a few cases in Patagonia and Tierra del Fuego, there has been a general retreat, with enhanced rates in recent decades. In the tropical Andes, the Quelccaya ice cap in Peru has had an accelerating and drastic loss over the last forty years (Thompson, 2000). López-Moreno et al. (2014) showed that from 1984 to 2011, glaciers of the Cordillera Huaytapallana range in Peru experienced a 55 percent decrease in surface area. In neighboring Bolivia, the Glacier Chacaltya lost no less than 40 percent of its average thickness, two-thirds of its volume, and more than 40 percent of its surface area between 1992 and 1998. In the Cordillera Real of the Bolivian Andes, between 1987 and 2010 glacier-covered areas were found to have diminished by more than 30 percent (Liu et al., 2013). In Patagonia, the calculated reduction in glacierized area between the LIA and 2011 was 4,131 km^2 (15.4 percent), with 660 km^2 (14.2 percent) being lost from the Northern Patagonia Icefield, 1,643 km^2 (11.4 percent) from the Southern Patagonia Icefield, and 306 km^2 (14.4 percent) from Cordillera Darwin (Davies and Glasser, 2012).

With regard to other mountainous areas, New Zealand glaciers lost between one-quarter and almost half of their volume between the LIA maximum and the 1970s. A further net ice volume loss between 1977 and 2005 has also been

Figure 10.2 The changing state of ice at Sukkertoppen, SW Greenland, between 1935 (top) and 2013 (bottom) (courtesy of NASA Earth Observatory).

reported. In the Pyrenees, since the first half of the nineteenth century, about two-thirds of the ice cover has been lost.

The recent response of Himalayan glaciers shows great regional variations depending on such factors as the amount of debris cover that occurs on their surfaces. As already noted, a thick cover insulates the glacier surface and so reduces ablation. Climatic conditions have also played a role. In the west of the region, around half of the westerlies-influenced Karakoram glaciers have been advancing or have been stable (Hewitt, 2005), while the great bulk of the monsoon-influenced glaciers in the central Himalaya have been retreating, especially those with a limited debris cover (Scherler et al., 2011). The work of Zhu et al. (2014) on the Naimona'Nyi glaciers in the western Himalaya shows that the glacier retreat rate between 2003 and 2013 was five times of that in the previous thirty years (1973–2003).

Since the early twentieth century, glaciers on the Tibetan Plateau have generally been retreating, and this trend has accelerated since the 1990s. In the central part, glaciers lost 22 percent of their coverage from 1977 to 2010 (Wang et al., 2013). In the Chinese Altai Mountains, analyses by Wang et al. (2015) showed that the collective area of all 201 glaciers investigated was reduced by 30.4 percent from 1959 to 2008. Fifty-five glaciers disappeared entirely. In China, the overall glacier

Figure 10.3 The lower portion of the Mer de Glace, French Alps. The exposed lateral moraines shows the great thinning of the glacier that has taken place since the Little Ice Age, and provide an illustration of what may happen in response to global warming.

loss is estimated at 20 percent since the LIA maximum. Glacier area is estimated to have decreased by 20–25 percent in the Tien Shan, 30–35 percent in the Pamirs, and 50 percent in northern Afghanistan during the twentieth century. For the second part of the century, glacier area loss for the Polar Urals was 22.3 percent (Khromava et al., 2014). Fast rates of retreat have also been described for the Caucasus glaciers (Tielidze et al., 2015).

In North America, glaciers have shown a general retreat after the LIA maximum, particularly at lower elevations and southern latitudes. In the Rocky Mountains the area lost since the LIA is about 25 percent. In the Sierra Nevadas, ice volume loss from 1903 to 2004 averaged no less than 55 percent and showed a rapid acceleration in the early 2000s (Basagic and Fountain, 2011). In Alaska (Arendt et al., 2002), glaciers are thinning at an accelerating rate, which in the late 1990s amounted to 1.8 m per year.

In the European Alps the glacier cover is estimated to have diminished (see Figure 10.3) by about 35 percent from the 1850s to the 1970s and by another 22 percent by 2000, though there were interruptions in the mid-1960s, late 1970s, and early 1980s. Indeed, some glaciers have advanced in recent warming decades. In Scandinavia, for example, glaciers close to the sea have seen a very strong mass gain since the 1970s, whereas mass losses have occurred with the more continental glaciers. The positive mass balance (and advance) of some Scandinavian glaciers, notwithstanding rising temperatures, has been attributed to increased storm activity and precipitation inputs coincident with a high index of the North Atlantic Oscillation (NAO) in winter months since 1980 (Nesje et al., 2000; Zeeberg and Forman, 2001). A positive mass balance phase in the Austrian Alps between 1965 and 1981 has been correlated with a negative NAO index. Indeed, the mass balances of glaciers in the north and south of Europe are inversely correlated (Six et al., 2001). However, since 2000, even the Norwegian glaciers, which were still advancing in the 1990s, have started to retreat rapidly (Nesje et al., 2008). Icelandic glaciers are also showing widespread retreat, especially since 1995. In the case of Virkisjökull–Falljökull, since 2005 retreat rates have increased considerably – averaging 35 m yr^{-1} – with the last 5 years representing the greatest amount of ice-front retreat ($\sim$190 m) in any five-year period since measurements began in 1932 (Bradwell et al., 2013).

Glaciers that calve into water can show especially fast rates of retreat, with rates greater than 1 km per year being possible (Venteris, 1999). In Patagonia, in the 1990s, rates of up to 500 m yr^{-1} were observed. This rapid retreat is accomplished by iceberg calving, with icebergs detaching from glacier termini when the ice connection is no longer able to resist the upward force of flotation and/or the downward force of gravity. It is favored by the thinning of ice near the termini, its flotation, and its weakening by bottom crevasses. The Columbia

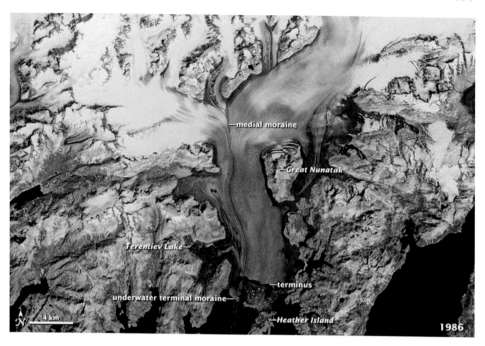

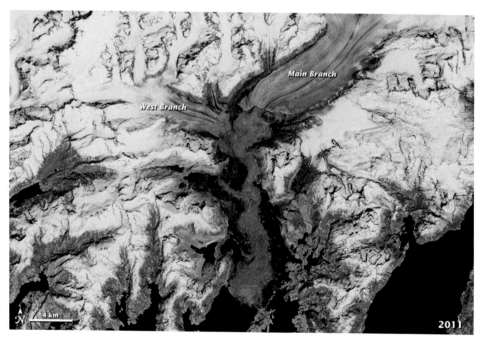

Figure 10.4 The Columbia tidewater glacier in 1986 (top) and 2011 (bottom). Source: http://earthobservatory.nasa.gov/Features/WorldOfChange/columbia_glacier.php. (accessed February 10, 2015).

tidewater glacier in Alaska retreated around 13 km between 1982 and 2000 (see Figure 10.4). Equally the Mendenhall Glacier, which calves into a proglacial lake, showed 3 km of terminus retreat in the twentieth century (Motyka et al., 2002). Certainly, calving permits much larger volumes of ice to be lost to the glacier than would be possible through surface ablation alone (van der Veen, 2002). The glaciers of the Antarctic Peninsula, all of which calve into the sea, also show a pattern of overall retreat, with 87 percent retreating in the last fifty years (Cook et al., 2005; Pritchard and Vaughan, 2007). In Novaya Zemlya, in the Russian Arctic, Carr et al. (2014) found that between 1992 and 2010, 90 percent of the glaciers they studied had retreated but that rates were an order of magnitude higher for marine-terminating outlets (52.1 m yr^{-1}) than for land-terminating glaciers (4.8 m yr^{-1}).

The WGMS (2008) suggests that globally the average annual mass loss of glaciers during the warming decade from 1996 to 2005 was twice that of the previous decade (1986–95) and over four times that from 1976 to 1985. Using WGMS data, Zemp et al. (2015) concluded that the rates of early twenty-first-century glacier mass loss are without precedent on a global scale.

This response to the warming of recent decades suggests that with even greater warming in coming decades, rates of recession and thinning may be very considerable. Radić and Hock (2011) tried to assess the glacier volume loss that will occur by 2100. They suggest that for the Earth as a whole, the total glacier volume will be reduced by about a fifth, but that in some locations, notably New Zealand and the European Alps, the loss will be about three-quarters. The largest glacier in the European Alps, the Grosser Aletsch, is likely to lose 90 percent of its volume by the end of the century (Jouvet et al., 2011). In a more recent analysis, Radić et al. (2014) suggested that glaciers in Central Europe, low-latitude South America, Caucasus, North Asia, Western Canada, and the United States will lose more than 80 percent of their volume by 2100. They suggested that the current global glacier volume will be reduced by between 29 and 41 percent. Finally, Clarke et al. (2015) predicted that by 2100, the volume of glacier ice in western Canada will shrink by 70 ± 10 percent relative to 2005. According to their simulations, few glaciers will remain in the Interior and Rockies regions, but maritime glaciers, in particular those in northwestern British Columbia, will survive in a diminished state.

Warming is not the only anthropogenic factor that may accelerate the retreat of glaciers and ice sheets. Also important, because of their effect on surface albedo, is the increasing role of such substances as black carbon (derived from fires and industrial sources) and aeolian dust. Their deposition has been implicated with forcing the rapid retreat of glaciers at the end of the LIA (Painter et al., 2013) and with current ice loss (Kaspari et al., 2015; Goelles et al., 2015).

10.3 Glacial Lakes

One consequence of the glacial downwasting outlined above, is that a variety of lakes have formed, which from time to time may drain catastrophically to give outburst floods (Harrison et al., 2006; Chen et al., 2010). There are many examples of accelerating lake formation. In western Greenland, Carrivick and Quincy (2014) used repeat satellite imagery for the period from 1987 to 2010. During that time there was a c. 44 percent net increase in the number of lakes, and a c. 20 percent expansion in total lake surface area on a 1,300 km length of ice margin. Similarly, in the Tibetan Plateau, in the fifteen years from 1986, the total number of lakes increased by 11 percent and the total lake surface areas increased by 47 percent. Glacial lakes also expanded in the neighboring Mount Everest area (Benn et al., 2012) and on the Bhutan-China border (Komori, 2008). In the mountains of Peru, Vilimek et al. (2005) showed that due to rapid deglaciation, the volume of Palcacocha Lake increased from 514,800 m^3 in 1972 to 3,690,000 m^3 in 2003. In the Patagonian Andes, Mölg (2014) found that between 1985 and 2011 the number of proglacial lakes increased from 223 to 327 and their area expanded by 59 percent.

10.4 Conclusions

Humans have caused great changes to the state of permafrost and have created a range of thermokarst features. Although humans have as yet had little or no direct influence on the state of glaciers, ice shelves, and ice sheets, the situation will change in the future as a consequence of global warming and of pollution, and this in turn will have an impact on river flows and on global sea levels. Most of the world's glaciers are in a state of retreat and downwasting, and, *inter alia*, this has had a major impact on the state of glacial lakes.

11

Conclusions on the Relationships Between Geomorphology and the Anthropocene

11.1 The Antiquity of Some Anthropogenic Geomorphological Changes

Without doubting the importance of the Industrial Era of the last three centuries and in particular the pace of change during the Great Acceleration since 1945, there is strong evidence that in some parts of the world humans have had much longer-term impacts on geomorphology. The following examples, drawn from earlier chapters in the book, summarize the key nature and dimensions of longer-term human impacts on landforms and earth surface processes. While these changes may have largely had only local or regional impacts, they set the scene for the wider changes that have occurred more recently.

Megafaunal extinctions. Megafaunal extinctions in the Late Pleistocene may have had important impacts on geomorphology. The extinction of large herbivores will have had an impact on vegetation cover, nutrient cycling, and trampling activity, all of which could have had geomorphological ramifications (see Chapter 2).

Fire effects. The use of fire goes back to Paleolithic times. In the eastern United States, human-induced fires started to affect vegetation early in the Holocene (Abrams and Nowacki, 2015). In New Zealand, Polynesians caused extensive burning of vegetation after their arrival in c. 1280 AD (McWethy et al., 2009; see Chapter 2). Fire can, *inter alia*, lead to increases in runoff and erosion (see Chapter 6).

Deforestation. Pollen analysis shows that deforestation of European forests started in Mesolithic and Neolithic times. Extensive deforestation took place in the Mediterranean lands in classical times and in Central Europe in medieval times. In equatorial rain forests, clearance for agriculture has been going on since at least 3000 BP in Africa, 7000 BP in South and Central America, and possibly since 9000 BP or earlier in India and New Guinea. It is possible that early to mid-Holocene deforestation and land cover change modified global climates by releasing carbon dioxide into the atmosphere (see Chapter 2).

Metal smelting. Copper smelting occurred in Turkey and Jordan from the sixth millennium BC and in Serbia from the fifth millennium BC. The spread of metal working into other areas was rapid particularly in the second half of the fifth millennium and by 2500 BC bronze products were in use from Britain in the West to northern China in the East. Metal smelting required fuel and so may have promoted vegetation clearance and subsequent erosion (see Chapter 2).

Tells and other mounds. In the Middle East and other areas of long-continued human urban settlement occupation mounds (*tells*) date back to the beginnings of urban settlement in the early Holocene (see Chapter 3). In Britain the burial mound at Silbury Hill, together with many long barrows, dates back to the Neolithic, while the pyramids of Central America, Egypt, and the Far East are even more spectacular early feats of landform creation. Likewise in the Americas, native Indians created large numbers of mounds for temples, burials, settlement, and effigies. In the Midwest and South of the United States, there may be as many as several hundred thousand artificial mounds that largely predate European colonization, and some are enormous (see Chapter 3).

Embankments. Some of the earliest examples of flood embankments on rivers date from c. 4,600 years ago on the Indus and c. 3,000 years ago in Egypt (see Chapter 3).

Coastal reclamation. Deliberate coastal reclamation to produce more available productive land started in Roman times in Britain, in the Fens, Romney Marsh, and the Somerset Levels. Many schemes were also developed in Europe in medieval times (see Chapter 3).

Terracing. In Jordan, terraced wadis and ancient floodwater farming possibly began in the Neolithic and Bronze Age. In the Negev, Israel terracing dates back to the Neolithic and Chalcolithic, but was also practiced in the Bronze Age, Iron Age, Nabataean, Roman, and Byzantine times (see Chapter 3).

Excavation. Landforms produced by excavation are widespread, and date back to the Neolithic in some places, for example in East Anglia, England, where people used antler picks and other means to dig deep pits in the chalk (see Chapter 3).

Qanat. These remarkable feats of hydraulic technology, in which underground tunnel wells are dug into dryland alluvial fans, originated 2,700–2,500 years ago in Iran (see Chapter 3).

Canals. The oldest-known canals are found in Mesopotamia and date back to c. 6,000 years BP. In Egypt, canals date back to c. 4300 BP, whereas in China large canals were constructed around 2,300 years ago (see Chapter 3).

Soil erosion. In the Mediterranean evidence of severe soil erosion has been found to date from the Neolithic in southern Greece (Fuchs et al., 2004), and from Roman times in Turkey (Casana, 2008). Changes in land use, such as the introduction of orchard crops, terracing, and cultivation, have had a profound and variable influence on the Mediterranean environment throughout the Holocene (see Chapter 6). In China, accelerated erosion from the Yellow River catchment has caused a tenfold increase in sediment accumulation on the continental shelf since c. 2300 BP in comparison with early Holocene rates (Milliman et al., 1987).

Lake sedimentation rates. Enhanced rates of lake sedimentation because of land cover changes causing higher catchment erosion rates have been found from the Neolithic onwards. For example, a thirteen-fold increase in sediment rates in Llangorse Lake, Wales, occurred rapidly after 5000 BP, related to forest clearance (Jones et al., 1985; see Chapter 6).

Landslides. Over a long time-span of 3,600 years, pulses of increased landslide activity in the Swiss Alps have been linked to phases of deforestation (Dapples et al., 2002; see Chapter 6).

Erosion and colluviation in Europe. Much evidence has been found of extensive erosion in many European catchments as a result of land-use changes in the Bronze Age, Iron Age, and later periods. For example, in British and Irish catchments, geomorphic instability linked with land-use changes was especially intense in the Bronze Age and the Iron Age (Foulds and Macklin, 2006). In Germany, soil erosion led to phases of accelerated erosion, colluviation, and alluvial sedimentation in the Bronze Age, the Iron Age/Roman period, and c. 1000 AD (Dreibrodt et al., 2010; see Chapter 7).

Postsettlement erosion rates. In Australia, New Zealand, and the United States, human-caused increases in erosion rate have occurred more recently than in Europe, but the trends are similar. For example, sediment fluxes in the Murrumbidgee river, Australia, have increased by more than 150 times since European settlers arrived 180 years ago (Olley and Wasson, 2003). In New Zealand, rates of sedimentation have increased from 0.1 mm yr^{-1} in pre-Polynesian times to 11 mm yr^{-1} since European land clearance in the 1880s (see Chapter 7).

Dune reactivation. There are numerous examples of dune systems being transformed quite early in the Holocene, such as in New Zealand, Canada, and California. In Europe, there is evidence of Mesolithic activation of dunes in Germany, sub-Atlantic modification of inland dunes in Hungary, and Neolithic changes to the European sand belt (see Chapter 8).

11.2 Highlights of Human Impacts on Geomorphology During the Great Acceleration

The geomorphological changes wrought by humans over the past few decades are impressive, and the following examples, drawn from earlier chapters, provide some quantification of the recent scale of human impacts.

Deliberate earth moving. Between 30 and 57 billion metric tons of earth is moved per year by humans on a global basis (Hooke, 1994; Douglas and Lawson, 2001). In comparison, the world's rivers transport between 8.3 and 51.1 billion tons per year (Walling, 2006). Thus, at the global scale humans move the same order of magnitude of earth as do rivers (Price et al., 2011).

Reservoir creation. In China, reservoir construction has dramatically increased since the 1950s. During the 1950s and 1960s nearly 72,000 reservoirs were built, and by 2013 there were 98,000 reservoirs in China (Yang and Lu, 2014). The total estimated storage capacity of Chinese reservoirs (794 km^3) is triple that of lakes (268 km^3).

Land subsidence. One of the most dramatic examples of subsidence occurred between 1928 and 1971 in Los Angeles, where 9.3 m of subsidence resulted from exploitation of the Wilmington oilfield.

Seismic activity. Ellsworth (2013) reports that within the central and eastern United States, the earthquake count has increased dramatically over the past few years as a result of drilling for hydrocarbons and other human activities (see Chapter 4). More than 300 earthquakes with $M \geq 3$ occurred in the three years from 2010 through 2012, compared with an average rate of twenty-one events per year observed from 1967 to 2000.

Soil erosion. In the eastern United States, Reusser et al. (2015) showed that by the early 1900s, averaged rates of hillslope erosion ($\sim$950 m per million years) exceeded ^{10}Be-derived background long-term erosion rates ($\sim$8 m per million years) by more than one hundred times.

Sediment accumulation rates in lakes. In a study of over 100 lakes in western Canada, using the median rate of sedimentation for the first half of the twentieth century as the background rate, many lakes showed a doubling of sedimentation rates in the second half of the twentieth century, and some showed a fourfold increase (Schiefer et al., 2013). In western Canada, sediment accumulation rates in recent decades are about 50 percent greater than the background levels of the first half of the twentieth century (Schiefer et al., 2013).

Landslides. In the mountainous terrain of the Pacific Northwest of the United States, the average landslide erosion rate for uncut forests is c. 3.8 times higher than for uncut forests.

Channelization. In east central Illinois, United States, humans are now seen to be the dominant agent of stream channel change, with human modification of streams in recent decades one to two orders of magnitude higher than natural geomorphological change (Urban and Rhodes, 2003).

River flows. In the United States as a whole human activities are estimated to have reduced the median annual flood in large rivers by an average of 29 percent, in medium rivers by 15 percent, and in small rivers by 7 percent (Fitzhugh and Vogel, 2011).

River sediment loads. On a global basis, large dams may retain 25–30 percent of global fluvial sediment flux (Vörösmarty et al., 2003). Syvitski et al. (2005) have calculated that humans have increased global fluvial sediment flux through soil erosion by c. 2.3 billion metric tons per year. They also calculated that sediment retention behind dams has reduced the net flux of sediment reaching the world's coasts by c. 1.4 billion metric tons per year. According to their calculations, therefore, global river sediment loads have increased more as a result of soil erosion than they have been reduced by dam retention. Wilkinson and McElroy (2007) calculated that "natural" sediment fluxes to the world's rivers are about 21 billion metric tons per year, and that "anthropogenic" losses may be around 75 billion metric tons per year.

Lake shrinkage. By 2008, because of irrigation and interbasin water transfers, the area of the Aral Sea was just 15.7 percent of that in 1961 (Kravstova and Tarasenko, 2010).

Dust emissions. Ginoux et al. (2012) estimated that anthropogenic sources account for 25 percent of global dust emissions. Several studies imply that dust storm activity has dramatically increased in recent years. For example, an ice core from the Antarctic Peninsula (McConnell et al., 2007) showed a doubling in dust deposition in the twentieth century, and this is explained by increasing temperatures, decreasing relative humidity, and widespread desertification in the source region, which lies in Patagonia and northern Argentina. Marx et al. (2014) used a core from an Australian alpine mire to show that since the 1880s rates of wind erosion have been ten times higher than background Holocene levels.

Coastal reclamation. Around the Yellow Sea, between China and Korea, coastal reclamation from the 1980s onwards led to the loss of 28 percent of tidal flats by the late 2000s (a loss rate of c. 1.2 percent annually, Murray et al., 2014). Over a five decade period up to 65 percent of tidal flats have been lost.

Coastal erosion. Sixty-eight percent of the coastline of New England and the Mid-Atlantic region of the United States has suffered erosion in recent decades (Hapke et al., 2013), while the total coastal wetland area in China has been reduced by >50 percent (Wang et al., 2014).

Coral cover. In the Indo-Pacific region as a whole, coral cover (which is a measure of the proportion of a reef surface covered by live stony coral rather than other organisms), declined from 42.5 percent in the early 1980s to only 22.1 percent by 2003, an average annual rate of 1 percent. In the Caribbean between 1977 and 2001 the annual rate was even higher – 1.5 percent (Bruno and Selig, 2007). Coral cover in the Caribbean has fallen slowly over the past 25–30 years from an average of roughly 50 percent to only 10 percent (Hughes et al., 2013).

Estuarine sedimentation. There are many examples of siltation of estuaries resulting from deforestation of the river catchments that feed them, and these are reviewed by Poirier et al. (2011), who find that in many of the works they cited, c. two- to six-fold increases in sedimentation rates were reported.

Loss of oyster reefs. Oyster reefs once dominated many estuaries, ecologically and economically, but centuries of resource extraction, exacerbated by coastal degradation, have pushed them to the brink of functional extinction worldwide. It is estimated that 85 percent of oyster reefs have been lost globally (Beck et al., 2011).

Loss of mangroves. Food and Agriculture Organization data suggest that the world's mangrove forests, which covered 19.8 million ha in 1980, have now been reduced to only 14.7 million ha, with the annual loss running at about 1 percent yr^{-1} (compared to 1.7 percent yr^{-1} from 1980 to 1990).

Valley glacier shrinkage. The World Glacier Monitoring Service (WGMS; 2008) suggests that the global average annual mass loss of glaciers during the warming decade from 1996 to 2005 was twice that of the previous decade (1986–95) and over four times that of the decade from 1976 to 1985.

Proglacial lake formation. In western Greenland, over the period 1987–2010 there was a c. 44 percent net increase in the number of lakes, and a c. 20 percent expansion in total lake surface area (Carrivick and Quincy, 2014).

11.3 The Future

In coming decades, combinations of land-use change and anthropogenic climate change will cause still further alterations in the geomorphological environment. With regards to climate change, some environments will change more than others – "geomorphological hotspots" – especially when crucial thresholds are crossed (Goudie, 1996). Feedbacks between geomorphological systems and the Earth System will be likely to cause further complex responses.

There are four types of reasons why some geomorphological processes, hazards, and landform assemblages will show substantial modification as climate changes: threshold reliance of some geomorphological processes on climatic conditions, the compound effects promoted by a combination of climate changes and other human pressures, environments that are susceptible because they are weak or in a dangerous location, and areas where there may be a risk of particularly severe climate change.

Threshold reliance. Some landforms and earth surface processes change across crucial climatic thresholds. For example, the melting of the cryosphere is strongly temperature dependent and permafrost can only exist where mean annual temperatures are negative. Thus as temperatures rise the limits of permafrost will move polewards and/or upwards in altitude, the depth of summer thaw will change and thermokarst depressions will form (see Chapter 10). Also, glacier mass balance is largely controlled by the relative significance of ablation and snow nourishment and these in turn depend on temperatures and precipitation amounts (see Chapter 10). Likewise, stream flow, especially in dry regions, can vary greatly with modest changes in moisture caused by changes in evapotranspiration. The Holocene history of valley fills in the American southwest (see Chapter 7) has shown how abruptly and greatly stream systems can switch between arroyo incision and aggradation. Aeolian activity is also strongly dependent on wind energy, sand supply, and the nature and extent of vegetation cover. If the last falls below a certain level, wind action is sharply intensified, as happened repeatedly during drought phases in the United States in the Holocene (see Chapter 8). Another example of threshold dependence of a landform type is the coral reef. These are highly sensitive to any changes in cyclone activity, to coral bleaching caused by elevated sea-surface temperatures, and to sea-level rise (see Chapter 9). There is also a strong temperature control of the distribution of mangrove swamps. Terminal lake basins, such as the Aral Sea or Lake Chad, are other landforms that have shown rapid and substantial variations in response to Holocene and twentieth-century climate changes (see Chapter 7).

Compound effects. There are numerous examples of landform change being promoted by a combination of climate changes and other human pressures. The US Dust Bowl of the 1930s, for example, was caused not only by a run of dry, hot years, but also coincided with a phase of land-use intensification (Chapter 8). This was also true of the desertification associated with the Sahel drought in West Africa from the mid-1960s. In coastal regions, the effects of rising sea levels are compounded by local subsidence caused by fluid abstraction (see Chapter 4) while beaches, marshes, and deltas starved of sediment by the damming of rivers and the construction of coastal defenses will be especially prone to erosion and inundation as global sea levels rise (see Chapter 9). Much of the Egyptian coast is now "undernourished" with sediment as a result of dam construction and so will be especially prone to erosion as sea levels rise. In permafrost regions, climate-induced thawing may exacerbate the effects of thawing produced by land cover changes.

Susceptible environments. Some landforms are robust, while others are not. Alluvial bottomlands such as the arroyos of the southwest United States, are intrinsically weak, being formed of relatively easily eroded material (see Chapter 7). The same is true of dongas developed in erodible colluvial aprons in southern Africa. Likewise, once their protective vegetation cover is removed, sand dunes in arid areas are easily reactivated (see Chapter 8). Muddy and sandy coastlines, such as those of the eastern seaboard of the United States, which are composed of features like barrier islands, will plainly be more prone to erosion than hard-rock coastlines. Slopes, shorelines, and river banks glued by permafrost will be more resistant that those features in the absence of permafrost. Some landforms are located in threatening areas – for example, at the margins of permafrost, in close proximity to retreating valley glaciers, or in low-lying areas subject to sea level rise. At the other extreme, there are some landscapes that appear to have remained relatively unperturbed by environmental change over millions of years (e.g. low-relief surfaces characteristic of shield areas in the interior of Australia).

Severity of climate change. The severity of climate change, and thus its likely impact, will also vary spatially. For example, the degree of temperature increase will be especially great in high northern latitudes (e.g. northern Canada and Siberia) so that the cryosphere in such regions may be subjected to particular pressures (Aalto et al., 2014). Similarly, reductions in soil moisture and stream flows may be especially great in some areas that are currently relatively dry, so that stream networks will shrink. Many dry areas will become drier, either because of less precipitation, or because of increased moisture losses resulting from evapo-transpiration. Warming will have an especially strong impact on river behavior in areas where winter precipitation currently falls as snow (see Chapter 7).

The possible consequences of global warming for geomorphology are numerous and are summarized in Table 11.1.

11.3.1 Not Just Climate Change

Global warming is not the only major human driver of future geomorphological change. Also of immense, and in many cases more immediate importance, are other aspects of global change, particularly those brought about by modifications of land use and land cover (Pelletier et al., 2015). In spite of the increasing pace of industrialization and urbanization, it is plowing and pastoralism which are responsible for many of our most serious environmental problems and which are still causing some of our most widespread changes in the landscape, as indeed they have through much of the Holocene. However, the complexity, frequency, and magnitude of impacts are increasing, partly because of steeply rising population levels and partly because of a general increase in *per capita* consumption. Human populations will increase, and will probably be greater by 2 – 4 billion people by 2050 (Cohen, 2003).

Deforestation and land cover change may cause climate changes at the regional scale which may be of comparable dimensions to those predicted to arise from global warming (e.g. Avila et al., 2012), changes in stream runoff and sediment loads caused by land cover change or dam construction may far exceed those caused by future changes in rainfall amounts (e.g. Ericson et al., 2006), loss of coastal wetlands due to direct human action may exceed those caused by sea-level rise (Nicholls et al., 1999), and landslide activity may owe more to changes in

Table 11.1 *Geomorphological consequences of global warming.*

Hydrological controls on rivers and lakes
Changes in state of peatbogs and wetlands
Consequential changes in river flows, drainage densities, and lake levels
Increased melting of glaciers and ice sheets may initially lead to higher river flows but in
    the longer term may lead to lower flows
Increased moisture loss due to evapotranspiration
Increased percentage of precipitation as rainfall at expense of winter snowfall
Increased precipitation as snowfall in very high latitudes
Less use of water by vegetation because of increased CO_2 effect on stomatal closure
More droughts in areas of precipitation deficit
Possible increased spread, frequency, and intensity of cyclones

Vegetational controls
Growth enhancement by CO_2 fertilization
Increasing fires in response to increasing droughts
Major changes in altitudinal distribution of vegetation types (c. 500 m for 3 °C)
Major changes in latitudinal extent of biomes
Reduction in boreal forest, increase in grassland, etc.

Aeolian
Increased dune movement and reactivation in areas of moisture deficit
Increased dust storm activity in areas of moisture deficit
Possible reduction in trade wind velocities

Coastal
Accelerated coast recession (particularly of sandy beaches) in response to sea level rise
Changes in rates of reef growth
Inundation of low-lying areas (including wetlands, sabkhas, deltas, reefs, lagoons, etc.)
    because of sea level rise
Loss of marshland in areas of coastal squeeze
More storm surges and tropical storms cause overtopping and erosion
Spread of mangrove swamp outwards from low latitudes

Weathering
Algal greening in response to wetter winters
Changes in number of freeze-thaw cycles
More salt weathering as climates dry
Possible increased rates of chemical weathering in response to temperature elevation
Spread of salt weathering in low-lying coastal areas as sea levels rise

Soil erosion and mass movements
Changes in response to changes in land use, fires, natural vegetation cover, rainfall
    erosivity, etc.
Changes resulting from soil erodibility modification (e.g. sodium and organic contents)
More intense rainfall events
Removal of glacial buttresses and of permafrost glue leading to increased slope instability
    in high latitudes and at high altitudes

Subsidence
Desiccation of clays under conditions of increased summer drought
Increased oxidation of organic soils under higher temperatures
Thermokarst development
Water shortages leading to more groundwater exploitation and thus to more subsidence

Cryosphere
Changes in rates of ablation and accumulation for glaciers and ice sheets leading to
  thinning and recession
Development of glacial lakes and associated outburst floods
Loss of ice shelves and of their buttressing activity on inflowing glaciers
Permafrost decay, thermokarst formation, increased thickness of active layer, instability of
  slopes, river banks, and shorelines
Sea ice melting leading to greater wave activity on high latitude coasts

human activity than to changes in climate (Crozier, 2010). Overexploitation of groundwater may be a more potent cause of saltwater intrusion, and possible increased amounts of salt weathering in coastal areas, than sea level rise caused by global warming (Ferguson and Gleeson, 2012).

11.4 Stage 3 of the Anthropocene: Stewardship

Humans are not merely passive observers of geomorphological change. There is now a huge array of techniques to control wind erosion and dune movement (see Chapter 8), water erosion of soils and landslides (see Chapter 6), and coastal retreat (see Chapter 9), while fluvial systems are being restored through restoration techniques and dam removals (see Chapter 7). Moreover, the world's network of protected areas has grown since the late nineteenth century and particularly since the early 1960s. Indeed, there are now over 120,000 terrestrial sites worldwide and these cover about 12.2 percent of the land area (comparable in size to the whole of South America). In many countries major developments in land use, construction, and industrialization now have to be preceded by the production of an Environmental Impact Assessment, in which geomorphological concerns may be important (Cavallin et al., 1994; Rivas et al., 1997; Veni, 1999). More specifically as regards landforms, a measure of protection is provided by the designation of World Heritage Sites (see Badman, 2010; Migoń, 2010; Migoń, 2014; May, 2015). UNESCO has commissioned reports on the need to designate areas of karst and caves (Williams, 2008), volcanic landforms (Wood, 2009) and deserts (Goudie and Seely, 2011). There are already a large number of essentially geomorphological World Heritage Sites (Table 11.2; Figure 11.1) in the natural category, as well as some cultural sites which may also have geomorphological value. The latter category includes the Aflaj Irrigation System of Oman (humanly made landforms associated with irrigation); the Al-Hijr Archaeological Site, Saudi Arabia (sandstone weathering); ancient Thebes and its Necropolis, Egypt (accelerated weathering, see Figure 5.4); archaeological ruins at Mohenjo-Daro, Pakistan (accelerated salt weathering, see Figure 5.3); the archaeological site at Volubilis, Morocco (calcrete development); rock art in the Hail Region of Saudi Arabia (rock coatings and a paleolake); the Champaner-Pavagadh Archaeological Park, India (fossil topographic dunes); Humberstone and Santa Laura Saltpeter Works, Chile (Caliche); incense route desert cities in the Negev (desert runoff processes at Avdat); rock art sites at Tadrart Acacus, Libya (rock rinds, desert varnish, sandstone weathering, and natural arches); the St. Catherine area, Egypt (granite weathering); Tsodilo Hills, Botswana (ancient linear dunes and pluvial lake deposits); Viñales Valley, Cuba (tropical karst); and Alto Douro, Portugal (granite terrain and terracing).

Table 11.2 *World Heritage Sites: selected geomorphological examples. (Data from UNESCO (http://whc.unesco.org/en/list/, accessed November 17, 2015).*

Iguazu National Park (Argentina)
Great Barrier Reef (Australia)
Purnululu National Park (Australia)
Sundarbans (Bangladesh)
Belize Barrier Reef Reserve System (Belize)
Okavango Delta (Botswana)
Nahanni National Park (Canada)
Lakes of Ounianga (Chad)
South China Karst (China)
Plitvice Lakes National Park (Croatia)
Pitons, cirques and ramparts of Reunion Island (France)
Western Ghats (India)
Great Himalayan National Park Conservation Area (India)
Dolomites (Italy)
Mount Etna (Italy)
Kenya Lake System in the Great Rift Valley (Kenya)
Tsingy de Bemaraha Strict Nature Reserve (Madagascar)
Gunung Mulu National Park (Malaysia)
Namib Sand Sea (Namibia)
Te Wahipounamu – Southwest New Zealand (New Zealand)
West Norwegian Fjords (Norway)
Aldabra Atoll (Seychelles)
Caves of Aggtelek Karst and Slovak Karst (Slovakia)
Teide National Park (Spain)
Swiss Alps Jungfrau-Aletsch (Switzerland)
Giant's Causeway and Causeway Coast (United Kingdom)
Dorset and East Devon Coast (United Kingdom)
Kilimanjaro National Park (Tanzania)
Grand Canyon National Park (United States)
Ha Long Bay (Vietnam)
Trang An Landscape Complex (Vietnam)
Mosi-oa-Tunya / Victoria Falls (Zambia and Zimbabwe)

Figure 11.1 The Namib Sand Sea in Namibia, with its superlative linear dunes, has been designated a World Heritage Site.

Individual countries have national parks which may exist primarily because of their beautiful landforms. Table 11.3 lists some of the superlative national parks of the United States and shows their importance in terms of visitor numbers and geotourism. In addition to national parks, there are also state parks, which may also be dominantly geomorphological. One example is shown in Figure 11.2.

Related to this is the establishment of geoparks and geomorphosites (Reynard, 2007; Joyce, 2010). Among the global geopark network are many of great geomorphological interest, such as the Alxa Desert and Zhangjiajie (Figure 11.3; Brierley et al., 2011) in China, Bohemian Paradise in the Czech Republic, the Rokua esker in Finland, the Burren and Cliff of Moher in Ireland, the Marble Arch caves of Northern Ireland, the Toya caldera and Usu volcano of Japan, and the Dong Van karst of Vietnam. In the United Kingdom, there is a network of regionally important geological/geomorphological sites (RIGS). Many sites have also been notified as Sites of Special Scientific Interest (SSSIs), and their importance recorded in a number of publications arising from the Geological Conservation Review of the Joint Nature Conservation Committee (e.g. Gregory, 1997;

Table 11.3 *Numbers of visitors to ten major National Parks United States (in millions) in 2012 (Data from National Park Service).*

Great Smoky Mountain, South Appalachians	9.7
Grand Canyon (Arizona)	4.4
Yosemite (California)	3.8
Olympic (Washington)	2.9
Yellowstone (Wyoming, Montana, and Idaho)	3.4
Cuyahoga Valley (Ohio)	2.3
Rocky Mountain (Colorado)	3.2
Zion (Utah)	3.0
Grand Teton (Wyoming)	2.7
Acadia (Maine)	2.4

Figure 11.2. Goosenecks State Park in Utah, United States, with its superlative incised meanders of the San Juan River.

May and Hansom, 2003). Less formal descriptions of nationally important sites are now available for a number of countries in the World Geomorphological Landscapes series, under the overall editorship of P. Migoń (e.g. Goudie and Viles, 2015).

Figure 11.3 Zhangjiajie, China, showing the remarkable forested sandstone towers.

The protection of geodiversity has become a major research and policy priority (Prosser et al., 2010; Gray 2013), while geomorphological tourism has become something that one must achieve before one dies (Bright, 2005). That said, tourism has many impacts on landforms and processes, including damage to caves, footpath erosion, trampling of coastal dunes, boat wake erosion of river channels, and disturbance of desert surfaces by off-road vehicles.

11.5 Geomorphological Changes and the Earth System

If, following Hamilton (2015), one accepts the purist definition of the Anthropocene which was introduced by Crutzen (2002) as a name for a new epoch in Earth's history – an epoch when human activities have "become so profound and pervasive that they rival, or exceed the great forces of Nature in influencing the functioning of the Earth System" (Steffen, 2010, p. 443), then it is important to ponder the influence that the geomorphological changes wrought by humans have had not merely on "the environment" but on "the Earth System." In reality, what we know about the operations and complex linkages in the Earth System is still insufficient to give definitive answers. Nonetheless, there are some examples of the highly important role of geomorphological changes which we can identify – notably the influence on dust, and the cycling of carbon, methane, and silica.

Goudie and Middleton (2006), Shao et al. (2011), and Mahowald et al. (2014; Figure 11.4) have considered the impact of wind erosion-related dust storm activity on the global system. Dust storms affect, *inter alia*, the albedo of ice bodies, radiation levels in the atmosphere, drought intensity, precipitation acidity, nutrient levels in the oceans, the drawdown of carbon dioxide by plankton, the state of coral reefs, the fertility of the Amazonian rainforest, and the transmission of pathogens. In turn, dust storm activity may be influenced by other geomorphological changes such as desiccation of lakes (Gill, 1996) and the winnowing of fine material from reactivated dunes (Bhattachan et al., 2013; 2014).

Another example of the effects of geomorphological change on the Earth System, touched upon in Chapter 5, relates to carbon budgets. Soil erosion by wind may play a significant role in these (Webb et al., 2012; Chappell et al., 2013), but so may water erosion from agricultural land (see Chapter 6), and the burning and subsidence of peat (see Chapter 4). Soil erosion is also implicated in budgets of other elements such as nitrogen and phosphorus (Qunnton et al., 2010). As Dadson (2010) has pointed out, there is a clear research need to quantify the effects on the carbon cycle of soil erosion and deposition, though approaches capable of doing this at appropriate scales have yet to be developed. The soil system is the third largest reservoir of carbon, next only to the lithosphere and oceans. Consequently, soil carbon represents a substantial and highly sensitive component within the

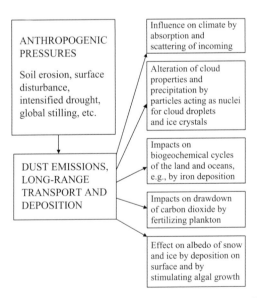

Figure 11.4 Some of the possible effects of changing amounts of aeolian dust on the Earth System.

global carbon cycle, and small changes in it may result in large changes of atmospheric CO_2 at timescales of 10^1–10^3 yr. Part of the Soil Organic Carbon (SOC) stored in the top few meters of soil, is redistributed across landscapes by soil erosion (Müller-Nedebock and Chaplot, 2015) and deposition. The erosion of peat bogs, described in Chapter 6, may release carbon (Evans et al., 2006). However, whether the combined effect of SOC redistribution and associated changes to ecosystem productivity result in a net C sink for or source to atmospheric CO_2 remains unresolved (Hoffmann et al., 2013; Kirkels et al., 2014), and land management techniques play a major role (Nadeu et al., 2015). However, rapid estuarine sedimentation of the type described in Chapter 9, could prove to be a sink, as could accumulation on floodplains and behind dams (Ran et al., 2014). The pervasive modification of river regimes, described in Chapter 7, is another influence on carbon budgets (Meybeck, 2003; Hoffmann, 2013). Recent studies have highlighted the role of rivers not only in transporting the carbon exported from terrestrial ecosystems but also in metabolizing and burying significant amounts of it.

Thermokarst development and the formation of thaw lakes, described in Chapter 10, may play a crucial role in the liberation of a major greenhouse gas, methane (Schaefer et al., 2014), and climate change is expected to increase the initiation and expansion of thaw features, thereby accelerating the release of methane (Schurr et al., 2015).

Land-use changes are, as shown in Chapter 5, modify silica transport rates in rivers (Struyf et al., 2010), and this can have an impact on the productivity of coastal waters. While controls on chemical weathering, such as bedrock geology, runoff, and temperature, are considered to be the primary drivers of Si transport from the continents to the oceans, recent work has highlighted terrestrial vegetation as an important control over Si cycling. Carey and Fulweiler (2012) have shown that at the regional scale (Southern New England, United States), land use and land cover are important variables controlling the net transport of Si from the land to the sea, accounting for at least 40 percent of dissolved Si fluxes. On the other hand, the postwar profusion of dams, a type of "artificial filter," means that reservoirs currently retain a third of the global sediment discharge (see Chapter 7). They trap biogenic and other forms of particulate, easily soluble silica, and thus starve downstream areas of dissolved Si, significantly reducing its flux to the ocean (Triplett et al., 2008; Treguer and De La Rocha, 2013). On the other hand, most "natural filters" have either been removed or greatly reduced in their functionality by human pressures (Meybeck and Vörösmarty, 2005): wetlands have been drained for agriculture and other land uses, floodplains have become more and more isolated from river channels by dykes and levees (see Chapter 3), and many estuaries are channelized for navigation and/or cultivated (see Chapter 9).

To conclude, this book has reviewed anthropogenic geomorphological changes and demonstrated their importance locally, regionally, and globally. Further research is urgently needed to provide quantification of the feedbacks between such geomorphological change and changes in the Earth System. If one accepts Johan Rockstrom's (2015) view that "The Anthropocene is up there with Copernicus's heliocentricity or Darwin's theory of evolution as one of the most profound shifts in worldview that has emerged from scientific endeavour," then it is important for the critical geomorphological contributions to the Anthropocene to be more fully researched.

References

Aagaard, T. and Sorensen, P., (2012). Coastal profile response to sea level rise: a process-based approach. *Earth Surface Processes and Landforms*, 37, 354–62.

Aalto, J., Venäläinen, A., Heikkinen, R. K., and Luoto, M. (2014). Potential for extreme loss in high-latitude Earth surface processes due to climate change. *Geophysical Research Letters*, 41, 3914–24.

Abidin, H. Z., Andreas, H., Gumilar, I., and Wibowo, I. R. R. (2015). On correlation between urban development, land subsidence and flooding phenomena in Jakarta. *Proceedings IAHS*, 370, 15–20.

Abrams, M. D. and Nowacki, G. J. (2015). Exploring the early Anthropocene burning hypothesis and climate-fire anomalies for the Eastern US. *Journal of Sustainable Forestry*, 34, 30–48.

Abudanh, F. and Twaissi, S. (2010). Innovation or technology immigration? The Qanat systems in the regions of Udhruh and Ma'an in Southern Jordan. *Bulletin of the American Schools of Oriental Research*, 360, 67–87.

Adam, J. C., Hamlet, A. F., and Lettenmaier, D. P. (2009). Implications of global climate change for snowmelt hydrology in the twenty-first century. *Hydrological Processes*, 23, 962–72.

Adams, H. D., Luce, C. H., Breshears, D. D., Allen, C. D., Weiler, M., Hale, V. C., Smith, A. M. S., and Huxman, T. E. (2012). Ecohydrological consequences of drought-and infestation-triggered tree die-off: insights and hypotheses. *Ecohydrology*, 5, 145–59.

Adams, W. M., Perrow, M. R., and Carpenter, A. (2004). Conservatives and champions: river managers and the river restoration discourse in the United Kingdom. *Environment and Planning A*, 36, 1929–42.

Aeschbach-Hertig, W. and Gleeson, T. (2012). Regional strategies for the accelerating global problem of groundwater depletion. *Nature Geoscience*, 5, 853–61.

Aikens, C. M. and Lee, G. A. (2014). Postglacial inception and growth of anthropogenic landscapes in China, Korea, Japan, and the Russian Far East. *Anthropocene*, 4, 46–56.

Akiner, S., Cooke, R. U., and French, R. A. (1992). Salt damage to Islamic monuments in Uzbekistan. *Geographical Journal*, 158, 257–72.

Albert, R. M. (2015). Anthropocene and early human behavior. *The Holocene*, 25, 1542–52 doi: 10.1177/0959683615588377.

Alcántara-Ayala, I., Esteban-Chávez, O., and Parrot, J. F. (2006). Landsliding related to land-cover change: a diachronic analysis of hillslope instability distribution in the Sierra Norte, Puebla, Mexico. *Catena*, 65, 152–65.

Alemayehu, T., Furi, W., and Legesse, D. (2007). Impact of water overexploitation on highland lakes of Eastern Ethiopia. *Environmental Geology*, 52, 147–54.

Algeet-Abarquero, N., Marchamalo, M., Bonatti, J., Fernández-Moya, J., and Moussa, R. (2015). Implications of land use change on runoff generation at the plot scale in the humid tropics of Costa Rica. *Catena*, 135, 263–70.

Al-Harthi, A. A. and Bankher, K. A. (1999). Collapsing loess-like soil in western Saudi Arabia. *Journal of Arid Environments*, 41, 383–9.

Allchin, B. (1976). Palaeolithic sites in the plains of Sind and their geographical implications. *Geographical Journal*, 142, 471–89.

Allen, B. (1971). Wet-field taro terraces on Mangaia, Cook Islands. *The Journal of the Polynesian Society*, 80, 371–8.

Allen, S. K., Cox, S. C., and Owens, I. F. (2011). Rock avalanches and other landslides in the central Southern Alps of New Zealand: a regional study considering possible climate change impacts. *Landslides*, 8, 33–48.

Allis, R. G. (2000). Review of subsidence at Wairakei field, New Zealand. *Geothermics*, 29, 455–78.

Allis, R. G., Bromley, C., and Currie, S. (2009). Update on subsidence at the Wairakei–Tauhara geothermal system, New Zealand. *Geothermics*, 38, 169–80.

Ambers, R. K., Druckenbrod, D. L., and Ambers, C. P. (2006). Geomorphic response to historical agriculture at Monument Hill in the Blue Ridge foothills of central Virginia. *Catena*, 65, 49–60.

Anderson, A. (2009). The rat and the octopus: initial human colonization and the prehistoric introduction of domestic animals to remote Oceania. *Biological Invasions*, 11, 1503–19.

Anderson, D. (1984). Depression, dust bowl, demography, and drought: the colonial state and soil conservation in East Africa during the 1930s. *African Affairs*, 83, 321–43.

Anderson, J. B., Wallace, D. J., Simms, A. R., Rodriguez, A. B., and Milliken, K. T. (2014). Variable response of coastal environments of the Northwestern Gulf of Mexico to sea-level rise and climate change: implications for future change. *Marine Geology*, 352, 348–66.

Andersson, A. J. and Gledhill, D. (2013). Ocean acidification and coral reefs: effects on reakdown, dissolution, and net ecosystem calcification. *Annual Review of Marine Science*, 5, 321–48.

André, M. F., Vautier, F., Voldoire, O., and Roussel, E. (2014). Accelerated stone deterioration induced by forest clearance around the Angkor temples. *Science of the Total Environment*, 493, 98–108.

Andréassian, V. (2004). Waters and forests: from historical controversy to scientific debate. *Journal of Hydrology*, 291, 1–27.

Angel, J. R. and Kunkel, K. E. (2010). The response of Great Lakes water levels to future climate scenarios with an emphasis on Lake Michigan-Huron. *Journal of Great Lakes Research*, 36, 51–8.

Anthony, E. J. (2014). The human influence on the Mediterranean coast over the last 200 years: a brief appraisal from a geomorphological perspective. *Géomorphologie: Relief, Processus, Environnement*, 3, 219–26.

(2015). Wave influence in the construction, shaping and destruction of river deltas: A review. *Marine Geology*, 361, 53–78.

Anthony, K., Maynard, J. A., Diaz-Pulido, G., Mumby, P. J., Marshall, P. A., Cao, L., and Hoegh-Guldberg, O. V. E. (2011). Ocean acidification and warming will lower coral reef resilience. *Global Change Biology*, 17, 1798–808.

Anthony, K. R. N., Kline, D. I., Diaz-Pulido, G., Dove, S., and Hoegh-Guldberg, O. (2008). Ocean acidification causes bleaching and productivity loss in coral reef builders. *Proceedings of the National Academy of Sciences*, 105, 17442–6.

April, R., Newton, R., and Coles, L. T. (1986). Chemical weathering in two Adirondack watersheds: past and present-day rates. *Geological Society of America Bulletin*, 97, 1232–8.

Arendt, A. A., Echelmeyer, K. A., Harriso, W. D., Kingle, C. S., and Valentine, V. B. (2002) Rapid wastage of Alaska glaciers and their contribution to rising sea level. *Science*, 297, 382–5.

Arens, S. M., Mulder, J. P., Slings, Q. L., Geelen, L. H., and Damsma, P. (2013). Dynamic dune management, integrating objectives of nature development and coastal safety: examples from the Netherlands. *Geomorphology*, 199, 205–13.

Arenson, L. U. and Jakob, M. (2015). Periglacial geohazard risks and ground temperature increases. In *Engineering Geology for Society and Territory: Vol. 1*, eds. G. Lollino, A. Manconi, J. Clague, W. Shan, and M. Chiarle. Springer International Publishing, pp. 233–7.

Armitage, A. R., Highfield, W. E., Brody, S. D., and Louchouarn, P. (2014). The contribution of mangrove expansion to salt marsh loss on the Texas Gulf Coast. *PloS one*, 10(5), e0125404.

Arnáez, J., Lana-Renault, N., Lasanta, T., Ruiz-Flaño, P., and Castroviejo, J. (2015). Effects of farming terraces on hydrological and geomorphological processes. A review. *Catena*, 128, 122–34.

Arnáez, J., Larrea, V., and Ortigosa, L. (2004). Surface runoff and soil erosion on unpaved forest roads from rainfall simulation tests in northeastern Spain. *Catena*, 57, 1–14.

Arnaud, F., Révillon, S., Debret, M., Revel, M., Chapron, E., Jacob, J., Giguet-Covexa, C., Poulenarda, J., and Magny, M. (2012). Lake Bourget regional erosion patterns reconstruction reveals Holocene NW European Alps soil evolution and paleohydrology. *Quaternary Science Reviews*, 51, 81–92.

Arnell, N. W. (2002). *Hydrology and Global Environmental Change*. Harlow: Prentice Hall.

Arnell, N. W. and Gosling, S. N. (2013). The impacts of climate change on river flow regimes at the global scale. *Journal of Hydrology*, 486, 351–64.

Arora, V. K. and Boer, G. J. (2001). Effects of simulated climate change on the hydrology of major river basins. *Journal of Geophysical Research*, 106, D4, 3335–48.

Ashkenazy, Y., Yizhaq, H., and Tsoar, H. (2012). Sand dune mobility under climate change in the Kalahari and Australian deserts. *Climatic Change*, 112, 901–23.

Ashton, A. D., Donnelly, J. P., and Evans, R. L. (2008). A discussion of the potential impacts of climate change on the shorelines of the northeastern USA. *Mitigation and Adaptive Strategies for Global Change*, 13, 719–43.

Ashton, A. D., Walkden, M. J. A., and Dickson, M. E. (2011). Equilibrium responses of cliffed coasts to changes in the rate of sea level rise. *Marine Geology*, 284, 217–29.

Ateweberhan, M., Feary, D. A., Keshavmurthy, S., Chen, A., Schleyer, M. H., and Sheppard, C. R. (2013). Climate change impacts on coral reefs: Synergies with local effects, possibilities for acclimation, and management implications. *Marine Pollution Bulletin*, 74, 526–39.

Atkinson, G., Assatourians, K., Cheadle, B., and Greig, W. (2015). Ground motions from three recent earthquakes in Western Alberta and Northeastern British Columbia and their implications for induced-seismicity hazard in Eastern Regions. *Seismological Research Letters*, 86, 1022–31.

Aubault, H., Webb, N. P., Strong, C. L., McTainsh, G. H., Leys, J. F., and Scanlan, J. C. (2015). Grazing impacts on the susceptibility of rangelands to wind erosion: the effects of stocking rate, stocking strategy and land condition. *Aeolian Research*, 17, 89–99.

Aukema, J. E., McCullough, D. G., Von Holle, B., Liebhold, A. M., Britton, K., and Frankel, S. J. (2010). Historical accumulation of nonindigenous forest pests in the continental United States. *BioScience*, 60, 886–97.

Avila, F. B., Pitman, A. J., Donat, M. G., Alexander, L. V. and Abramowitz, G. (2012). Climate model simulated changes in temperature extremes due to land cover change. *Journal of Geophysical Research*, 117, D04108, doi: 10.1029/2011JD016382.

Ayenew, T. and Legesse, D. (2007). The changing face of the Ethiopian rift lakes and their environs: call of the time. *Lakes & Reservoirs: Research and Management*, 12, 149–65.

Azorin-Molina, C., Vicente-Serrano, S. M., McVicar, T. R., Jerez, S., Sanchez-Lorenzo, A., López-Moreno, J., Revuelto, J., Trigo, R. M., Lopez-Bustins, J. A., and Espírito-Santo, F. (2014). Homogenization and assessment of observed near-surface wind speed trends over Spain and Portugal, 1961–2011. *Journal of Climate*, 27, 3692–712.

Baartman, J. E., Temme, A. J., Schoorl, J. M., Braakhekke, M. H., and Veldkamp, T. (2012). Did tillage erosion play a role in millennial scale landscape development? *Earth Surface Processes and Landforms*, 37, 1615–26.

Baccari, N., Boussema, M. R., Lamachère, J. M., and Nasri, S. (2008). Efficiency of contour benches, filling-in and silting-up of a hillside reservoir in a semi-arid climate in Tunisia. *C. R. Geoscience*, 340, 38–48.

Badescu, V. and Cathcart, R. B. (2011). Aral Sea partial restoration.1. A Caspian importation macroproject. *International Journal of Environment and Waste Water Management*, 7, 161–82.

Badman, T. (2010). World Heritage and geomorphology. In *Geomorphological Landscapes of the World*, ed. P. Migoń. Dordrecht: Springer, pp. 357–68.

Bailey, D. W., Tringham, R., Bass, J., Stevanović, M., Hamilton, M., Neumann, H., and Raduncheva, A. (1998). Expanding the dimensions of early agricultural tells: the Podgoritsa Archaeological Project, Bulgaria. *Journal of Field Archaeology*, 25, 373–96.

Bain, D. J., Green, M. B., Campbell, J. L., Chamblee, J. F., Chaoka, S., Fraterrigo, J. M., and Leigh, D. S. (2012). Legacy effects in material flux: structural catchment changes predate long-term studies. *Bioscience*, 62, 575–84.

Baine, M. (2001). Artificial reefs: a review of their design, application, management and performance. *Ocean & Coastal Management*, 44, 241–59.

Baisch, S., Vörös, R., Rothert, E., Stang, H., Jung, R., and Schellschmidt, R. (2010). A numerical model for fluid injection induced seismicity at Soultz-sous-Forêts. *International Journal of Rock Mechanics and Mining Sciences*, 47, 405–13.

Baker, A. and Simms, M. J. (1998). Active deposition of calcareous tufa in Wessex, UK, and its implications for the "late-Holocene tufa decline." *The Holocene*, 8, 359–365.

Baker, A. C., Glynn, P. W., and Riegl, B. (2008). Climate change and coral reef bleaching: an ecological assessment of long-term impacts, recovery trends and future outlook. *Estuarine, Coastal and Shelf Science*, 8, 435–71.

Baker, V. R. (2014). Uniformitarianism, earth system science, and geology. *Anthropocene*, 5, 76–9.

Bakker, M. M., Govers, G., van Doorn, A., Quetier, F., Chouvardas, D., and Rounsevell, M. (2008). The response of soil erosion and sediment export to land-use change in four areas of Europe: the importance of landscape pattern. *Geomorphology*, 98, 213–26.

Bakoariniaina, L. N., Kusky, T., and Raharimahefa, T. (2006). Disappearing Lake Alaotra: monitoring catastrophic erosion, waterway silting, and land degradation hazards in Madagascar using Landsat imagery. *Journal of African Earth Sciences*, 44, 241–52.

Balcerak, E. (2011). Thermokarst lakes expand and drain laterally as permafrost degrades. *Eos, Transactions American Geophysical Union*, 92 (45), 408.

Balke, T. and Friess, D. A. (2015). Geomorphic knowledge for mangrove restoration: a pan-tropical categorization. *Earth Surface Processes and Landforms*. doi: 10.1002/esp.3841.

Balling, R. C. and Wells, S. G. (1990). Historical rainfall patterns and arroyo activity within the Zuni river drainage basin, New Mexico. *Annals of the Association of American Geographers*, 80, 603–17.

Balme, J. (2013). On boats and string: the maritime colonisation of Australia. *Quaternary International* 285, 68–75.

Balter, M. (2010). The tangled roots of agriculture. *Science*, 327, 404–6.
  (2013). Farming's tangled European roots. *Science*, 342, 181–82.

Balthazar, V., Vanacker, V., Girma, A., Poesen, J., and Golla, S. (2013). Human impact on sediment fluxes within the Blue Nile and Atbara River basins. *Geomorphology*, 180, 231–41.

Bamber, J. L. and Aspinall, W. P. (2013). An expert judgement assessment of future sea level rise from the ice sheets. *Nature Climate Change*, 3, 424–7.

Banks, V. J., Jones, P. F., Lowe, D. J., Lee, J. R., Rushton, J., and Ellis, M. A. (2012). Review of tufa deposition and palaeohydrological conditions in the White Peak, Derbyshire, UK: implications for Quaternary landscape evolution. *Proceedings of the Geologists' Association*, 123, 117–29.

Barca, D., Comite, V., Belfiore, C. M., Bonazza, A., La Russa, M. F., Ruffolo, S. A., and Sabbioni, C. (2014). Impact of air pollution in deterioration of carbonate building materials in Italian urban environments. *Applied Geochemistry*, 48, 122–31.

Barker, G. (2006). *The Agricultural Revolution in Prehistory; Why Did Foragers Become Farmers?* Cambridge: Cambridge University Press.

Barker, G. W. and Hunt, C. O. (1995). *Quaternary valley floor erosion and aluviation in the Biferno Valley*, Molise, Italy: the role of tectonics, climate, sea level change, and human activity. In *Mediterranean Quaternary River Environments*, ed. J. Lewin, M. G. Macklin and J. C. Woodward, pp. 145–57.

Barnard, P. L., Owen, L. A., Sharma, M. C., and Finkel, R. C. (2001). Natural and human-induced landsliding in the Garhwal Himalaya of northern India. *Geomorphology*, 40, 21–35.

Barry, R. G. (1985). The cryosphere and climate change. In *Detecting the Climatic Effects of Increasing Carbon Dioxide*, ed. M. C. MacCracken and F. M. Luther, Washington, DC: US Deptarment of Energy, pp. 111–48.

Bartley, R., Bainbridge, Z. T., Lewis, S. E., Kroon, F. J., Wilkinson, S. N., Brodie, J. E., and Silburn, D. M. (2014). Relating sediment impacts on coral reefs to watershed sources, processes and management: a review. *Science of the Total Environment*, 468, 1138–53.

Bar-Yosef, O. (1998). On the nature of transitions: the Middle to Upper Palaeolithic and the Neolithic Revolution. *Cambridge Archaeological Journal*, 8, 141–63.

Basagic, H. J. and Fountain, A. G. (2011). Quantifying 20th century glacier change in the Sierra Nevada, California. *Arctic, Antarctic and Alpine Research*, 43, 317–30.

Battany, M. C. and Grismer, M. E. (2000). Rainfall runoff and erosion in Napa Valley vineyards: effects of slope, cover and surface roughness. *Hydrological Processes*, 14, 1289–304.

Battarbee, R. W., Appleby, P. G., Odel, K., and Flower, R. J. (1985). [210]Pb dating of Scottish Lake sediments, afforestation and accelerated soil erosion. *Earth Surface Processes and Landforms*, 10, 137–42.

Baú, D., Gambolati, G., and Teatini, P. (2000). Residual land subsidence near abandoned gas fields raises concern over Northern Adriatic coastland. *Eos, Transactions American Geophysical Union*, 81(22), 245–9.

Bauer, J. E., Cai, W. J., Raymond, P. A., Bianchi, T. S., Hopkinson, C. S., and Regnier, P. A. (2013). The changing carbon cycle of the coastal ocean. *Nature*, 504, 61–70.

Bautista S., Bellot, J., and Ramón Vallejo, V. (1996). Mulching treatment for postfire soil conservation in a semi-arid ecosystem. *Arid Soil Research and Rehabilitation*, 10, 235–42.

Bayliss, A., McAvoy, F., and Whittle, A. (2007). The world recreated: redating Silbury Hill in its monumental landscape. *Antiquity*, 81, 26.

Bayon, G., Dennielou, B., Etoubleau, J., Ponzevera, E., Toucanne, S., and Bermell, S. (2012). Intensifying weathering and land use in iron age Central Africa. *Science*, 335, 1219–22.

Beach, T., Dunning, N., Luzzadder-Beach, S., Cook, D. E., and Lohse, J. (2006). Impacts of the ancient Maya on soils and soil erosion in the central Maya Lowlands. *Catena*, 65, 166–78.

Beach, T., Luzzadder-Beach, S., Cook, D., Dunning, N., Kennett, D. J., Krause, S., Terry, R., Trein, D., and Valdez, F. (2015). Ancient Maya impacts on the Earth's surface: an early Anthropocene analog? *Quaternary Science Reviews*, 124, 1–30.

Beaulieu, E., Goddéris, Y., Donnadieu, Y., Labat, D., and Roelandt, C. (2012). High sensitivity of the continental-weathering carbon dioxide sink to future climate change. *Nature Climate Change*, 2, 346–9.

Beaumont, P. B. (2011). The edge: more on fire-making by about 1.7 million years ago at Wonderwerk Cave in South Africa. *Current Anthropology*, 52, 585–94.

Beck, M. W., Brumbaugh, R. D., Airoldi, L., Carranza, A., Coen, L. D., Crawford, C., Defeo, O., Edgar, G. J., Hancock, B., Kay, M. C., and Guo, X. (2011). Oyster reefs at risk and recommendations for conservation, restoration, and management. *Bioscience*, 61, 107–16.

Beckers, B., Berking, J., and Schütt, B. (2013). Ancient water harvesting methods in the drylands of the Mediterranean and Western Asia. *Journal for Ancient Studies.* 2, 145–64.

Beerten, K., Vandersmissen, N., Deforce, K., and Vandenberghe, N. (2014). Late Quaternary (15 ka to present) development of a sandy landscape in the Mol area, Campine region, north-east Belgium. *Journal of Quaternary Science*, 29, 433–44.

Bégin, C., Brooks, G., Larson, R. A., Dragićević, S., Ramos Scharrón, C. E., and Côté, I. M. (2014). Increased sediment loads over coral reefs in Saint Lucia in relation to land use change in contributing watersheds. *Ocean & Coastal Management*, 95, 35–45.

Beinart, W. (1984). Soil erosion, conservationism and ideas about development: a southern African exploration, 1900–1960. *Journal of Southern African Studies*, 11, 52–83.

Bekendam, R. F. and Pöttgens, J. J. (1995). Ground movements over the coal mines of southern Limburg, The Netherlands, and their relation to rising mine waters. Land Subsidence. *IAHS Publication*, 234, 3–12.

Bell, F. G., Stacey, T. R., and Genske, D. D. (2000). Mining subsidence and its effect on the environment: some differing examples. *Environmental Geology*, 40, 135–52.

Bell, F. G., Donnelly, L. J., Genske, D. D., and Ojeda, J. (2005). Unusual cases of mining subsidence from Great Britain, Germany and Colombia. *Environmental Geology*, 47, 620–31.

Bell, M. L. (1982). The effect of land-use and climate on valley sedimentation. In *Climatic Change in Later Prehistory*, ed. A. F. Harding. Edinburgh: Edinburgh University Press, pp. 127–42.

Bellard, C., Leclerc, C., and Courchamp, F. (2013). Potential impact of sea level rise on French islands worldwide. *Nature Conservation*, 5, 75–86.

Bellin, N., van Wesemael, B., Meerkerk, A., Vanacker, V., and Barbera, G. G. (2009). Abandonment of soil and water conservation structures in Mediterranean ecosystems: a case study from south east Spain. *Catena*, 76, 114–21.

Belnap, J. and Gillette, D. A. (1997). Disturbance of biological soil crusts: impacts on potential wind erodibility of sandy desert soils in southeastern Utah. *Land Degradation & Development*, 8, 355–62.

Belski, A. J. (1996). Viewpoint: western juniper expansion: is it a threat to arid northwest ecosystems? *Journal of Range Management*, 49, 53–9.

Benda, L., Miller, D., Bigelow, P., and Andras, K. (2003). Effects of post-wildfire erosion on channel environments, Boise River, Idaho. *Forest Ecology and Management*, 178, 105–19.

Benito, G., Del Campo, P. P., Gutiérrez-Elorza, M., and Sancho, C. (1995). Natural and human-induced sinkholes in gypsum terrain and associated environmental problems in NE Spain. *Environmental Geology*, 25, 156–64.

Benn, D. I., Bolch, T., Hands, K., Gulley, J., Luckman, A., Nicholson, L. I,, Quincey, D., Thompson, S., Toumi, R., and Wiseman, S. (2012). Response of debris-covered glaciers in the Mount Everest region to recent warming, and implications for outburst flood hazards. *Earth-Science Reviews*, 114, 156–74.

Bennett, H. H. (1938). *Soil Conservation*. New York: McGraw-Hill.

Bennett, M. R. (2003). Ice streams as the arteries of an ice sheet: their mechanics, stability and significance. *Earth-Science Reviews*, 61, 309–39.

Berg, R. D. and Carter, D. L. (1980). Furrow erosion and sediment losses on irrigated cropland. *Journal of Soil and Water Conservation*, 35, 267–70.

Berhe, A. A., Harte, J., Harden, J. W., and Torn, M. S. (2007). The significance of the erosion-induced terrestrial carbon sink. *BioScience*, 57, 337–46.

Bern, C. R., Clark, M. L., Schmidt, T. S., Holloway, J. M., and McDougal, R. R. (2015). Soil disturbance as a driver of increased stream salinity in a semiarid watershed undergoing energy development. *Journal of Hydrology*, 524, 123–36.

Berna, F., Goldberg, P., Horwitz, L. K., Brink, J., Holt, S., Bamford, M., and Chazan, M. (2012). Microstratigraphic evidence of in situ fire in the Acheulean strata of Wonderwerk Cave, Northern Cape Province, South Africa. *Proceedings of the National Academy of Sciences*, 109, E1215–20.

Berner, R. A. (1998). The carbon cycle and CO_2 over Phanerozoic time: the role of land plants. *Philosophical Transactions of the Royal Society B*, 353, 75–82.

Bertness, M. D. and Coverdale, T. C. (2013). An invasive species facilitates the recovery of salt marsh ecosystems on Cape Cod. *Ecology*, 94, 1937–43.

Bertness, M. D. and Silliman, B. R. (2008). Consumer control of salt marshes driven by human disturbance. *Conservation Biology*, 22, 618–23.

Bertran, P. (2004). Soil erosion in small catchments of the Quercy region (southwestern France) during the Holocene. *The Holocene*, 14, 597–606.

Beschta, R. L. (1978). Long-term patterns of sediment production following road construction and logging in the Oregon Coast Range. *Water Resources Research*, 14, 1011–6.

Bhagwat, S., Kettle, C. J., and Koh, L. P. (2014). The history of deforestation and forest fragmentation: a global perspective. In *Global Forest Fragmentation*, ed. C. J. Kettle and L. P. Koh. Wallingford: CABI, pp. 5–19.

Bhark, E. W. and Small, E. E. (2003). Association between plant canopies and the spatial patterns of infiltration in shrubland and grassland of the Chihuahuan Desert, New Mexico. *Ecosystems*, 6, 185–96.

Bhattachan, A., D'Odorico, P., Okin, G. S., and Dintwe, K. (2013). Potential dust emissions from the southern Kalahari's dunelands. *Journal of Geophysical Research: Earth Surface*, 118, 307–14.

Bhattachan, A., D'Odorico, P., Dintwe, K., Okin, G. S., and Collins, S. L. (2014). Resilience and recovery potential of duneland vegetation in the southern Kalahari. *Ecosphere*, 5(1), art2.

Biemans, H., Haddeland, I., Kabat, P., Ludwig, F., Hutjes, R. W. A., Heinke, J., and Gerten, D. (2011). Impact of reservoirs on river discharge and irrigation water supply during the 20th century. *Water Resources Research*, 47, doi: 10.1029/2009WR008929.

Bigelow, G. F., Ferrante, S. M., Hall, S. T., Kimball, L. M., Proctor, R. E., and Remington, S. L. (2005). Researching catastrophic environmental changes on northern coastlines: a geoarchaeological case study from the Shetland Islands. *Arctic Anthropology*, 42, 88–102.

Binford, M. W., Brenner, M., Whitmore, T. J., Higuera-Grundy, A., Deevey, E. S., and Leyden, B. (1987). Ecosystems, palaeoecology, and human disturbance in subtropical and tropical America. *Quaternary Science Review*, 6, 115–28.

Bini, C., Gemignani, S., and Zilocchi, L. (2006). Effect of different land use on soil erosion in the pre-alpine fringe (North-East Italy): Ion budget and sediment yield. *Science of the Total Environment*, 369, 433–40.

Bird, E. C. F. (1979). Coastal processes. In *Man and Environmental Processes*, ed. K. J. Gregory and D. E. Walling. Folkestone: Dawson, pp. 82–101.

(1993). *Submerging Coasts*. Chichester: Wiley.

(1996). *Beach Management*. Chichester: Wiley.

Bird, E. C. F. and Lewis, N. (2015). Causes of beach erosion. In *Beach Renourishment* ed. E. Bird and N. Lewis. Springer International Publishing. pp. 7–28.

Birkeland, C. (ed.) (2015). *Coral Reefs in the Anthropocene*. Dordrecht: Springer.

Birken, A. S. and Cooper, D. J. (2006). Processes of Tamarix invasion and floodplain development along the lower Green River, Utah. *Ecological Applications*, 16, 1103–20.

Birkinshaw, S. J., Bathurst, J. C., and Robinson, M. (2014). 45 years of non-stationary hydrology over a forest plantation growth cycle, Coalburn catchment, Northern England. *Journal of Hydrology*, 519, 559–73.

Bischoff, M., Cete, A., Fritschen, R., and Meier, T. (2010). Coal mining induced seismicity in the Ruhr area, Germany. *Pure and Applied Geophysics*, 167, 63–75.

Bishop, P. and Muñoz-Salinas, E. (2013). Tectonics, geomorphology and water mill location in Scotland, and the potential impacts of mill dam failure. *Applied Geography*, 42, 195–205.

Black, K. and Mead, S. (2009). Design of surfing reefs. *Reef Journal*, 1, 177–91.

Blais-Stevens, A., Kremer, M., Bonnaventure, P. P., Smith, S. L., Lipovsky, P., and Lewkowicz, A. G. (2015). Active layer detachment slides and retrogressive thaw slumps susceptibility mapping for current and future permafrost distribution, Yukon Alaska Highway Corridor. In *Engineering Geology for Society and Territory-Volume 1* (pp. 449–53), ed. G. Lollino, A. Manconi, J. Clague, W. Shan, and M. Chiarle, Springer International Publishing, pp. 449–53.

Blanco P. D., Rostagno, C. M., del Valle, H. F., Beeskow, A. M., and Wiegand, T. (2008). Grazing impacts in vegetated dune fields: predictions from spatial pattern analysis. *Rangeland Ecology and Management*, 61, 194–203.

Blankespoor, B., Dasgupta, S., and Laplante, B. (2014). Sea-level rise and coastal wetlands. *Ambio*, 43, 996–1005.

Blanton, P. and Marcus, W. A. (2013). Transportation infrastructure, river confinement, and impacts on floodplain and channel habitat, Yakima and Chehalis rivers, Washington, USA. *Geomorphology*, 189, 55–65.

Blavet, D., De Noni, G., Le Bissonnais, Y., Leonard, M., Maillo, L., Laurent, J. Y., and Roose, E. (2009). Effect of land use and management on the early stages of soil water erosion in French Mediterranean vineyards. *Soil and Tillage Research*, 106, 124–36.

Bliss, A., Hock, R., and Radić, V. (2014). Global response of glacier runoff to twenty-first century climate change. *Journal of Geophysical Research: Earth Surface*, 119, 717–30.

Blott, S. J., Pye, K., Van der Wal, D., and Neal, A. (2006). Long-term morphological change and its causes in the Mersey Estuary, NW England. *Geomorphology*, 81, 185–206.

Boardman, J. (1998). An average soil erosion rate for Europe: myth or reality. *Journal of Soil and Water Conservation*, 53, 46–50.

(2013). The hydrological role of "sunken lanes" with respect to sediment mobilization and delivery to watercourses with particular reference to West Sussex, southern England. *Journal of Soils and Sediments*, 13, 1636–44.

(2014). How old are the gullies (dongas) of the Sneeuberg uplands, Eastern Karoo, South Africa? *Catena*, 113, 79–85.

Böckh, A. (1973). Consequences of uncontrolled human activities in the Valencia lake basin. In *The Careless Technology*, ed. M. T. Farvar and J. P. Milton. London: Tom Stacey, pp. 301–17.

Boenzi, F., Caldara, M., Capolongo, D., Dellino, P., Piccarreta, M., and Simone, O. (2008). Late Pleistocene–Holocene landscape evolution in Fossa Bradanica, Basilicata (southern Italy). *Geomorphology*, 102, 297–306.

Boix-Fayos, C., Barberá, G. G., López-Bermúdez, F., and Castillo, V. M. (2007). Effects of check dams, reforestation and land-use changes on river channel morphology: case study of the Rogativa catchment (Murcia, Spain). *Geomorphology*, 91, 103–23.

Boix-Fayos, C., de Vente, J., Martínez-Mena, M., Barberá, G. G., and Castillo, V. (2008). The impact of land use change and check-dams on catchment sediment yield. *Hydrological Processes*, 22, 4922–35.

Bonazza, A., Messina, P., Sabbioni, C., Grossi, C. M., and Brimblecombe, P. (2009a). Mapping the impact of climate change on surface recession of carbonate buildings in Europe. *Science of the Total Environment*, 407, 2039–50.

Bonazza, A., Sabbioni, C., Messina, P., Guaraldi, C., and De Nuntiis, P. (2009b). Climate change impact: mapping thermal stress on Carrara marble in Europe. *Science of the Total Environment*, 407, 4506–12.

Bonhomme, D., Boudouresque, C. F., Astruch, P., Bonhomme, J., Bonhomme, P., Goujard, A., and Thibaut, T. (2015). Typology of the reef formations of the Mediterranean seagrass

Posidonia oceanica, and the discovery of extensive reefs in the Gulf of Hyères (Provence, Mediterranean). *Scientific Report of Port-Cros National Park*, 29, 41–73.

Borrelli, P., Ballabio, C., Panagos, P., and Montanarella, L. (2014). Wind erosion susceptibility of European soils. *Geoderma*, 232, 471–8.

Bostock, H. C., Lowe, D. J., Gillespie, R., Priestley, R., Newnham, R. M., and Mooney, S. D. (2015). The advent of the Anthropocene in Australasia. *Quaternary Australasia*, 32(1), 7.

Boulanger, M. T. and Lyman, R. L. (2014). Northeastern North American Pleistocene megafauna chronologically overlapped minimally with Paleoindians. *Quaternary Science Reviews*, 85, 35–46.

Bountry, J. A., Lai, Y. G., and Randle, T. J. (2013). Sediment impacts from the savage rapids dam removal, Rogue River, Oregon. *Reviews in Engineering Geology*, 21, 93–104.

Boussingault, J. B., 1845, *Rural Economy* (2nd edn). London: Baillière.

Bowden, W. B. (2010). Climate change in the Arctic – Permafrost, thermokarst, and why they matter to the non-Arctic world. *Geography Compass*, 4, 1553–66.

Bowman, D. M. (2014). What is the relevance of pyrogeography to the Anthropocene? *The Anthropocene Review*, 2, 73–6.

Bozec, Y. M. and Mumby, P. J. (2015). Synergistic impacts of global warming on the resilience of coral reefs. *Philosophical Transactions of the Royal Society B: Biological Sciences*, 370 (1659), 20130267.

Bradbury, J., Cullen, P., Dixon, G., and Pemberton, M. (1995). Monitoring and management of streambank erosion and natural revegetation on the lower Gordon River, Tasmanian Wilderness World Heritage Area, Australia. *Environmental Management*, 19, 259–72.

Bradley, R. and Fraser, E. (2010). Bronze Age barrows on the heathlands of southern England: construction, forms and interpretations. *Oxford Journal of Archaeology*, 29, 15–33.

Bradley, R. S., Vuille, M., Diaz, H. F., and Vergara, W. (2006). Threats to water supplies in the tropical Andes. *Science*, 312, 1755–6.

Bradwell, T., Sigurðsson, O., and Everest, J. (2013). Recent, very rapid retreat of a temperate glacier in SE Iceland. *Boreas*, 42, 959–73.

Bragg, O. M. and Tallis, J. H. (2001). The sensitivity of peat-covered upland landscapes. *Catena*, 42, 345–60.

Braje, T., Erlandson, J., Aikens, C. M., Beach, T., Fitzpatrick, S., Gonzalez, S., and Zeder, M. A. (2014). An Anthropocene without Archaeology —Should we care? *The SAA Archaeological Record*, 14, 26–9.

Braje, T. J. (2015). Earth Systems, Human Agency, and the Anthropocene: Planet Earth in the Human Age. *Journal of Archaeological Research*, doi: 10.1007/s10814-015-9087-y.

Braje, T. J. and Erlandson, J. M. (2014). Looking forward, looking back: humans, anthropogenic change, and the Anthropocene. *Anthropocene*, 4, 116–21.

Brander, K. E., Owen, K. E. and Potter, K. W. (2004). Modelled impacts of development type on runoff volume and infiltration performance. *Journal of the American Water Resources Association*, 40, 961–9.

Brantley, S. L., Megonigal, J. P., Scatena, F. N, Zalogh-Brunstad, Z, Barnes, R. T., Bruns, M. A., and 21 others. (2011). Twelve testable hypotheses on the geobiology of weathering. *Geobiology*, 9, 140–65.

Brattebo, B. O. and Booth, D. B. (2003). Long-term stormwater quantity and quality performance of permeable pavement systems. *Water Research*, 37, 4369–76.

Breckle S. W., Yair, A., and Veste, M. (2008). General conclusions – sand dune deserts, desertification, rehabilitation and conservation. *Ecological Studies* 200, 441–59.

Brenot, J., Quiquerez, A., Petit, C., and Garcia, J. P. (2008). Erosion rates and sediment budgets in vineyards at 1-m resolution based on stock unearthing (Burgundy, France). *Geomorphology*, 100, 345–55.

Breslow, P. B. and Sailor, D. J. (2002). Vulnerability of wind power resources to climate change in the continental United States. *Renewable Energy*, 27, 585–98.

Brestolani, F., Solari, L., Rinaldi, M., and Lollino, G. (2015). On the morphological impacts of gravel mining: The case of the Orco River. In *Engineering Geology for Society and Territory-Volume 3*, ed. G. Lollino, M. Arattano, M. Rinaldi, O. Giustolisi, J. C. Marechal and G. E. Grant. Springer International Publishing, pp. 319–22.

Brierley, G., Huang, H. Q., Chen, A., Aiken, S., Crozier, M., Eder, W., Goudie, A., Ma, Y., May, J. H., Migon, P., and Nanson, G. A. (2011). Naming conventions in geomorphology: contributions and controversies in the sandstone landscape of Zhangjiajie Geopark, China. *Earth Surface Processes and Landforms*, 36, 1981–4.

Bright, M. (2005). *1001 Natural Wonders You Must See Before You Die*. London: Cassell Illustrated.

Brimblecombe, P. and Camuffo, D. (2003). Long term damage to the built environment. In *The Effects of Air Pollution on the Built Environment*, ed. P. Brimblecombe. London: Imperial College Press, pp. 1–30.

Brookes, A. (1985). River channelization: traditional engineering methods, physical consequences, and alternative practices. *Progress in Physical Geography*, 9, 44–73.

(1987). River channel adjustments downstream from channelization works in England and Wales.

Earth Surface Processes and Landforms, 12, 337–51.

Brookes, A. and Brierley, G. J. (1997). Geomorphic responses of lower Bega River to catchment disturbance, 1851–1926. *Geomorphology*, 18, 291–304.

Brooks, N. and Legrand, M. (2000). Dust variability over northern Africa and rainfall in the Sahel. In *Linking Land Surface Change to Climate Change*, ed. S. J. McLaren, and D. Kniverton. Dordrecht: Kluwer, pp. 1–25.

Brooks, S. M. and Spencer, T. (2012). Shoreline retreat and sediment release in response to accelerating sea level rise: measuring and modelling cliffloine dynamics on the Suffolk coast, UK. *Global and Planetary Change*, 80–81, 165–79.

Brooks, S. S. and Lake, P. S. (2007). River restoration in Victoria, Australia: change is in the wind, and none too soon. *Restoration Ecology*, 15, 584–91.

Broothaerts, N., Kissi, E., Poesen, J., Van Rompaey, A., Getahun, K., Van Ranst, E., and Diels, J. (2012). Spatial patterns, causes and consequences of landslides in the Gilgel Gibe catchment, SW Ethiopia. *Catena*, 97, 127–36.

Broothaerts, N., Notebaert, B., Verstraeten, G., Kasse, C., Bohncke, S., and Vandenberghe, J. (2014). Non-uniform and diachronous Holocene floodplain evolution: a case study from the Dijle catchment, Belgium. *Journal of Quaternary Science*, 29, 351–60.

Brouns, K., Eikelboom, T., Jansen, P. C., Janssen, R., Kwakernaak, C., van den Akker, J. J., and Verhoeven, J. T. (2015). Spatial analysis of soil subsidence in peat meadow areas in Friesland in relation to land and water management, climate change, and adaptation. *Environmental Management*, 55, 360–72.

Brown, A. E., Zhang, L., McMahon, T. A., Western, A. W., and Vertessy, R. A. (2005). A review of paired catchment studies for determining changes in water yield resulting from alterations in vegetation. *Journal of Hydrology*, 310, 28–61.

Brown, A. G. (2014). The Anthropocene: a geomorphological and sedimentary view. *Strati 2013*, 909–14.

Brown, A. G., Tooth, S., Chiverrell, R. C., Rose, J., Thomas, D. S., Wainwright, J., Bullard, J. E., Thorndycraft, V. R., Aalto, R., and Downs, P. (2013). The Anthropocene: is there a geomorphological case? *Earth Surface Processes and Landforms* 38, 431–4.

Brown, A. V., Lyttle, M. M., and Brown, K. B. (1998). Impacts of gravel mining on gravel bed streams. *Transactions of the American Fisheries Society*, 127, 979–94.

Brown, D., Jorgenson, M. T., Douglas, T. A., Romanovsky, V. E., Kielland, K., Hiemstra, C., Euskirchen, E. S., and Ruess, R. W. (2015). Interactions of fire and climate exacerbate permafrost degradation in Alaskan lowland forests. *Journal*

of Geophysical Research: Biogeosciences. 120, 1619–37, doi: 10.1002/2015JG003033.

Brown, E. H. (1970). Man shapes the earth. *Geographical Journal* 136, 74–85.

Brown, S. and Nicholls, R. J. (2015). Subsidence and human influences in mega deltas: the case of the Ganges–Brahmaputra–Meghna. *Science of the Total Environment*, 527, 362–74.

Brown, S., Barton, M., and Nicholls, R. (2011). Coastal retreat and/or advance adjacent to defences in England and Wales. *Journal of Coastal Conservation*, 15, 659–70.

Bruijnzeel, L. A. (1990). *Hydrology of Moist Tropical Forests and Effects of Conversion: a State of Knowledge Review*. Amsterdam: Free University for UNESCO International Hydrological Programme.

Bruins, H. J. (2012). Ancient desert agriculture in the Negev and climate-zone boundary changes during average, wet and drought years. *Journal of Arid Environments*, 86, 28–42.

Brunbjerg, A. K., Svenning, J. C., and Ejrnæs, R. (2014). Experimental evidence for disturbance as key to the conservation of dune grassland. *Biological Conservation*, 174, 101–10.

Brunel, C. and Sabatier, F. (2009). Potential influence of sea-level rise in controlling shoreline position on the French Mediterranean Coast. *Geomorphology*, 107, 47–57.

Brunet, F., Potot, C., Probst, A., and Probst, J. L. (2011). Stable carbon isotope evidence for nitrogenous fertilizer impact on carbonate weathering in a small agricultural watershed. *Rapid Communications in Mass Spectrometry*, 25, 2682–90.

Brungard, C. W., Boettinger, J. L., and Hipps, L. E. (2015). Wind erosion potential of lacustrine and alluvial soils before and after disturbance in the eastern Great Basin, USA: estimating threshold friction velocity using easier-to-measure soil properties. *Aeolian Research*, 18, 185–203.

Brunier, G., Anthony, E. J., Goichot, M., Provansal, M., and Dussouillez, P. (2014). Recent morphological changes in the Mekong and Bassac river channels, Mekong delta: the marked impact of river-bed mining and implications for delta destabilisation. *Geomorphology*, 224, 177–91.

Bruno, J. F. and Selig, E. R. (2007). Regional decline of coral cover in the Indo-Pacific: timing, extent and subregional comparisons. *PlosOne*, 8, e711, doi: 10.1371/journal.pone.0000711.

Brunsden, D., Coombe, K., Goudie, A. S., and Parker, A. G. (1996). The structural geomorphology of the Isle of Portland, southern England. *Proceedings of the Geologists' Association*, 107, 209–30.

Bruun, P. (1962). Sea level rise as a cause of shore erosion. *American Society of Civil Engineers Proceedings: Journal of Waterways and Harbors Division*, 88, 117–30.

Bryan, K. (1928). Historic evidence of changes in the channel of Rio Puerco, a tributary of the Rio

Grande in New Mexico. *Journal of Geology*, 36, 265–82.

Buendia, C., Batalla, R. J., Sabater, S., Palau, A., and Marcé, R. (2015). Runoff trends driven by climate and afforestation in a Pyrenean Basin. *Land Degradation and Development*, doi: 10.1002/ldr.2384.

Buijse, A. D., Coops, H., Staras, M., Jans, L. H., Van Geest, G. J., Grift, R. E., Ibelings, B. W., Oosterberg, W., and Roozen, F. C. (2002). Restoration strategies for river floodplains along large lowland rivers in Europe. *Freshwater Biology*, 47, 889–907.

Bull, W. B. (1997). Discontinuous ephemeral streams. *Geomorphology*, 19, 227–76.

Burrin, P. J. (1985). Holocene alluviation in southeast England and some implications for palaeohydrological studies. *Earth Surface Processes and Landforms*, 10, 257–71.

Burroughs, B. A., Hayes, D. B., Klomp, K. D., Hansen, J. F., and Mistak. J. (2009). Effects of Stronach Dam removal on fluvial geomorphology in the Pine River, Michigan, United States. *Geomorphology*, 110, 96–107.

Burt, J. A. (2014). The environmental costs of coastal urbanization in the Arabian Gulf. *City*, 18, 760–70.

Burt, J. A., Bartholomew, A., and Feary, D. A. (2012). Man-made structures as artificial reefs in the Gulf. In *Coral Reefs of the Gulf*, ed. B. M. Riegel and S. J. Purkis. Netherlands: Springer, pp. 171–86.

Burt, J. A., Bartholomew, A., Usseglio, P., Bauman, A., and Sale, P. F. (2009). Are artificial reefs surrogates of natural habitats for corals and fish in Dubai, United Arab Emirates? *Coral Reefs*, 28, 663–75.

Burt, T. P. (1994). Long-term study of the natural environment – perceptive science or mindless monitoring? *Progress in Physical Geography*, 18, 475–96.

Butler, D. R. (2006). Human-induced changes in animal populations and distributions, and the subsequent effects on fluvial systems. *Geomorphology*, 79, 448–59.

Buttrick, D. and Van Schalkwyk, A. (1998). Hazard and risk assessment for sinkhole formation on dolomite land in South Africa. *Environmental Geology*, 36, 170–8.

Butzer, K. W. (1974). Accelerated soil erosion: a problem of man-land relationships. In *Perspectives on Environments*, ed. I. R. Manners and M. W. Mikesell. Washington, DC: Association of American Geographers. pp. 57–77.

(1976). *Early Hydraulic Civilization in Egypt*. Chicago: University of Chicago Press.

(2005). Environmental history in the Mediterranean world: cross-disciplinary investigation of cause-and-effect for degradation and soil erosion. *Journal of Archaeological Science*, 32, 1773–800.

Buynevich, I., Bitinas, A., and Pupienis, D. (2007). Reactivation of coastal dunes documented by subsurface imaging of the Great Dune Ridge, Lithuania. *Journal of Coastal Research*, 50, 226–30.

Cadol, D., Rathburn, S. L., and Cooper, D. J. (2011). Aerial photographic analysis of channel narrowing and vegetation expansion in Canyon de Chelly National Monument, Arizona, USA, 1935–2004. *River Research and Applications*, 27, 841–56.

Caldararo, N. (2002). Human ecological intervention and the role of forest fires in human ecology. *The Science of the Total Environment*, 292, 141–65.

Caldiera, K. and Wickett, M. G. (2003). Anthropogenic carbon and ocean pH. *Nature*, 425, 365.

Caldwell, P. V., Sun, G., McNulty, S. G., Cohen, E. C., and Moore Myers, J. A. (2012). Impacts of impervious cover, water withdrawals, and climate change on river flows in the conterminous US. *Hydrology and Earth System Sciences*, 16, 2839–57.

Calmels, D., Gaillardet, J., and François, L. (2014). Sensitivity of carbonate weathering to soil CO_2 production by biological activity along a temperate climate transect. *Chemical Geology*, 390, 74–86.

Cambi, M., Certini, G., Neri, F., and Marchi, E. (2015). The impact of heavy traffic on forest soils: a review. *Forest Ecology and Management*, 338, 124–38.

Cannon, S. H., Bigio, E. R., and Mine, E. (2001a,). A process for fire-related debris flow initiation, Cerro Grande fire, New Mexico. *Hydrological Processes*, 15, 3011–23.

Cannon, S. H., Kirkham, R. M., and Parise, M. (2001b). Wildfire-related debris-flow initiation processes, Storm King Mountains, Colorado. *Geomorphology*, 39, 171–88.

Cao, S., Tian, T., Chen, L., Dong, X., Yu, X., and Wang, G. (2010). Damage caused to the environment by reforestation policies in arid and semi-arid areas of China. *Ambio*, 39, 279–83.

Capezzuoli, E., Gandin, A., and Sandrelli, F. (2010). Calcareous tufa as indicators of climatic variability: a case study from southern Tuscany (Italy). *Geological Society, London, Special Publications*, 336, 263–81.

Carbognin, L., Teatini, P., Tomasin, A., and Tosi, L. (2010). Global change and relative sea level rise at Venice: what impact in term of flooding. *Climate Dynamics*, 35, 1039–47.

Carbonell, E. and 29 others (2008). The first hominim of Europe. *Nature*, 452, 465–9.

Carey, J. C. and Fulweiler, R. W. (2012). Human activities directly alter watershed dissolved silica fluxes. *Biogeochemistry*, 111, 125–38.

Carminati, E. and Martinelli, G. (2002). Subsidence rates in the Po Plain, northern Italy: the relative impact of natural and anthropogenic causation. *Engineering Geology*, 66, 241–55.

Carmo J. A., Reis, C. S., and Freitas, H. (2010).
Working with nature by protecting sand dunes:
lessons learned. *Journal of Coastal Research* 26,
1068–78.

Carmona, P. and Ruiz, J. M. (2011). Historical mor-
phogenesis of the Turia River coastal flood plain
in the Mediterranean littoral of Spain. *Catena*,
86, 139–49.

Carnec, C. and Fabriol, H. (1999). Monitoring and
modeling land subsidence at the Cerro Prieto
geothermal field, Baja California, Mexico, using
SAR interferometry. *Geophysical Research
Letters*, 26, 1211–4.

Carr, J. R., Stokes, C., and Vieli, A. (2014). Recent
retreat of major outlet glaciers on Novaya Zemlya, Russian Arctic, influenced by fjord geometry and sea-ice conditions. *Journal of
Glaciology*, 60, 155–70.

Carrara, P. E. and Carroll, T. R. (1979). The determination of erosion rates from exposed tree roots in
the Piceance Basin, Colorado. *Earth Surface
Processes*, 4, 407–17.

Carrivick, J. L. and Quincey, D. J. (2014). Progressive
increase in number and volume of ice-marginal
lakes on the western margin of the Greenland Ice
Sheet. *Global and Planetary Change*, 116,
156–63.

Casalí, J., Giménez, R., De Santisteban, L., Álvarez-
Mozos, J., Mena, J., and Del Valle de Lersundi,
J. (2009). Determination of long-term erosion
rates in vineyards of Navarre (Spain) using
botanical benchmarks. *Catena*, 78, 12–9.

Casana, J. (2008). Mediterranean valleys revisited:
Linking soil erosion, land use and climate variability in the Northern Levant. *Geomorphology*,
101, 429–42.

Castaneda, C. and Herrero, J. (2008). Assessing the
degradation of saline wetlands in an arid agricultural region in Spain. *Catena*, 72, 205–13.

Castillo, V. M., Mosch, W. M., Garcia, C. C., Barberá,
G. G., Cano, J. A., and López-Bermúdez, F.
(2007). Effectiveness and geomorphological
impacts of check dams for soil erosion control
in a semiarid Mediterranean catchment: El Cár-
cavo (Murcia, Spain). *Catena*, 70, 416–27.

Castree, N. (2014a). the Anthropocene and geography I:
the back story. *Geography Compass*, 8, 436–49.
    (2014b). Geography and the Anthropocene II: cur-
    rent contributions. *Geography Compass*, 8,
    450–63.
    (2014c). The Anthropocene and geography III:
    future directions. *Geography Compass*, 8,
    464–76.
    (2015). The Anthropocene: a primer for geograpers.
    Geography, 100, 66–75.

Cathcart, R. B. (1983). Mediterranean Basin – Sahara
reclamation. *Speculations in Science and Technology*, 6, 150–2.

Catto, N. (2002). Anthropogenic pressures on coastal
dunes, southwestern Newfoundland. *The Canadian Geographer*, 46, 17–32.

Cavallin, A., Marchetti, M., Panizza, M., and Soldati,
M. (1994). The role of geomorphology in environmental impact assessment. *Geomorphology*,
9, 143–53.

Cavanaugh, K. C., Kellner, J. R., Forde, A. J., Gruner,
D. S., Parker, J. D., Rodriguez, W., and Feller,
I. C. (2014). Poleward expansion of mangroves
is a threshold response to decreased frequency of
extreme cold events. *Proceedings of the
National Academy of Sciences*, 111, 723–7.

Cazenave, A. and Cozannet, G. L. (2014). Sea level
rise and its coastal impacts. *Earth's Future*, 2(2),
15–34.

Cazenave, A. and Llovel, W. (2010). Contemporary
sea level rise. *Annual Review of Marine Science*,
2, 145–73.

Cazenave, A., Dieng, H. B., Meyssignac, B., von
Schuckmann, K., Decharme, B., and Berthier,
E. (2014). The rate of sea-level rise. *Nature
Climate Change*, 4, 358–61.

Cerdà, A. (2007). Soil water erosion on road embankments in eastern Spain. *Science of the Total
Environment*, 378, 151–5.

Cerdà, A. and Doerr, S. H. (2008). The effect of ash
and needle cover on surface runoff and erosion
in the immediate post-fire period. *Catena*, 74,
256–63.

Cerdan, O., Govers, G., Le Bissonnais, Y., Van Oost,
K., Poesen, J., Saby, N., and Dostal, T. (2010).
Rates and spatial variations of soil erosion in
Europe: a study based on erosion plot data.
Geomorphology, 122, 167–77.

Cerney, D. L. (2010). The use of repeat photography in
contemporary geomorphic studies: an evolving
approach to understanding landscape change.
Geography Compass, 4, 1339–57.

Certini, G. and Scalenghe, R. (2011). Anthropogenic
soils are the golden spikes for the Anthropocene.
The Holocene, 21, 1269–74.

Chai J-C, Shen S-L. Zhu, H. H., and Zhang, X. L.
(2004). Land subsidence due to groundwater
drawdown in Shanghai. *Géotechnique*, 54,
143–7.

Chan, N. and Connolly, S. R. (2013). Sensitivity
of coral calcification to ocean acidification: a
meta-analysis. *Global Change Biology*, 19,
282–90.

Chang, J. C. and Slaymaker, O. (2002). Frequency and
spatial distribution of landslides in a mountainous drainage basin: Western Foothills, Taiwan.
Catena, 46, 285–307.

Chappell, A., Webb, N. P., Butler, H. J., Strong, C. L.,
McTainsh, G. H., Leys, J. F., and Viscarra Rossel,
R. A. (2013). Soil organic carbon dust emission:
an omitted global source of atmospheric CO2.
Global Change Biology, 19, 3238–44.

Charlier, R. H., Chaineux, M. C. P., and Morcos, S.
(2005). Panorama of the history of coastal protection. *Journal of Coastal Research*, 21, 79–111.

Chaussard, E., Amelung, F., Abidin, H., and Hong,
S. H. (2013). Sinking cities in Indonesia: Alos

Palsar detects rapid subsidence due to groundwater and gas extraction. *Remote Sensing of Environment*, 128, 150–61.

Chen, C., Pei, S., and Jiao, J. (2003). Land subsidence caused by groundwater exploitation in Suzhou City, China. *Hydrogeology Journal*, 11, 275–87.

Chen, C. T., Hu, J. C., Lu, C. Y., Lee, J. C., and Chan, Y. C. (2007). Thirty-year land elevation change from subsidence to uplift following the termination of groundwater pumping and its geological implications in the Metropolitan Taipei Basin, Northern Taiwan. *Engineering Geology*, 95, 30–47.

Chen, J. L., Wilson, C. R., and Tapley, B. D. (2013). Contribution of ice sheet and mountain glacier melt to recent sea level rise. *Nature Geoscience*, 6, 549–52.

Chen, N., Wu, Y., Wu, J., Yan, X., and Hong, H. (2014). Natural and human influences on dissolved silica export from watershed to coast in Southeast China. *Journal of Geophysical Research: Biogeosciences*, 119, 95–109.

Chen, N., Chen, M., Li, J., He, N., Deng, M., Tanoli, J. I., and Cai, M. (2015). Effects of human activity on erosion, sedimentation and debris flow activity– a case study of the Qionghai Lake watershed, southeastern Tibetan Plateau, China. *The Holocene*, 25, 973–88.

Chen, X. Q., Cui, P., Li, Y., Yang, Z., and Qi, Y. Q. (2007). Changes in glacial lakes and glaciers of post-1986 in the Poiqu River basin, Nyalam, Xizang (Tibet). *Geomorphology*, 88, 298–311.

Chen, Y., Xu, C., Chen, Y., Li, W., and Liu, J. (2010). Response of glacial-lake outburst floods to climate change in the Yarkant River basin on northern slope of Karakoram Mountains, China. *Quaternary International*, 226, 75–81.

Chen, Z. and Saito, Y. (2011). The Megadeltas of Asia: interlinkage of land and sea, and human development. *Earth Surface Processes and Landforms*, 36, 1703–4.

Cheng, F. and Granata, T. (2007). Sediment transport and channel adjustments associated with dam removal: field observations. *Water Resources Research*, 43, W03444, doi:10.1029/2005WR004271.

Cheng, J. D., Lin, J. P., Lu, S. Y., Huang, L. S., and Wu, H. L. (2008). Hydrological characteristics of betel nut plantations on slopelands in central Taiwan. *Hydrological Sciences Journal*, 53, 1208–20.

Chepil, W. S. (1945). Dynamics of wind erosion. *Soil Science*, 60, 305–20; 397–411; 475–80.

Chepil, W. S. and Woodruff, N. P. (1963). The physics of wind erosion and its control. *Advances in Agronomy*, 15, 211–302.

Chevallier, P., Pouyaud, B., Suarez, W., and Condom, T. (2011). Climate change threats to environment in the tropical Andes: glaciers and water resources. *Regional Environmental Change*, 11, 179–87.

Chi, S. C. and Reilinger, R. E. (1984). Geodetic evidence for subsidence due to groundwater withdrawal in many parts of the United States of America. *Journal of Hydrology*, 67, 155–82.

Chiang, S. H. and Chang, K. T. (2011). The potential impact of climate change on typhoon-triggered landslides in Taiwan, 2010–2099. *Geomorphology*, 133, 143–51.

Chin, A. (2006). Urban transformation of river landscapes in a global context. *Geomorphology*, 79, 460–87.

Chin, A., Florsheim, J. L., Wohl, E., and Collins, B. D. (2014). Feedbacks in human–landscape systems. *Environmental Management*, 53, 28–41.

Chin, M., Diehl, T., Tan, Q., Prospero, J. M., Kahn, R. A., Remer, L. A., Yu, H., Sayer, A. M., Bian, H., Geogdzhayev, I. V., Holben, B. N., Howell, S. G., Huebert, B. J., Hsu, N. C., Kim, D., Kucsera, T. L., Levy, R. C., Mishchenko, M. I., Pan, X., Quinn, P. K., Schuster, G. L., Streets, D. G., Strode, S. A., Torres, O., and Zhao, X.-P. (2014). Multi-decadal aerosol variations from 1980 to 2009: a perspective from observations and a global model. *Atmospheric Chemistry and Physics*, 14, 3657–90.

Chiverrell, R. C., Harvey, A. M., and Foster, G. C. (2007). Hillslope gullying in the Solway Firth—Morecambe Bay region, Great Britain: Responses to human impact and/or climatic deterioration? *Geomorphology*, 84, 317–43.

Chu, Z. X., Zhai, S. K., Lu, X. X., Liu, J. P., Xu, J. X., and Xu, K.H. (2009). A quantitative assessment of human impacts on decrease in sediment flux from major Chinese rivers entering the western Pacific Ocean. *Geophysical Research Letters*, 36, L19603, doi: 10.1029/2009GL039513.

Church, J. A. and 13 others (2013). Sea level change. (2013). *In Climate Change 2013: the Physical Science Basis*, ed. T. F. Stocker, D. Qin, G. K. Plattner, M. Tignor, S. K. Allen, J. Boschung, and P. M. Midgley. Cambridge: Cambridge University Press, pp. 1137–216.

Church, M. J., Burt, T., Galay, V. J., and Kondolf, G. M. (2009). Rivers. In *Geomorphology and Global Environmental Change*, ed. O. Slaymaker, T. Spencer, and C. Embleton-Hamann. Cambridge: Cambridge University Press, pp. 98–129.

Ciccarelli, D. (2014). Mediterranean coastal sand dune vegetation: influence of natural and anthropogenic Factors. *Environmental Management*, 54, 194–204.

Clarke, D. and Ayutthaya, S. S. N. (2010). Predicted effects of climate change, vegetation and tree cover on dune slack habitats at Ainsdale on the Sefton Coast, UK. *Journal of Coastal Conservation*, 14, 115–25.

Clark, J. M., Bottrell, S. H., Evans, C. D., Monteith, D. T., Bartlett, R., Rose, R., and Chapman, P. J. (2010). The importance of the relationship between scale and process in understanding long-term DOC dynamics. *Science of the Total Environment*, 408, 2768–75.

Clark, P. U., Church, J. A., Gregory, J. M., and Payne, A. J. (2015). Recent progress in understanding and projecting regional and global mean sea level change. *Current Climate Change Reports*, 1, 1–23.

Clark, S. and Edwards, A. J. (1999). An evaluation of artificial reef structures as tools for marine habitat rehabilitation in the Maldives. *Aquatic Conservation: Marine and Freshwater Ecosystems*, 9, 5–21.

Clarke, G. K., Jarosch, A. H., Anslow, F. S., Radić, V., and Menounos, B. (2015). Projected deglaciation of western Canada in the twenty-first century. *Nature Geoscience*, doi:10.1038/NGE02407.

Clarke, M. A. and Walsh, R. P. D. (2006). Long-term erosion and surface roughness change of rainforest terrain following selective logging, Danum Valley, Sabah, Malaysia. *Catena*, 68, 109–23.

Clarke, M. L. and Rendell, H. M. (2009). The impact of North Atlantic storminess on western European coasts: a review. *Quaternary International*, 195, 31–41.

Clarke, M. L. and Rendell, H. M. (2015). "This restless enemy of all fertility": exploring paradigms of coastal dune management in Western Europe over the last 700 years. *Transactions of the Institute of British Geographers*, 40, 414–29.

Clay, G. D., Dixon, S., Evans, M. G., Rowson, J. G., and Worrall, F. (2012). Carbon dioxide fluxes and DOC concentrations of eroding blanket peat gullies. *Earth Surface Processes and Landforms*, 37, 562–71.

Clements, R., Sodhi, N. S., Schilthuizen, M., and Ng, P. K. (2006). Limestone karsts of Southeast Asia: imperiled arks of biodiversity. *Bioscience*, 56, 733–42.

Clemmensen, L. B., Bjornsen, M., Murray, A., and Pedersen, K. (2007). Formation of aeolian dunes on Anholt, Denmark since AD 1560: a record of deforestation and increased storminess. *Sedimentary Geology*, 199, 171–87.

Closson, D., LaMoreaux, P. E., Karaki, N. A. and Al-Fugha, H. (2007). Karst system developed in salt layers of the Lisan Peninsula, Dead Sea, Jordan. *Environmental Geology*, 52, 155–72.

Clymans, W., Struyf, E., Govers, G., Vandevenne, F., and Conley, D. J. (2011). Anthropogenic impact on amorphous silica pools in temperate soils. *Biogeosciences*, 8, 2281–93.

Coates, D. R. (1977). Landslide perspective. *Reviews in Engineering Geology*, 3, 3–28.

Cocco, S., Brecciaroli, G., Agnelli, A., Weindorf, D., and Corti, G. (2015). Soil genesis and evolution on calanchi (badland-like landform) of central Italy. *Geomorphology*, 248, 33–45.

Cohen, J. E. (2003). Human population: the next half century. *Science*, 302, 1172–5.

Coleman, R. (1981). Footpath erosion in the English Lake District. *Applied Geography*, 1, 121–31

Collison, A., Wade, S., Griffiths, J., and Dehn, M. (2000). Modelling the impact of predicted climate change on landslide frequency and magnitude in S. E. England. *Engineering Geology*, 55, 205–18.

Comeaux, R. S., Allison, M. A., and Bianchi, T. S. (2012). Mangrove expansion in the Gulf of Mexico with climate change: implications for wetland health and resistance to rising sea levels. *Estuarine, Coastal and Shelf Science*, 96, 81–95.

Comiti, F., Da Canal, M., Surian, N., Mao, L., Picco, L., and Lenzi, M. A. (2011). Channel adjustments and vegetation cover dynamics in a large gravel bed river over the last 200 years. *Geomorphology*, 125, 147–59.

Compton J. S., Herbert, C. T., Hoffman, M. T., Schneider, R. R., and Dtuut, J. B. (2010). A tenfold increase in the Orange River mean Holocene mud flux: implications for soil erosion in South Africa. *The Holocene*, 20, 115–22.

Conley, D. J., Likens, G. E., Buso, D. C., Saccone, L., Bailey, S. W., and Johnson, C. E. (2008). Deforestation causes increased dissolved silicate losses in the Hubbard Brook Experimental Forest. *Global Change Biology*, 14, 2548–54.

Connaughton, C. A. (1935). Forest fires and accelerated erosion. *Journal of Forestry*, 33, 751–2.

Conway, V. M. (1954). Stratigraphy and pollen analysis of southern Pennine blanket peats. *Journal of Ecology*, 42, 117–47.

Cook, A. J. and Vaughan, D. G. (2010). Overview of areal changes of the ice shelves on the Antarctic Peninsula over the past 50 years. *The Cryosphere*, 4, 77–98.

Cook, A. J., Fox, A. J., Vaughan, D. G., and Ferrigno, J. G. (2005). Retreating glacier fronts on the Antarctic Peninsula over the past half-century. *Science*, 308, 541–4.

Cook, B. I., Miller, R. L., and Seager, R. (2009). Amplification of the North American "Dust Bowl" drought through human-induced land degradation. *Proceedings of the National Academy of Sciences*, 106, 4997–5001.

Cooke, R. U. and Reeves, R. W. (1976). *Arroyos and Environmental Change in the American Southwest*. Oxford: Clarendon Press.

Coones, P. and Patten, J. H. C. (1986). *The landscape of England and Wales*. Harmondsworth: Penguin Books.

Cooper, A. H. (2002). Halite karst geohazards (natural and man-made) in the United Kingdom. *Environmental Geology*, 42, 505–12.

Cooper, J. A. G. and Pilkey, O. H. (2004). Sea-level rise and shoreline retreat: time to abandon the Bruun Rule. *Global and Planetary Change*, 43, 157–71.

Cooper, S. D., Lake, P. S., Sabater, S., Melack, J. M., and Sabo, J. L. (2013). The effects of land use changes on streams and rivers in Mediterranean climates. *Hydrobiologia*, 719, 383–425.

Cordova, C. E. (2008). Floodplain degradation and settlement history in Wadi al-Wala and Wadi ash-Shallalah, Jordan. *Geomorphology*, 101, 443–57.

Corella, J. P., El Amrani, A., Sigró, J., Morellón, M., Rico, E., and Valero-Garcés, B. L. (2011). Recent evolution of Lake Arreo, northern Spain: influences of land use change and climate. *Journal of Paleolimnology*, 46, 469–85.

Cornell, J. D. and Miller, M. (2007). Slash and burn. *Encyclopedia of Earth*, 31, www.eoearth.org/view/article/156045. (Accessed October 16, 2015).

Corona, M. G., Vincente, A. M., and Novo, F. G. (1988). Long-term vegetation changes on the stabilized dunes of Doñana National Park (SW Spain). *Vegetatio* 75, 73–80.

Corti, G., Cavallo, E., Cocco, S., Biddoccu, M., Brecciaroli, G., and Agnelli, A. (2011). Evaluation of erosion intensity and some of its consequences in vineyards from two hilly environments under a Mediterranean type of climate, Italy. In *Soil Erosion in Agriculture*, ed. D. Godone and S. Stranchi. Rijeka: Intech, pp. 113–60.

Cosandey, C., Andréassian, V., Martin, C., Didon-Lescot, J. F., Lavabre, J., Folton, N., and Richard, D. (2005). The hydrological impact of the Mediterranean forest: a review of French research. *Journal of Hydrology*, 301, 235–49.

Costa-Cabral, M., Roy, S. B., Maurer, E. P., Mills, W. B., and Chen, L. (2013). Snowpack and runoff response to climate change in Owens Valley and Mono Lake watersheds. *Climatic Change*, 116, 97–109.

Cosyns, E., Degezelle, T., Demeulenaere, E., and Hoffmann, M. (2001). Feeding ecology of Konik horses and donkeys in Belgian coastal dunes and its implications for nature management. *Belgian Journal of Zoology*, 131, Supplement 2, 111–8.

Cotton, J. M., Jeffery, M. L., and Sheldon, N. D. (2013). Climate controls on soil respired CO_2 in the United States: implications for 21st century chemical weathering rates in temperate and arid ecosystems. *Chemical Geology*, 358, 37–45.

Coverdale, T. C., Herrmann, N. C., Altieri, A. H., and Bertness, M. D. (2013). Latent impacts: the role of historical human activity in coastal habitat loss. *Frontiers in Ecology and the Environment*, 11, 69–74.

Cowie, S. M., Knippertz, P., and Marsham, J. H. (2013). Are vegetation-related roughness changes the cause of the recent decrease in dust emission from the Sahel? *Geophysical Research Letters*, 40, 1868–72.

Cox, R., Bierman, P., Jungers, M. C., and Rakotondrazafy, A. M. (2009). Erosion rates and sediment sources in Madagascar inferred from ^{10}Be analysis of lavaka, slope, and river sediment. *Journal of Geology*, 117, 363–76.

Cox, R., Zentner, D. B., Rakotondrazafy, A. F. M., and Rasoazanamparany, C. F. (2010). Shakedown in Madagascar: occurrence of lavakas (erosional gullies) associated with seismic activity. *Geology*, 38, 179–82.

Cremaschi, M. (2014). When did the Anthropocene begin? A geoarchaeological approach to deciphering the consequences of human activity in pre-protohistoric times: selected cases from the Po Plain (northern Italy). *Rendiconti Lincei*, 25, 101–12.

Croke, J. and Nethery, M. (2006). Modelling runoff and soil erosion in logged forests: scope and application of some existing models. *Catena*, 67, 35–49.

Croke, J., Hairsine, P., and Fogarty, P. (1999). Sediment transport, redistribution and storage on logged forest hillslopes in south-eastern Australia. *Hydrological Processes*, 13, 2705–20.

(2001). Soil recovery from track construction and harvesting changes in surface infiltration, erosion and delivery rates with time. *Forest Ecology and Management*, 143, 3–12.

Crooks, J. A. (2002). Characterizing ecosystem-level consequences of biological invasions: the role of ecosystem engineers. *Oikos*, 97, 153–66.

Crozier, M. J. (2010). Deciphering the effect of climate change on landslide activity: a review. *Geomorphology*, 124, 260–7.

Crozier, M. J., Marx, S. L., and Grant, I. J., 1978, Impact of off-road recreational vehicles on soil and vegetation. *Proceedings of the 9th New Zealand Geography Conference*, Dunedin, 76–79.

Crutzen, P. J. (2002). Geology of mankind. *Nature*, 415, 23.

Csiki, S. J. and Rhoads, B. L. (2014). Influence of four run-of-river dams on channel morphology and sediment characteristics in Illinois, USA. *Geomorphology*, 206, 215–29.

Cui, Y., Wooster, J. K., Braudrick, C. A., and Orr, B. K. (2014). Lessons learned from sediment transport model predictions and long-term postremoval monitoring: Marmot Dam removal project on the Sandy River in Oregon. *Journal of Hydraulic Engineering*, 140, 04014044.

Curreli, A., Wallace, H., Freeman, C., Hollingham, M., Stratford, C., Johnson, H., and Jones, L. (2013). Eco-hydrological requirements of dune slack vegetation and the implications of climate change. *Science of the Total Environment*, 443, 910–9.

Cushman, G. T. (2011). Humboldtian science, creole meteorology, and the discovery of human-caused climate change in South America. *Osiris*, 26, 19–44.

Cuvilliez, A., Deloffre, J., Lafite, R., and Bessineton, C. (2009). Morphological responses of an estuarine intertidal mudflat to constructions since 1978 to 2005: the Seine estuary (France). *Geomorphology*, 104, 165–74.

Cypser, D. A. and Davis, S. D., 1998, Induced seismicity and the potential for liability under U.S. law. *Tectonophysics*, 289, 239–55.

Dabkowski, J., Brou, L., and Naton, H. G. (2015). New stratigraphic and geochemical data on the Holocene environment and climate from a tufa deposit at Direndall (Mamer Valley, Luxembourg). *The Holocene*, 25, 1153–64.

Dadson, S. (2010). Geomorphology and Earth system science. *Progress in Physical Geography*, 34, 385–98.

Dai, A. (2011). Drought under global warming: a review. *Interdisciplinary Reviews: Climate Change*, 2, 45–65.

Dai, S. B. and Lu, X. X. (2014). Sediment load change in the Yangtze River (Changjiang): a review. *Geomorphology*, 215, 60–73.

Dai, Z. and Liu, J. T. (2013). Impacts of large dams on downstream fluvial sedimentation: an example of the Three Gorges Dam (TGD) on the Changjiang (Yangtze River). *Journal of Hydrology*, 480, 10–8.

Dale, T. and Carter V. G. (1955) *Topsoil and Civilization*. Norman, Oklahoma: University of Oklahoma Press.

da Luz, R. A. and Rodrigues, C. (2015). Anthropogenic changes in urbanised hydromorphological systems in a humid tropical environment: River Pinheiros, Sao Paulo, Brazil. *Zeitschrift für Geomorphologie*, Supplementary Issues, 59, 109–35.

Daniel, D. W., Smith, L. M., and McMurry, S. T. (2015). Land use effects on sedimentation and water storage volume in playas of the rainwater basin of Nebraska. *Land Use Policy*, 42, 426–31.

Daniel, T. C., McGuire, P. E., Stoffel, D., and Millfe, B. (1979). Sediment and nutrient yield from residential construction sites. *Journal of Environmental Quality*, 8, 304–8.

Dapples, F., Lotter, A. F., van Leeuwen, J. F., van der Knaap, W. O., Dimitriadis, S., and Oswald, D. (2002). Paleolimnological evidence for increased landslide activity due to forest clearing and land-use since 3600 cal BP in the western Swiss Alps. *Journal of Paleolimnology*, 27, 239–48.

DasGupta, R. and Shaw, R. (2013). Cumulative impacts of human interventions and climate change on mangrove ecosystems of South and Southeast Asia: an overview. *Journal of Ecosystems, article* 379429.

Davies, B. J and Glasser, N. F. (2012). Accelerating shrinkage of Patagonian glaciers from the Little Ice Age (~AD 1870) to 2011. *Journal of Glaciology*, 58, 1063–84.

Davies, R., Foulger, G., Bindley, A., and Styles, P. (2013). Induced seismicity and hydraulic fracturing for the recovery of hydrocarbons. *Marine and Petroleum Geology*, 45, 171–85.

Dawson, Q., Kechavarzi, C., Leeds-Harrison, P. B., and Burton, R. G. O. (2010). Subsidence and degradation of agricultural peatlands in the Fenlands of Norfolk, U.K. *Geoderma*, 154, 181–7.

Day, M. (1996). Conservation of karst in Belize. *Journal of Caves and Karst Studies*, 58, 139–44.

De Alba, S., Lindstrom, M., Schumacher, T. E., and Malo, D. D. (2004). Soil landscape evolution due to soil redistribution by tillage: a new conceptual model of soil catena evolution in agricultural landscapes. *Catena*, 58, 77–100.

De Angelis, H. and Skvarca, P. (2003). Glacier surge after Ice Shelf collapse. *Science*, 299, 1560–2.

De Bruyn, I. A. and Bell, F. G. (2001). The occurrence of sinkholes and subsidence depressions in the Far West Rand and Gauteng Province, South Africa, and their engineering implications. *Environmental & Engineering Geoscience*, 7, 281–95.

de Jong, C., Carletti, G., and Previtali, F. (2015). Assessing impacts of climate change, ski slope, snow and hydraulic engineering on slope stability in ski resorts (French and Italian Alps). In *Engineering Geology for Society and Territory*, 1, ed. G. Lollino, A. Manconi, J. Clague, W. Shan, and M. Chiarle. Springer International Publishing, pp. 51–5.

De Meyer, A., Poesen, J., Isabirye, M., Deckers, J., and Raes, D. (2011). Soil erosion rates in tropical villages: a case study from Lake Victoria Basin, Uganda. *Catena*, 84, 89–98.

De Santisteban, L. M., Casalí, J., and López, J. J. (2006). Assessing soil erosion rates in cultivated areas of Navarre (Spain). *Earth Surface Processes and Landforms*, 31, 487–506.

De Wit, M. and Stanciewicz J. (2006). Changes in surface water supply across Africa with predicted climate change. *Science*, 311, 1917–21.

Dean, J. R., Leng, M. J., and Mackay, A. W. (2014). Is there an isotopic signature of the Anthropocene? *The Anthropocene Review*, 1, 276–87.

Death, R. G., Fuller, I. C., and Macklin, M. G. (2015). Resetting the river template: the potential for climate-related extreme floods to transform river geomorphology and ecology. *Freshwater Biology*, 60, 2477–96, doi: 10.1111/fwb.12639.

DeBano, L. F. (2000). The role of fire and soil heating on water repellency in wildland environments: a review. *Journal of Hydrology*, 231/2, 195–206.

DeCarlo, T. M., Cohen, A. L., Barkley, H. C., Cobban, Q., Young, C., Shamberger, K. E., and Golbuu, Y. (2014). Coral macrobioerosion is accelerated by ocean acidification and nutrients. *Geology*, 43, 7–10.

DeGeorges, A., Goreau, T. J., and Reilly, B. (2010). Land-sourced pollution with an emphasis on domestic sewage: lessons from the Caribbean and implications for coastal development on Indian Ocean and Pacific coral reefs. *Sustainability*, 2, 2919–49.

Dehn, M. and Buma, J. (1999). Modelling future landslide activity based on general circulation models. *Geomorphology*, 30, 175–87.

Dehn, M., Gurger, G., Buma, J., and Gasparetto, P. (2000). Impact of climate change on slope stability using expanded downscaling. *Engineering Geology*, 55, 193–204.

Deline, P., Gardent, M., Magnin, F,. and Ravanel, L. (2012). The morphodynamics of the Mont Blanc massif in a changing cryosphere: a comprehensive review. *Geografiska Annaler*, 94 A, 265–83.

Del Vecchio, S., Acosta, A., and Stanisci, A. (2013). The impact of *Acacia saligna* invasion on Italian coastal dune EC habitats. *Comptes Rendus Biologies*, 336, 364–9.

DeLong, S. B., Johnson, J. P., and Whipple, K. X. (2014). Arroyo channel head evolution in a flash-flood–dominated discontinuous ephemeral stream system. *Geological Society of America Bulletin*, 126, 1683–701.

DeLong, S. B., Pelletier, J. D., and Arnold, L. J. (2011). Late Holocene alluvial history of the Cuyama River, California, *USA. Geological Society of America Bulletin*, B30312–1.

Delworth, T. L. and Zeng, F. (2014). Regional rainfall decline in Australia attributed to anthropogenic greenhouse gases and ozone levels. *Nature Geoscience*, 7, 583–7.

Denevan, W. M. (1992). The pristine myth: the landscape of the Americas in 1492. *Annals of the Association of American Geographers*, 82, 369–85.

(2001). *Cultivated Landscapes of Native Amazonia and the Andes*. Oxford: Oxford University Press.

Deo, R. C., Syktus, J. I., McAlpine, C. A., Lawrence, P. J., McGowan, H. A., and Phinn, S.R., (2009). Impact of historical land cover change on daily indices of climate extremes including droughts in eastern Australia. *Geophysical Research Letters*, 36, L08705, doi:10.1029/2009GL037666.

Dickinson, W. R. (1999). Holocene sea-level record on Funafuti and potential impact of global warming on Central Pacific atolls. *Quaternary Research*, 51, 124–32.

Dillehay, T. D. (2003). Tracking the first Americans. *Nature*, 425, 23–4.

Dolan, R., Godfrey, P. J., and Odum, W. E. (1973). Man's impact on the barrier islands of North Carolina. *American Scientist*, 61, 152–62.

Domínguez-Villar, D., Vázquez-Navarro, J. A., and Carrasco, R. M. (2012). Mid-Holocene erosive episodes in tufa deposits from Trabaque Canyon, central Spain, as a result of abrupt arid climate transitions. *Geomorphology*, 161, 15–25.

Doney, S. C. (2006). The dangers of ocean acidification. *Scientific American*, 294 (3), 38–45.

Dong, Z., Chen, G., He, X., Han, Z., and Wang, X. (2004). Controlling blown sand along the highway crossing the Taklimakan Desert. *Journal of Arid Environments*, 57, 329–44.

Donkin, R. A. (1979). *Agricultural Terracing in the Aboriginal New World*. Tucson: University of Arizona Press.

Donnelly, L. J. (2009). A review of international cases of fault reactivation during mining subsidence and fluid abstraction. *Quarterly Journal of Engineering Geology and Hydrogeology*, 42, 73–94.

Donner, S. D. (2009). Coping with commitment: projected thermal stress on coral reefs under different future scenarios. *Plos One*, 4, e5712, 1–9.

Doody, J. P. (2013). Coastal squeeze and managed realignment in southeast England, does it tell us anything about the future? *Ocean and Coastal Management* 79, 34–41.

Doody, P. (ed.) (1984). *Spartina anglica in Great Britain*. Shrewsbury: Nature Conservancy Council.

Doody, P. and Barnett, B. (eds). (1987). *The Wash and its Environment*. Peterborough: Nature Conservancy Council.

Doren, R. F., Richards, J. H., and Volin, J. C. (2009). A conceptual ecological model to facilitate understanding the role of invasive species in large-scale ecosystem restoration. *Ecological Indicators*, 9, S150–60.

Dotterweich, M. (2008).The history of soil erosion and fluvial deposits in small catchments of central Europe: deciphering the long-term interaction between humans and the environment – a review. *Geomorphology*, 101, 192–208.

(2013). The history of human-induced soil erosion: geomorphic legacies, early descriptions and research, and the development of soil conservation—a global synopsis. *Geomorphology*, 201, 1–34.

Dotterweich, M., Ivester, A. H., Hanson, P. R., Larsen, D., and Dye, D. H. (2015). Natural and human induced prehistoric and historical soil erosion and landscape development in Southwestern Tennessee, USA. *Anthropocene*, 8, 6–24.

Doughty, C. E., Wolf, A., and Field, C. B. (2010). Biophysical feedbacks between the Pleistocene megafauna extinction and climate: the first human-induced global warming? *Geophysical Research Letters*, 37, doi: 10.1029/2010GL043985.

Douglas, I. (1983). *The Urban Environment*. London: Arnold.

(1996). The impact of land-use changes, especially logging, shifting cultivation, mining and urbanization on sediment yields in humid tropical Southeast Asia: a review with special reference to Borneo. *IAHS Publications-Series of Proceedings and Reports-Intern Assoc Hydrological Sciences*, 236, 463–72.

Douglas, I., and Lawson, N. (2001). The human dimensions of geomorphological work in Britain. *Journal of Industrial Ecology*, 4, 9–33.

(2003). Airport construction: materials use and geomorphic change. *Journal of Air Transport Management*, 9, 177–85.

Douglas-Mankin, K. R., Srinivasan, R., and Arnold, J. G. (2010). Soil and Water Assessment Tool (SWAT) model: current developments and applications. *Transactions of the American Society of Agricultural and Biological Engineers*, 53, 1423–31.

Douville H., Chauvin, F., Planton, S., Royer, J. F., Salas-Mélia, D., and Tyteca, S. (2002).

Sensitivity of the hydrological cycle to increasing amounts of greenhouse gases and aerosols. *Climate Dynamics*, 20, 45–68.

Downs, P. W. and Gregory, K. J. (2004). *River Channel Management*. Arnold: London.

Downs, P. W., Dusterhoff, S. R., and Sears, W. A. (2013). Reach-scale channel sensitivity to multiple human activities and natural events: lower Santa Clara River, California, USA. *Geomorphology*, 189, 121–34.

Downward, S. and Skinner, K. (2005). Working rivers: the geomorphological legacy of English freshwater mills. *Area*, 37, 138–47.

Doyle, M. W., Stanley, E. H., and Harbor, J. M. (2003). Channel adjustments following two dam removals in Wisconsin. *Water Resources Research*, 39, W1011, doi:10.1029/2002/WR001714.

Draut, A. E. (2012). Effects of river regulation on aeolian landscapes, Colorado River, southwestern USA. *Journal of Geophysical Research: Earth Surface*, 117(F2), doi: 10.1029/2011JF002329.

Draut, A. E. and Ritchie, A. C. (2013). Sedimentology of new fluvial deposits on the Elwha River, Washington, USA, formed during large-scale dam removal. *River Research and Applications*, doi: 10.1002/rra.2724.

Dreibrodt, S., Lubos, C., Terhorst, B., Damm, B., and Bork, H. R. (2010). Historical soil erosion by water in Germany: scales and archives, chronology, research perspectives. *Quaternary International*, 222, 80–95.

Drew, D. P. (1983). Accelerated soil erosion in a karst area: the Burren, western Ireland. *Journal of Hydrology*, 61, 113–24.

Drewry, D. J. (1991). The response of the Antarctic ice sheet to climate change. In *Antarctica and Global Climatic Change*, ed. C. M. Harris and B. Stonehouse. London: Belhaven Press, pp. 90–106.

Duarte, F., Jones, N., and Fleskens, L. (2008). Traditional olive orchards on sloping land: sustainability or abandonment? *Journal of Environmental Management*, 89, 86–98.

Dubois, R. N. (2013). How does a barrier shoreface respond to a sea-level rise? *Journal of Coastal Research*, 18, 612–28.

Dufour, S. and Piégay, H. (2009). From the myth of a lost paradise to targeted river restoration: forget natural references and focus on human benefits. *River Research and Applications*, 25, 568–81.

Dukes, J. S. and Mooney, H. A. (2004). Disruption of ecosystem processes in western North America by invasive species. *Revista Chilena de Historia Natural*, 77, 411–37.

Dunning, N. P. and Beach, T. (1994). Soil erosion, slope management, and ancient terracing in the Maya lowlands. *Latin American Antiquity*, 5, 51–69.

Du Toit, G. van N., Snyman, H. A., and Malan, P. J. (2009). Physical impact of grazing by sheep on soil parameters in the Nama Karoo subshrub/grass rangeland of South Africa. *Journal of Arid Environments*, 73, 804–10.

Dupin, B., De Rouw, A., Phantahvong, K. B., and Valentin, C. (2009). Assessment of tillage erosion rates on steep slopes in northern Laos. *Soil and Tillage Research*, 103, 119–26.

Durá-Gómez, I. and Talwani, P. (2010). Reservoir-induced seismicity associated with the Itoiz Reservoir, Spain: a case study. *Geophysical Journal International*, 181, 343–56.

Durán Zuazo, V. H., Francia Martínez, J. R., and Martínez Raya, A. (2004). Impact of vegetative cover on runoff and soil erosion at hillslope scale in Lanjaron, Spain. *The Environmentalist*, 24, 39–48.

Dusar, B., Verstraeten, G., Notebaert, B., and Bakker, J. (2011). Holocene environmental change and its impact on sediment dynamics in the Eastern Mediterranean. *Earth-Science Reviews*, 108, 137–57.

East, A. E., Pess, G. R., Bountry, J. A., Magirl, C. S., Ritchie, A. C., Logan, J. B., and Shafroth, P. B. (2014). Large-scale dam removal on the Elwha River, Washington, USA: river channel and floodplain geomorphic change. *Geomorphology*, 228, 765–86.

Edeso, J. M., Merino, A., Gonzalez, M. J., and Marauri, P. (1999). Soil erosion under different harvesting managements in steep forestlands from northern Spain. *Land Degradation & Development*, 10, 79–88.

Edgeworth, M., deB Richter, D., Waters, C., Haff, P., Neal, C., and Price, S. J. (2015). Diachronous beginnings of the Anthropocene: the lower bounding surface of anthropogenic deposits. *The Anthropocene Review*, 2, 33–58.

Edinger, E. N. and Risk, M. J. (2013). Effect of land-based pollution on Central Java coral reefs. *Journal of Coastal Development*, 3, 593–613.

Edwards, K. J. and Whittington, G. (2001). Lake sediments, erosion and landscape change during the Holocene in Britain and Ireland. *Catena*, 42, 143–73.

Ehrenfeld, J. G. (2010). Ecosystem consequences of biological invasions. *Annual Review of Ecology, Evolution, and Systematics*, 41, 59–80.

Eitner, V. (1996). Geomorphological response to the East Frisian barrier islands to sea level rise: an investigation of past and future evolution. *Geomorphology*, 15, 57–65.

El Banna, M. M. and Frihy, O. E. (2009). Human-induced changes in the geomorphology of the northeastern coast of the Nile delta, Egypt. *Geomorphology*, 107, 72–8.

Eldridge, D. J. (1998). Trampling of microphytic crusts on calcareous soils, and its impact on erosion under rain-impacted flow. *Catena*, 33, 221–39.

Eldridge, D. J. and Robson, A. D. (1997). Blade-ploughing and exclosure influence soil properties in a semi-arid Australian woodland. *Journal of Range Management*, 50, 191–8.

Elguindi, N. and Giorgi, F. (2006). Projected changes in the Caspian Sea level for the 21st century based on the latest AOGCM simulations. *Geophysical Research Letters*, 33, L08706, doi: 10.1029/2006GL025943.

El Hariri, M., Abercrombie, R. E., Rowe, C. A., and Do Nascimento, A. F. (2010). The role of fluids in triggering earthquakes: observations from reservoir induced seismicity in Brazil. *Geophysical Journal International*, 181, 1566–74.

Elliot, W. J., Hall, D. E., and Graves, S. R. (1999). Predicting sedimentation from forest roads. *Journal of Forestry*, 97, 23–9.

Ellis, E. C., Fuller, D. Q., Kaplan, J. O., and Lutters, W. G. (2013b). Dating the Anthropocene: towards an empirical global history of human transformation of the terrestrial biosphere. *Elementa: Science of the Anthropocene*, doi: 10.12952/journal.elementa.000018.

Ellis, E. C., Kaplan, J. O., Fuller, D. Q., Vavrus, S., Goldewijk, K. K., and Verburg, P. H. (2013a). Used planet: a global history. *Proceedings of the National Academy of Sciences*, 110, 7978–85.

Ellison, J. C. (2015). Vulnerability assessment of mangroves to climate change and sea-level rise impacts. *Wetlands Ecology and Management*, 23, 115–37.

Ellsworth, W. L. (2013). Injection-induced earthquakes. *Science*, 341, doi: 10.1126/science.1225942.

Elschot, K., Bouma, T. J., Temmerman, S., and Bakker, J. P. (2013). Effects of long-term grazing on sediment deposition and salt-marsh accretion rates. *Estuarine, Coastal and Shelf Science*, 133, 109–15.

Elsner. J. B., Kossin, J. P., and Jagger, T. H. (2008). The increasing intensity of the strongest tropical cyclones. *Nature*, 455, 92–5.

Emadodin, I., Reiss, S., and Bork, H. R. (2011). Colluviation and soil formation as geoindicators to study long-term environmental changes. *Environmental Earth Sciences*, 62, 1695–706.

Emery, F. V. (1962). Moated settlements in England. *Geography*, 47, 378–88.

Engelstaedter, S., Tegen, I., and Washington, R. (2006). North African dust emissions and transport. *Earth-Science Reviews*, 79, 73–100.

Engstrom, D. R., Almendinger, J. E., and Wolin, J. A. (2009). Historical changes in sediment and phosphorus loading to the upper Mississippi River: mass-balance reconstructions from the sediments of Lake Pepin. *Journal of Paleolimnology*, 41, 563–88.

Erban, L. E., Gorelick, S. M., and Zebker, H. A. (2014). Groundwater extraction, land subsidence, and sea-level rise in the Mekong Delta, Vietnam. *Environmental Research Letters*, 9, 084010.

Ericson, J. P., Vorosmarty, C. J., Dingham, L., Ward, L. G., and Meybeck, M. (2006). Effective sea-level rise and deltas: causes of change and human dimension implications. *Global and Planetary Change*, 50, 63–82.

Eriksson, M. G., Olley, J. M., and Payton, R. W. (2000). Soil erosion history in central Tanzania based on OSL dating of colluvial and alluvial hillslope deposits. *Geomorphology*, 36, 107–128.

Erlandson, J. M. (2013). Shell middens and other anthropogenic soils as global stratigraphic signatures of the Anthropocene. *Anthropocene*, 4, 24–32.

Erlandson, J. M. and Braje, T. J. (2014). Archaeology and the Anthropocene. *Anthropocene*, 4, 1–7.

Erlandson J. M., Rick, T. C., and Peterson, C. (2005). A geoarchaeological chronology of Holocene dune building on San Miguel Island, California. *The Holocene*, 15, 1227–35.

Erskine, W. D., Saynor, M. J., Chalmers, A., and Riley, S. J. (2012). Water, wind, wood, and trees: interactions, spatial variations, temporal dynamics, and their potential role in river rehabilitation. *Geographical Research*, 50, 60–74.

Escalante, S. A. and Pimentel, A .S. (2008). Coastal dune stabilization using geotextile tubes at Las Colorados. *Geosynthetics*, 26, 16–24.

Esteves, L. S. (2014). *Managed Realignment: a Viable Long-Term Coastal Management Strategy?* Dordrecht: Springer.

Etienne, D., Ruffaldi, P., Goepp, S., Ritz, F., Georges-Leroy, M., Pollier, B., and Dambrine, E. (2011). The origin of closed depressions in Northeastern France: a new assessment. *Geomorphology*, 126, 121–31.

Evans, D. J., Ewertowski, M., Jamieson, S. S., and Orton, C. (2015). Surficial geology and geomorphology of the Kumtor Gold Mine, Kyrgyzstan: human impacts on mountain glacier landsystems. *Journal of Maps*, doi: 10.1080/17445647.2015.1071720.

Evans, G. (2008). Man's impact on the coastline. *Journal of Iberian Geology*, 34, 167–90.

Evans, K. G., Saynor, M. J., Willgoose, G. R., and Riley, S. J. (2000). Post-mining landform evolution modelling: 1. Derivation of sediment transport model and rainfall–runoff model parameters. *Earth Surface Processes and Landforms*, 25, 743–63.

Evans, M. and Lindsay, J. (2010). High resolution quantification of gully erosion in upland peatlands at the landscape scale. *Earth Surface Processes and Landforms*, 35, 876–86.

Evans, M. and Warburton, J. (2007). *Geomorphology of Upland Peat*. Oxford: Blackwell.

Evans, M., Warburton, J., and Yang, J. (2006). Eroding blanket peat catchments: global and local implications of upland organic sediment budgets. *Geomorphology* 79, 45–57.

Evans, R. (1997). Soil erosion in the UK initiated by grazing animals: a need for a national survey. *Applied Geography*, 17, 127–41.

(1998). The erosional impacts of grazing animals. *Progress in Physical Geography*, 22, 251–68.

(2005). Curtailing grazing-induced erosion in a small catchment and its environs, the Peak

District, Central England. *Applied Geography*, 25, 81–95.

Evenari, M., Shanan, L., and Tadmor, N. (1982). *The Negev: The Challenge of a Desert*. Cambridge, Mass.: Harvard University Press.

Fabi, G., Spagnolo, A., Bellan-Santini, D., Charbonnel, E., Çiçek, B. A., García, J. J. G., and Santos, M. N. D. (2011). Overview on artificial reefs in Europe. *Brazilian Journal of Oceanography*, 59 (SPE1), 155–66.

Fabricius, K. E. (2005). Effects of terrestrial runoff on the ecology of corals and coral reefs: review and synthesis. *Marine Pollution Bulletin*, 50, 125–46.

Faith, J. T. (2014). Late Pleistocene and Holocene mammal extinctions on continental Africa. *Earth-Science Reviews*, 128, 105–21.

Fan, B., Guo, L., Li, N., Chen, J., Lin, H., Zhang, X. and Ma, L. (2014). Earlier vegetation green-up has reduced spring dust storms. *Scientific Reports*, 4, 6749, doi: 10.1038/srep06749.

Farley, K. A., Jobbágy, E. G., and Jackson, R. B. (2005). Effects of afforestation on water yield: a global synthesis with implications for policy. *Global Change Biology*, 11, 1565–76.

Faulkner, H. (1995). Gully erosion associated with the expansion of unterraced almond cultivation in the coastal Sierra de Lujar, S. Spain. *Land Degradation & Development*, 6, 179–200.

Faulkner, H., Ruiz, J., Zukowskyj, P., and Downward, S. (2003). Erosion risk associated with rapid and extensive agricultural clearances on dispersive materials in southeast Spain. *Environmental Science & Policy*, 6, 115–27.

Favis-Mortlock, D. and Boardman, J. (1995). Nonlinear responses of soil erosion to climate change: modelling study on the UK South Downs. *Catena*, 25, 365–87.

Favis-Mortlock, D. T. and Guerra, A. J. T. (1999). The implications of general circulation model estimates of rainfall for future erosion: a case study from Brazil. *Catena*, 37, 329–54.

Fedick, S. L. (1994). Ancient Maya agricultural terracing in the upper Belize River area. *Ancient Mesoamerica*, 5, 107–27.

Feeser, I. and O'Connell, M. (2009). Fresh insights into long-term changes in flora, vegetation, land use and soil erosion in the karstic environment of the Burren, western Ireland. *Journal of Ecology*, 97, 1083–100.

Fei, S., Phillips, J., and Shouse, M. (2014). Biogeomorphic impacts of invasive species. *Annual Review of Ecology, Evolution, and Systematics*, 45, 69–87.

Feld, C. K., Birk, S., Bradley, D. C., Hering, D., Kail, J., Marzin, A., and Friberg, N. (2011). From natural to degraded rivers and back again: a test of restoration ecology theory and practice. *Advances in Ecological Research*, 44, 119–209.

Fencl, J. S., Mather, M. E., Costigan, K. H., and Daniels, M. D. (2014). How big of an effect do small dams have? Using geomorphological footprints to quantify spatial impact of low-head dams and identify patterns of across-dam variation. *PloS one*, 10(11), e0141210.

Feng, Q. Y., Liu, G. J., Meng, L., Fu, E. J., Zhang, H. R., and Zhang, K. F. (2008). Land subsidence induced by groundwater extraction and building damage level assessment—a case study of Datun, China. *Journal of China University of Mining and Technology*, 18, 556–60.

Feola, S., Carranza, M. L., Schaminée, J. H. J., Janssen, J. A. M., and Acosta, A. T. R. (2011). EU habitats of interest: an insight into Atlantic and Mediterranean beach and foredunes. *Biodiversity and Conservation*, 20, 1457–68.

Ferguson, G. and Gleeson, T. (2012). Vulnerability of coastal aquifers to groundwater use and climate change. *Nature Climate Change*, 2, 342–5.

Fernandez, D. P., Neff, J. C., and Reynolds, R. L. (2008). Biogeochemical and ecological impacts of livestock grazing in semi-arid southeastern Utah, USA. *Journal of Arid Environments*, 72, 777–91.

Fernández-Moya, J., Alvarado, A., Forsythe, W., Ramírez, L., Algeet-Abarquero, N., and Marchamalo-Sacristán, M. (2014). Soil erosion under teak (*Tectona grandis* Lf) plantations: general patterns, assumptions and controversies. *Catena*, 123, 236–42.

Ferris, T. M. C., Lowther, K. A., and Smith, B. J. (1993). Changes in footpath degradation 1983–1992: a study of the Brandy Pad, Mourne Mountains. *Irish Geography*, 26, 133–40.

Fidelibus, M. D., Gutiérrez, F., and Spilotro, G. (2011). Human-induced hydrogeological changes and sinkholes in the coastal gypsum karst of Lesina Marina area (Foggia Province, Italy). *Engineering Geology*, 118, 1–19.

Field, M. E., Ogston, A. S., and Storlazzi, C. D. (2011). Rising sea level may cause decline of fringing coral reefs. *Eos*, 92, 273–80.

Filin, S., Avni, Y., Baruch, A., Morik, S., Arav, R., and Marco, S. (2014). Characterization of land degradation along the receding Dead Sea coastal zone using airborne laser scanning. *Geomorphology*, 206, 403–20.

Filip, F. and Giosan, L. (2014). Evolution of Chilia lobes of the Danube delta: reorganization of deltaic processes under cultural pressures. *Anthropocene*, 5, 65–70.

Fillenham, L. F. (1963). Holme Fen Post. *Geographical Journal*, 129, 502–3.

Fisher, C. T., Pollard, H. P., Israde-Alcántara, I., Garduño-Monroy, V. H., and Banerjee, S. K. (2003). A reexamination of human-induced environmental change within the Lake Patzcuaro Basin, Michoacan, Mexico. *Proceedings of the National Academy of Sciences*, 100, 4957–62.

Fitzhugh, T. W. and Vogel, R. M. (2011). The impact of dams on flood flows in the United States. *River Research and Applications*, 27, 1192–215.

Fitzner, B., Heinrichs, K., and La Bouchardiere, D. (2002). Limestone weathering on historical monuments in Cairo, Egypt. *Special Publication of the Geological Society of London*, 205, 217–40.

Flannigan, M., Cantin, A. S., de Groot, W. J., Wotton, M., Newbery, A., and Gowman, L. M. (2013). Global wildland fire season severity in the 21st century. *Forest Ecology and Management*, 294, 54–61.

Fleitmann, D., Dunbar, R. B., McCulloch, M., Mudelsee, M., Vuille, M., McClanahan, T. R., and Eggins, S. (2007). East African soil erosion recorded in a 300 year old coral colony from Kenya. *Geophysical Research Letters*, 34, doi: 10.1029/2006GL028525.

Flenley, J. R. (1979). *The Equatorial Rain Forest: A Geological History*. London: Butterworth.

Fleskens, L. and Stroosnijder, L. (2007). Is soil erosion in olive groves as bad as often claimed? *Geoderma*, 141, 260–71.

Fletcher, S., Bateman, P., and Emery, A. (2011). The governance of the Boscombe artificial surf reef, UK. *Land Use Policy*, 28, 395–401.

Flor-Blanco, G., Pando, L., Morales, J. A., and Flor, G. (2015). Evolution of beach–dune fields systems following the construction of jetties in estuarine mouths (Cantabrian coast, NW Spain). *Environmental Earth Sciences*, 73, 1317–30.

Foley, R. A. and Lahr, M. M. (2015). Lithic landscapes: early human impact from stone tool production on the central Saharan environment. *PloS one*, 10(3), e0116482.

Foley, S. F., Gronenborn, D., Andreae, M. O., Kadereit, J. W., Esper, J., Scholz, D., and Crutzen, P. J. (2013). The Palaeoanthropocene – The beginnings of anthropogenic environmental change. *Anthropocene*, 3, 83–8.

Foltz, R. B., Copeland, N. S., and Elliot, W. J. (2009). Reopening abandoned forest roads in northern Idaho, USA: quantification of runoff, sediment concentration, infiltration, and interrill erosion parameters. *Journal of Environmental Management*, 90, 2542–50.

Ford, D. C. and Williams, P. W. (1989). *Karst Geomorphology and Hydrology*. London: Unwin Hyman.

Ford, J. R., Price, S. J., Cooper, A. H., and Waters, C. N. (2014). An assessment of lithostratigraphy for anthropogenic deposits. *Geological Society, London, Special Publications*, 395, 55–89.

Fornasiero, A., Putti, M.., Teatini, P., Ferraris, S., Rizzetto, F., and Tosi, L. (2003). Monitoring of hydrological parameters related to peat oxidation in a subsiding coastal basin south of Venice, Italy. *International Association of Hydrological Sciences, Publication*, 278, 458–62.

Foster, G. C., Chiverrell, R. C., Thomas, G. S. P., Marshall, P., and Hamilton, D. (2009). Fluvial development and the sediment regime of the lower Calder, Ribble catchment, northwest England. *Catena*, 77, 81–95.

Foster, I. D. L., Dearing, J. A., and Appleby, R. G. (1986). Historical trends in catchment sediment yields: a case study in reconstruction from lake-sediment records in Warwickshire, UK. *Hydrological Science Journal*, 31, 427–43.

Foster, I. D. L., Collins, A. L., Naden, P. S., Sear, D. A., Jones, J. I., and Zhang, Y. (2011). The potential for paleolimnology to determine historic sediment delivery to rivers. *Journal of Paleolimnology*, 45, 287–306.

Foulds, S. A. and Macklin, M. G. (2006). Holocene land-use change and its impact on river basin dynamics in Great Britain and Ireland. *Progress in Physical Geography*, 30, 589–604.

Fox, D. M., Martin, N., Carrega, P., Andrieu, J., Adnès, C., Emsellem, K., and Fox, E. A. (2015). Increases in fire risk due to warmer summer temperatures and wildland urban interface changes do not necessarily lead to more fires. *Applied Geography*, 56, 1–12.

Fox, H. L. (1976). The urbanizing river: a case study in the Maryland piedmont. In *Geomorphology and Engineering*, ed. D. R. Coates. Stroudsburg: Dowden, Hutchinson and Ross, pp. 245–71.

Francia Martínez, J. R., Durán Zuazo, V. H., and Martínez Raya, A. (2006). Environmental impact from mountainous olive orchards under different soil-management systems (SE Spain). *Science of the Total Environment*, 358, 46–60.

Frankl, A., Nyssen, J., De Dapper, M., Haile, M., Billi, P., Munro, R. N., and Poesen, J. (2011). Linking long-term gully and river channel dynamics to environmental change using repeat photography (Northern Ethiopia). *Geomorphology*, 129, 238–51.

Frans, C., Istanbulluoglu, E., Mishra, V., Munoz-Arriola, F., and Lettenmaier, D. P. (2013). Are climatic or land cover changes the dominant cause of runoff trends in the Upper Mississippi River Basin? *Geophysical Research Letters*, 40, 1104–10.

Frazier, T. G., Wood, N., Yarnal, B., and Bauer D. H. (2010). Influence of potential sea level rise on societal vulnerability to hurricane storm-surge hazards, Sarasota County, Florida. *Applied Geography*, 30, 490–505.

French, C. A. and Whitelaw, T. M. (1999). Soil erosion, agricultural terracing and site formation processes at Markiani, Amorgos, Greece: the micromorphological perspective. *Geoarchaeology*, 14, 151–89.

French, C., Periman, R., Cummings, L. S., Hall, S., Goodman-Elgar, M., and Boreham, J. (2009). Holocene alluvial sequences, cumulic soils and fire signatures in the Middle Rio Puerco Basin at Guadelupe Ruin, New Mexico. *Geoarchaeology*, 24, 638–76.

French, H. M. (1976). *The Periglacial Environment*. London: Longman.

French, J. R., Spencer, T., and Reed, D. J. (eds). (1995). Geomorphic responses to sea level rise: existing evidence and future impacts. *Earth Surface Processes and Landforms*, 20, 1–6.

Friedman, J. M., Auble, G. T., Shafroth, P. B., Scott, M. L., Merigliano, M. F., Freehling, M. D., and Griffin, E. R. (2005). Dominance of non-native riparian trees in western USA. *Biological Invasions*, 7, 747–51.

Friedman, J. M., Vincent, K. R., Griffin, E. R., Scott, M. L., Shafroth, P. B., and Auble, G. T. (2014). Processes of arroyo filling in northern New Mexico, USA. *Geological Society of America Bulletin*, 127, 621–40.

Friess, D. A., Möller, I., Spencer, T., Smith, G. M., Thomson, A. G., and Hill, R. A. (2014). Coastal saltmarsh managed realignment drives rapid breach inlet and external creek evolution, Freiston Shore (UK). *Geomorphology*, 208, 22–33.

Fuchs, M., Lang, A., and Wagner, G. A. (2004). The history of Holocene soil erosion in the Phlious Basin, NE Peloponnese, Greece, based on optical dating. *The Holocene*, 14, 334–45.

Fuller, I. C., Macklin, M. G., and Richardson, J. M. (2015). The geography of the Anthropocene in New Zealand: differential river catchment response to human impact. *Geographical Research*, doi: 10.1111/1745-5871.12121.

Gabet, E. J. (2014). Fire increases dust production from chaparral soils. *Geomorphology* 217, 182–92.

Gale, S. J. and Hoare, P. G. (2012). The stratigraphic status of the Anthropocene. *The Holocene*, 22, 1491–4.

Gallardo, A. H., Marui, A., Takeda, S., and Okuda, F. (2009). Groundwater supply under land subsidence constrains in the Nobi Plain. *Geosciences Journal*, 13, 151–9.

Gallego-Sala, A. V. and Prentice, I. C. (2012). Blanket peat biome endangered by climate change. *Nature Climate Change*, 3, 152–5.

Galloway, D. L. and Burbey, T. J. (2011). Review: regional land subsidence accompanying groundwater extraction. *Hydrogeology Journal*, 19, 1459–86.

Gambolati, G., Ricceri, G., Bertoni, W., Brighenti, G., and Vuillermin, E. (1991). Mathematical simulation of the subsidence of Ravenna. *Water Resources Research*, 27, 2899–918.

Gambolati, G., Putti, M., Teatini, P., Camporese, M., Ferraris, S., Stori, G. G., and Tosi, L. (2005). Peat land oxidation enhances subsidence in the Venice watershed. *Eos, Transactions American Geophysical Union*, 86(23), 217–20.

Gambolati, G., Putti, M., Teatini, P., and Stori, G. G. (2006). Subsidence due to peat oxidation and impact on drainage infrastructures in a farmland catchment south of the Venice Lagoon. *Environmental Geology*, 49, 814–20.

Gao, J. H., Xu, X., Jia, J., Kettner, A. J., Xing, F., Wang, Y. P., and Gao, S. (2015). A numerical investigation of freshwater and sediment discharge variations of Poyang Lake catchment, China over the last 1000 years. *The Holocene*, 25, 1470–82.

Garbutt, R. A., Reading, C. J., Wolters, M., Gray, A. J., and Rothery, P. (2006). Monitoring the development of intertidal habitats on former agricultural land after the managed realignment of coastal defences at Tollesbury, Essex, UK. *Marine Pollution Bulletin*, 53, 155–64.

García-Moreno, I. and Mateos, R. M. (2011). Sinkholes related to discontinuous pumping: susceptibility mapping based on geophysical studies. The case of Crestatx (Majorca, Spain). *Environmental Earth Sciences*, 64, 523–37.

García-Ruiz, J., Lasanta, T., and Alberto, F. (1997). Soil erosion by piping in irrigated fields. *Geomorphology*, 20, 269–78.

García-Ruiz, J. M. (2010). The effects of land uses on soil erosion in Spain: a review. *Catena*, 81, 1–11.

García-Ruiz, J. M. and Valero-Garcés, B. L. (1998). Historical geomorphic processes and human activities in the Central Spanish Pyrenees. *Mountain Research and Development*, 10, 267–79.

García-Ruiz, J. M., Beguería, S., Alatorre, L. C., and Puigdefábregas, J. (2010). Land cover changes and shallow landsliding in the flysch sector of the Spanish Pyrenees. *Geomorphology*, 124, 250–9.

García-Ruiz, J. M., López-Moreno, J. I., Vicente-Serrano, S. M., Lasanta–Martínez, T., and Beguería, S. (2011). Mediterranean water resources in a global change scenario. *Earth-Science Reviews*, 105, 121–39.

Gardner, R. (2009). Trees as technology: planting shelterbelts on the Great Plains. *History and Technology*, 25, 325–41.

Garland, G. G., Hudson, C., and Blackshaw, J. (1985). An approach to the study of path erosion in the Natal Drakensberg, a mountain wilderness area. *Environmental Conservation*, 12, 337–42.

Gautney, J. R. and Holliday, T. W. (2015). New estimations of habitable land area and human population size at the last glacial maximum. *Journal of Archaeological Science*, 58, 103–12.

Gedan, K. B., Altieri, A. H., and Bertness, M. D. (2011). Uncertain future of New England salt marshes. *Marine Ecology Progress Series*, 434, 229–37.

Gedan, K. B., Silliman, B. R., and Bertness, M. D. (2009). Centuries of human-driven change in salt marsh ecosystems. *Annual Review of Marine Science*, 1, 117–41.

Gedney, N., Cox, P. M., Betts, R. A., Boucher, O. Huntingford, C., and Stott, P. A. (2006). Detection of a direct carbon dioxide effect in continental river runoff records. *Nature*, 439, 835–8.

Geertsema, M., Clague, J. J., Schwab, J. W., and Evans, S. G. (2006). An overview of recent large catastrophic landslides in northern British Columbia, Canada. *Engineering Geology*, 83, 120–43.

Gelfenbaum, G., Stevens, A. W., Miller, I., Warrick, J. A., Ogston, A. S., and Eidam, E. (2015). Large-scale dam removal on the Elwha River, Washington, USA: coastal geomorphic change. *Geomorphology*, 246, 649–68.

Gellatly, A. F., Whalley, W. B., and Gordon, J. E. (1986). Footpath deterioration in the Lyngen Peninsula, north Norway. *Mountain Research and Development*, 6, 167–76.

Gellis, A. C., Pavich, M. J., Ellwein, A. L., Aby, S., Clark, I., Wieczorek, M. E., and Viger, R. (2012). Erosion, storage, and transport of sediment in two subbasins of the Rio Puerco, New Mexico. *Geological Society of America Bulletin*, 124, 817–41.

Genevois, R. and Ghirotti, M. (2005). The 1963 Vaiont landslide. *Giornale di Geologia Applicata*, 1, 41–52.

Gerasimov, I. P. (1979). Anthropogene and its major problem. *Boreas*, 8, 23–30.

Germer, S., Neill, C., Krusche, A. V., and Elsenbeer, H. (2010). Influence of land-use on near-surface hydrological processes: undisturbed forest to pasture. *Journal of Hydrology*, 380, 473–80.

Ghoneim, E., Mashaly, J., Gamble, D., Halls, J., and AbuBakr, M. (2015). Nile Delta exhibited a spatial reversal in the rates of shoreline retreat on the Rosetta promontory comparing pre-and post-beach protection. *Geomorphology*, 228, 1–14.

Gifford, G. F. and Hawkins, R. H. (1978). Hydrologic impact of grazing on infiltration: a critical review. *Water Resources Research*, 14, 305–13.

Giguet-Covex, C., Pansu, J., Arnaud, F., Rey, P. J., Griggo, C., Gielly, L., and Taberlet, P. (2014). Long livestock farming history and human landscape shaping revealed by lake sediment DNA. *Nature communications*, 5, doi:10.1038/ncomms4211.

Gilbert, G. K. (1917). Hydraulic-mining débris in the Sierra Nevada. *United States Geological Survey Professional Paper*, 105.

Gilbertson D. D., Schwenninger, Kemp, R. A., and Rhodes, E. J. (1999). Sand-drift and soil formation along an exposed North Atlantic coastline: 14,000 years of diverse geomorphological, climatic and human impacts. *Journal of Archaeological Science*, 26, 439–69.

Gill, R. A. (2007). Influence of 90 years of protection from grazing on plant and soil processes in the subalpine of the Wasatch Plateau, USA. *Rangeland Ecology & Management*, 60, 88–98.

Gill, T. E. (1996). Eolian sediments generated by anthropogenic disturbance of playas: human impacts on the geomorphic system and geomorphic impacts on the human system. *Geomorphology*, 17, 207–28.

Gillies, J. A. (2013). Fundamentals of aeolian sediment transport: dust emissions and transport near surface. In *Treatise on Geomorphology*, ed. J. Shroder (Editor in Chief), N. Lancaster, D. J. Sherman, and A. C. W. Baas. Academic Press, San Diego, CA, 11, Aeolian Geomorphology, pp. 43–63.

Gillies, J. A., Etyemezian, V., Kuhns, H., Nikolic, D., and Gillette, D. A. (2005). Effect of vehicle characteristics on unpaved road dust emissions. *Atmospheric Environment*, 39, 2341–7.

Gillies, J. A., Kuhns, H., Engelbrecht, J. P., Uppapalli, S., Etyemezian, V., and Nikolich, G. (2007). Particulate emissions from US Department of Defense artillery backblast testing. *Journal of the Air & Waste Management Association*, 57, 551–60.

Gilvear, D. J., Waters, T. M., and Milner, A. M. (1995). Image analysis of aerial photography to quantify changes in channel morphology and instream habitat following placer mining in interior Alaska. *Freshwater Biology*, 34, 389–98.

Ginoux, P., Prospero, J. M., Gill, T. E., Hsu, N. C., and Zhao, M. (2012). Global-scale attribution of anthropogenic and natural dust sources and their emission rates based on MODIS deep blue aerosol products. *Reviews of Geophysics*, 50, RG3005, doi: 10.1029/2012RG000388.

Giosan, L., Syvitski, J., Constantinescu, S., and Day, J. (2014). Climate change: protect the world's deltas. *Nature*, 516, 31–3.

Giraldéz, J. V., Ayuso, J. L., Garcia, A., Lopez, J. G., and Roldán, J. (1988). Water harvesting strategies in the semiarid climate of southeastern Spain. *Agricultural Water Management*, 14, 253–63.

Gislason, S. R., Oelkers, E. H., Eiriksdottir, E. S., Kardjilov, M. I., Gisladottir, G., Sigfusson, B., and Oskarsson, N. (2009). Direct evidence of the feedback between climate and weathering. *Earth and Planetary Science Letters*, 277, 213–22.

Glacken, C. (1967). *Traces on the Rhodian Shore: Nature and Culture in Western Thought from Ancient Times to the End of the Eighteenth Century*. Berkeley: University of California Press.

Glade, T. (2003). Landslide occurrence as a response to land use change: a review of evidence from New Zealand. *Catena*, 51, 297–314.

Glikson, A. (2013). Fire and human evolution: the deep-time blueprints of the Anthropocene. *Anthropocene*, 3, 89–92.

Glowacka, E., Gonzalez, J., and Nava, F. A. (2000). Subsidence in Cerro Prieto geothermal field, Baja California, Mexico. In *Proceedings World Geothermal Congress*, Kyushu, Japan: pp. 591–6.

Goddéris, Y. and Brantley, S. L. (2013). Earthcasting the future Critical Zone. *Elementa: Science of the Anthropocene*, 1(1), 000019.

Godoy, J. M., Padovani, C. R., Guimarães, J. R., Pereira, J. C., Vieira, L. M., Carvalho, Z. L., and Galdino, S. (2002). Evaluation of the siltation of River Taquari, Pantanal, Brazil, through 210Pb geochronology of floodplain lake sediments. *Journal of the Brazilian Chemical Society*, 13, 71–7.

Godoy, M. D. P. and de Lacerda, L. D. (2013). River-island morphological response to basin land-use change within the Jaguaribe River Estuary, NE Brazil. *Journal of Coastal Research*, 30, 399–410.

Goebel, T., Waters, M. R. and O'Rourke, D. H. (2008). The late Pleistocene dispersal of modern humans in the Americas. *Science*, 319, 1497–502.

Goelles, T., Bøggild, C. E., and Greve, R. (2015). Ice sheet mass loss caused by dust and black carbon accumulation. *The Cryosphere Discussions*, 9, 2563–96.

Goldewijk, K. K. (2001). Estimating global land use change over the past 300 years: the HYDE database, *Global Biogeochemical Cycles*, 15, 417–33.

Golomb, B. and Eder, H. M. (1964). Landforms made by man. *Landscape*, 14, 4–7.

Golosov, V. N., Sosin, P. M., Belyaev, V. R., Wolfgramm, B., and Khodzhaev, S. I. (2015). Effect of irrigation-induced erosion on the degradation of soils in river valleys of the alpine Pamir. *Eurasian Soil Science*, 48, 325–35.

Gómez, J. A., Giráldez, J. V., and Fereres, E. (2005). Water erosion in olive orchards in Andalusia (Southern Spain): a review. *Geophysical Research Abstracts*, 7, 08406.

Gómez, J. A., Battany, M., Renschler, C. S., and Fereres, E. (2003). Evaluating the impact of soil management on soil loss in olive orchards. *Soil Use and Management*, 19, 127–34.

Gómez, J. A., Romero, P., Giráldez, J. V., and Fereres, E. (2004). Experimental assessment of runoff and soil erosion in an olive grove on a Vertic soil in southern Spain as affected by soil management. *Soil Use and Management*, 20, 426–31.

Gómez, J. A., Guzmán, M. G., Giráldez, J. V., and Fereres, E. (2009). The influence of cover crops and tillage on water and sediment yield, and on nutrient, and organic matter losses in an olive orchard on a sandy loam soil. *Soil and Tillage Research*, 106, 137–44.

Gómez, J. A., Llewellyn, C., Basch, G., Sutton, P. B., Dyson, J. S., and Jones, C. A. (2011). The effects of cover crops and conventional tillage on soil and runoff loss in vineyards and olive groves in several Mediterranean countries. *Soil Use and Management*, 27, 502–14.

Gómez-Pina, G., Munoz-Pérez. J. J., Ramírez, J. L., and Ley, C. (2002). Sand dune management problems and techniques, *Spain. Journal of Coastal Research* S1 36, 325–32.

Gonzalez, M. A. (2001). Recent formation of arroyos in the Little Missouri Badlands of southwestern Dakota. *Geomorphology*, 38, 63–84.

González-Amuchastegui, M. J. and Serrano, E. (2015). Tufa buildups, landscape evolution and human impact during the Holocene in the Upper Ebro Basin. *Quaternary International*, 364, 54–64.

Goode, J. R., Luce, C. H., and Buffington, J. M. (2012). Enhanced sediment delivery in a changing climate in semi-arid mountain basins: implications for water resource management and aquatic habitat in the northern Rocky Mountains. *Geomorphology*, 139–140, 1–15.

Goossens, D., Buck, B., and McLaurin, B. (2012). Contributions to atmospheric dust production of natural and anthropogenic emissions in a recreational area designated for off-road vehicular activity (Nellis Dunes, Nevada, USA). *Journal of Arid Environments*, 78, 80–99.

Gornitz, V., Couch, S., and Hartig, E. K. (2002). Impacts of sea level rise in the New York City metropolitan area. *Global and Planetary Change*, 32, 61–88.

Gornitz, V., Rosenzweig, C., and Hillel, D. (1997). Effects of anthropogenic intervention in the land hydrologic cycle on global sea level rise. *Global and Planetary Change*, 14, 147–61.

Gottschalk, L. C. (1945). Effects of soil erosion on navigation in Upper Chesapeake Bay. *Geographical Review*, 35, 219–38.

Goudie, A. S. (1973). *Duricrusts of Tropical and Subtropical Landscapes*. Oxford: Clarendon Press.

(1977). Sodium sulphate weathering and the disintegration of Mohenjo-Daro, Pakistan. *Earth Surface Processes*, 2, 75–86.

(1978). Dust storms and their geomorphological implications. *Journal of Arid Environments*, 1, 291–310.

(1983). Dust storms in space and time. *Progress in Physical Geography*, 7, 502–30.

(1996). Geomorphological 'hotspots' and global warming. *Interdisciplinary Science Reviews*, 21, 253–9.

(2013a). *Arid and Semi-arid Geomorphology*. Cambridge: Cambridge University Press.

(2013b). *The Human Impact on the Natural Environment* (7th edn). Oxford: Wiley-Blackwell.

(2014). Desert dust and human health disorders. *Environment International*, 63, 101–13.

Goudie, A. and Seely, M. (2011). World Heritage desert landscapes: potential priorities for the recognition of desert landscapes and geomorphological sites on the World Heritage List. *IUCN World Heritage Studies*, 9.

Goudie, A. S. and Middleton, N. J. (1992). The changing frequency of dust storms through time. *Climate Change*, 20, 197–225.

(2006) *Desert Dust in the Global System*. Berlin and Heidelberg: Springer.

Goudie, A. S. and Viles, H. A. (1997). *Salt Weathering Hazards*. Chichester: Wiley.

(2012). Weathering and the global carbon cycle: geomorphological perspectives. *Earth-Science Reviews*, 113, 59–71.

(2015). *Landscapes and Landforms of Namibia.* Springer: Dordrecht.

Goudie, A. S., Viles, H. A., and Pentecost, A. (1993). The late-Holocene tufa decline in Europe. *The Holocene*, 3, 181–6.

Gourou, P. (1961). *The Tropical World* (3rd edn). London: Longman.

Goutal, N., Keller, T., Défossez, P., and Ranger, J. (2013). Soil compaction due to heavy forest traffic: measurements and simulations using an analytical soil compaction model. *Annals of Forest Science*, 70, 545–56.

Gowlett, J. A. J., Harris, J. W. K., Walton, D., and Wood, B. A. (1981). Early archaeological sites, hominid remains and traces of fire from Cheso-wanja, Kenya. *Nature*, 284, 125–9.

Graber, E. R., Fine, P., and Levy, G. J. (2006). Soil stabilization in semiarid and arid land agriculture. *Journal of Materials in Civil Engineering*, 18, 190–205.

Gradziński, M., Hercman, H., Jaśkiewicz, M., and Szczurek, S. (2013). Holocene tufa in the Slovak Karst: facies, sedimentary environments and depositional history. *Geological Quarterly*, 57, 769–88.

Graf, J. B., Webb, R. H., and Hereford, R. (1991). Relation of sediment load and flood-plain formation to climatic variability, Paria River drainage basin, Utah and Arizona. *Bulletin of the Geological Society of America*, 103, 1405–15.

Graf, W. L. (1977). Network characteristics in suburbanizing streams. *Water Resources Research*, 13, 459–63.

(1979). Mining and channel response. *Annals of the Association of American Geographers*, 69, 262–75.

Grainger, S. and Conway, D. (2014). Climate change and international river boundaries: fixed points in shifting sands. *Wiley Interdisciplinary Reviews: Climate Change*, 5, 835–48.

Grant, G. E. and Lewis, S. L. (2015). The remains of the dam: What have we learned from 15 years of US dam removals? *Engineering Geology for Society and Territory*, 3, 31–5.

Grantham, T. E., Figueroa, R., and Prat, N. (2013). Water management in mediterranean river basins: a comparison of management frameworks, physical impacts, and ecological responses. *Hydrobiologia*, 719, 451–82.

Grattan, J. P., Gilbertson, D. D., and Hunt, C. O. (2007). The local and global dimensions of metalliferous pollution derived from a reconstruction of an eight thousand year record of copper smelting and mining at a desert-mountain frontier in southern Jordan. *Journal of Archaeological Science*, 34, 83–110.

Gray, M. (2013). *Geodiversity: Valuing and Conserving Abiotic Nature* (2nd edn). Chichester: John Wiley & Sons.

Grayson, R., Holden, J., and Rose, R. (2010). Long-term change in storm hydrographs in response to peatland vegetation change. *Journal of Hydrology*, 389, 336–43.

Greenfield, H. J. (2010). The secondary products revolution: the past, the present and the future. *World Archaeology*, 42, 29–54.

Greenwood, P. and Kuhn, N. J. (2014). Does the invasive plant, *Impatiens glandulifera*, promote soil erosion along the riparian zone? An investigation on a small watercourse in northwest Switzerland. *Journal of Soils and Sediments*, 14, 637–50.

Gregory, K. J. (1997). *Fluvial Geomorphology of Great Britain.* London: Chapman and Hall.

Gregory, J. M., White, N. J., Church, J. A., Bierkens, M. F. P., Box, J. E., Van den Broeke, M. R., and Van de Wal, R. S. W. (2013). Twentieth-Century global-mean sea level rise: is the whole greater than the sum of the parts? *Journal of Climate*, 26, 4476–99.

Griffith, M. B., Norton, S. B., Alexander, L. C., Pollard, A. L., and LeDuc, S. D. (2012). The effects of mountaintop mines and valley fills on the physicochemical quality of stream ecosystems in the central Appalachians: a review. *Science of the Total Environment*, 417, 1–12.

Grimm, N. B., Chacon, A., Dahm, C. N., Hostetler, S. W., Lind, O. T., Starkweather, P. L., and Wertsbaugh, W. W. (1997). Sensitivity of aquatic ecosystems to climatic and anthropogenic changes: the Basin and Range, American south-west and Mexico. In *Freshwater Ecosystems and Climate Change in North America: a Regional Assessment*, ed. C. E. Cushing. Chichester: Wiley, pp. 205–23.

Grinsted, A., Moore, J. C., and Jevrejeva, S. (2013). Projected Atlantic hurricane surge threat from rising temperatures. *Proceedings of the National Academy of Sciences*, 110, 5369–73.

Grossi, C. M., Brimblecombe, P., and Harris, I. (2007). Predicting long term freeze–thaw risks on Europe built heritage and archaeological sites in a changing climate. *Science of the Total Environment*, 377, 273–81.

Grossi, C. M., Bonazza, A., Brimblecombe, P., Harris, I., and Sabbioni, C. (2008). Predicting twenty-first century recession of architectural limestone in European cities. *Environmental Geology*, 56, 455–61.

Grossi, C. M., Brimblecombe, P., Menéndez, B., Benavente, D., Harris, I., and Déqué, M. (2011). Climatology of salt transitions and implications for stone weathering. *Science of the Total Environment*, 409, 2577–85.

Grove, A. T. and Rackham, O. (2001). *The Nature of Mediterranean Europe: an Ecological History.* New Haven and London: Yale University Press.

Grove, A. T. and Sutton, J. E. G. (1989). Agricultural terracing south of the Sahara. *Azania: Journal of the British Institute in Eastern Africa*, 24, 113–22.

Grove, R. H. (1996). *Green Imperialism: Colonial Expansion, Tropical Island Edens and the Origins of Environmentalism, 1600–1860*. Cambridge: Cambridge University Press.

Grove, R. H. and Damodaran, V. (2006). Imperialism, intellectual networks and environmental change. *Economic and Social Weekly*, October 14th, 4345–54.

Grünthal, G. (2014). Induced seismicity related to geothermal projects versus natural tectonic earthquakes and other types of induced seismic events in central Europe. *Geothermics*. 52, 22–35.

Guha, S. K. (ed.) (2000). *Induced Earthquakes*. Dordrecht: Kluwer.

Gumilar, I., Abidin, H. Z., Hutasoit, L. M., Hakim, D. M., Sidiq, T. P., and Andreas, H. (2015). Land subsidence in Bandung Basin and its possible caused factors. *Procedia Earth and Planetary Science*, 12, 47–62.

Gunnell, Y. and Krishnamurthy, A. (2003). Past and present status of runoff harvesting systems in dryland peninsular India: a critical review. *Ambio*, 32, 320–4.

Guns, M. and Vanacker, V. (2014). Shifts in landslide frequency–area distribution after forest conversion in the tropical Andes. *Anthropocene*, 6, 75–85.

Guo, M., Wu, W., Zhou, X., Chen, Y., and Li, J. (2015). Investigation of the dramatic changes in lake level of the Bosten Lake in northwestern China. *Theoretical and Applied Climatology*, 119, 341–51.

Gupta, H. K. (2002). A review of recent studies of triggered earthquakes by artificial water reservoirs with special emphasis on earthquakes in Koyna, India. *Earth-Science Reviews*, 58, 279–310.

Gupta, H., Kao, S. J., and Dai, M. (2012). The role of mega dams in reducing sediment fluxes: a case study of large Asian rivers. *Journal of Hydrology*, 464, 447–58.

Gutiérrez, F., Parise, M., De Waele, J., and Jourde, H. (2014). A review on natural and human-induced geohazards and impacts in karst. *Earth-Science Reviews*, 138, 61–88.

Habersack, H. and Piégay, H. (2007). River restoration in the Alps and their surroundings: past experience and future challenges. *Developments in Earth Surface Processes*, 11, 703–35.

Habersack, H., Haspel, D., and Kondolf, M. (2014). Large rivers in the Anthropocene: insights and tools for understanding climatic, land use, and reservoir influences. *Water Resources Research*, 50, 3641–6.

Haddock, K. (1998). *Giant Earthmovers: an Illustrated History*. Osceola, WI: MotorBooks International.

Haeberli, W. and Burn, C. R. (2002). Natural hazards in forests: glacier and permafrost effects as related to climate change. In *Environmental Change and Geomorphic Effects in Forests*, ed. R. C. Sidle. Wallingford: CABI, pp. 167–202.

Haeberli, W., Clague, J. J., Huggel, C., and Kääb, A. (2008). Hazards from lakes in high-mountain glacier and permafrost regions: climate change effects and process interactions. *Avances de la Geomorphología en España, 2010*, 439–46.

Haff, P. (2014). Humans and technology in the Anthropocene: six rules. *The Anthropocene Review*, 1, 126–36.

Haff, P. K. (2010). Hillslopes, rivers, plows and trucks: mass transport on Earth's surface by natural and technological processes. *Earth Surface Processes and Landforms*, 35, 1157–66.

Haig, J., Nott, J., and Reichart, G. J. (2014). Australian tropical cyclone activity lower than at any time over the past 550–1,500 years. *Nature*, 505, 667–71.

Hall, B. A. (1994). Formation processes of large earthen residential mounds in La Mixtequilla, Veracruz, Mexico. *Latin American Antiquity*, 5, 31–50.

Hall, C., Hamilton, A., Hoff, W. D., Viles, H. A., and Eklund, J. A. (2011). Moisture dynamics in walls: response to micro-environment and climate change. *Proceedings of the Royal Society A: Mathematical, Physical and Engineering Science*, 467, 194–211.

Hall, R. (2015). Population and the future. *Geography*, 100, 36–44.

Hallegatte, S., Green, C., Nicholls, R. J., and Corfee-Morlot, J. (2013). Future flood losses in major coastal cities. *Nature Climate Change*, 3, 802–6.

Hamilton, C. (2015). Getting the Anthropocene so wrong. *The Anthropocene Review*, 2, 102–7.

Hamilton, C., Bonneuil, C., and Gemenne, F. (eds.) (2015). *The Anthropocene and the Global Environmental Crisis*. London and New York: Routledge.

Hamza, M. A. and Anderson, W. K. (2005). Soil compaction in cropping systems: a review of the nature, causes and possible solutions. *Soil and Tillage Research*, 82, 121–45.

Han, Z., Wang, T., Dong, Z., Hu, Y., and Yao, Z. (2007). Chemical stabilization of mobile dunefields along a highway in the Taklimakan Desert of China. *Journal of Arid Environments*, 68, 260–70.

Hanley, M. E., Hoggart, S. P. G., Simmonds, D. J., Bichot, A., Colangelo, M. A., Bozzeda, F., and Thompson, R. C. (2014). Shifting sands? Coastal protection by sand banks, beaches and dunes. *Coastal Engineering*, 87, 136–46.

Hanson, P. R., Joeckle, R. M., Young, A. R., and Horn, J. (2009). Late Holocene dune activity in the eastern Platte River Valley, Nebraska. *Geomorphology*, 103, 555–61.

Hapke, C. J., Kratzmann, M. G., and Himmelstoss, E. A. (2013). Geomorphic and human influence on large-scale coastal change. *Geomorphology*, 199, 160–70.

Happ, S. C., Rittenhouse, G., and Dobson, G. C. (1940). *Some Principles of Accelerated Stream and Valley Sedimentation* (No. 695). US Department of Agriculture.

Harbor, J. (1999). Engineering geomorphology at the cutting edge of land disturbance: erosion and sediment control on construction sites. *Geomorphology*, 31, 247–63.

Harden, C. P. (2006). Human impacts on headwater fluvial systems in the northern and central Andes. *Geomorphology*, 79, 249–63.

Harley, G. L., Polk, J. S., North, L. A., and Reeder, P. P. (2011). Application of a cave inventory system to stimulate development of management strategies: the case of west-central Florida, USA. *Journal of Environmental Management*, 92, 2547–57.

Harmand, S., Lewis, J. E., Feibel, C. S., Lepre, C. J., Prat, S., Lenoble, A., and Roche, H. (2015). 3.3-million-year-old stone tools from Lomekwi 3, West Turkana, Kenya. *Nature*, 521, 310–5.

Harnischmacher, S. (2010). Quantification of mining subsidence in the Ruhr District (Germany). *Géomorphologie*, 3, 261–74.

Harnischmacher, S. and Zepp, H. (2014). Mining and its impact on the earth surface in the Ruhr District (Germany). *Zeitschrift für Geomorphologie*, 58, Supplement, 3–22.

Harris, N. and Evans, J. E. (2014). Channel evolution of sandy reservoir sediments following low-head dam removal, Ottawa River, Northwestern Ohio, USA. *Open Journal of Modern Hydrology, 2014*, doi:10.4236/ojmh.2014.42004.

Harrison, S., Glasser, N., Winchester, V., Haresign, E., Warren, C., and Jansson, K. (2006). A glacial lake outburst flood associated with recent mountain glacier retreat, Patagonian Andes. *The Holocene*, 16, 611–20.

Hartemink, A. (2006). Soil erosion: perennial crop plantations. In *Encyclopedia of Soil Science*, ed. R. Lal. Taylor and Francis: New York, pp. 1613–7.

Hartmann, H. C. (1990). Climate change impacts on Laurentian Great Lakes levels. *Climatic Change*, 17, 49–67.

Harvey, A. M. and Renwick, W. H. (1987). Holocene alluvial fan and terrace formation in the Bowland Fells, Northwest England. *Earth Surface Processes and Landforms*, 12, 249–57.

Harvey, A. M., Oldfield, F., Baron, A. F., and Pearson, G. W., (1981). Dating of post-glacial landforms in the central Howgills. *Earth Surface Processes and Landforms*, 6, 401–12.

Harvey, G. L., Moorhouse, T. P., Clifford, N. J., Henshaw, A. J., Johnson, M. F., Macdonald, D. W., and Rice, S. (2011). Evaluating the role of invasive aquatic species as drivers of fine sediment-related river management problems: the case of the signal crayfish (*Pacifastacus leniusculus*). *Progress in Physical Geography*, 35, 517–33.

Harvey, J. E. and Pederson, J. L. (2011). Reconciling arroyo cycle and paleoflood approaches to late Holocene alluvial records in dryland streams. *Quaternary Science Reviews*, 30, 855–66.

Harvey, J. E., Pederson, J. L., and Rittenour, T. M. (2011). Exploring relations between arroyo cycles and canyon paleoflood records in Buckskin Wash, Utah: reconciling scientific paradigms. *Geological Society of America Bulletin*, 123, 2266–76.

Hawley, R. J., MacMannis, K. R., and Wooten, M. S. (2013). Bed coarsening, riffle shortening, and channel enlargement in urbanizing watersheds, northern Kentucky, USA. *Geomorphology*, 201, 111–26.

Haycraft, W. R. (2002). *Yellow Steel: the Story of the Earth Moving Equipment Industry*. Urbana and Chicago: University of Illinois Press.

Hayhoe, K., VanDorn, J., Croley II, T., Schlegal, N., and Wuebbles, D. (2010). Regional climate change projections for Chicago and the US Great Lakes. *Journal of Great Lakes Research*, 36, 7–21.

Hayhoe, S. J., Neill, C., Porder, S., McHorney, R., LeFebvre, P., Coe, M. T., and Krusche, A. V. (2011). Conversion to soy on the Amazonian agricultural frontier increases streamflow without affecting stormflow dynamics. *Global Change Biology*, 17, 1821–33.

Haynes, G. (2012). Elephants (and extinct relatives) as earth-movers and ecosystem engineers. *Geomorphology*, 157, 99–107.

He, F., Vavrus, S. J., Kutzbach, J. E., Ruddiman, W. F., Kaplan, J. O., and Krumhardt, K. M. (2014). Simulating global and local surface temperature changes due to Holocene anthropogenic land cover change. *Geophysical Research Letters*, 41, 623–31.

He, K., Bin, W., and Dunyun, Z. (2004). Mechanism and mechanical model of karst collapse in an over-pumping area. *Environmental Geology*, 46, 1102–7.

He, K., Liu, C. and Wang, S. (2003). Karst collapse related to over-pumping and a criterion for its stability. *Environmental Geology*, 43, 720–4.

Healy, T. (1996). Sea level rise and impacts on near-shore sedimentation. *Geologische Rundschau*, 85, 546–53.

Heberger, M., Cooley, H., Herrera, P., Gleick, P. H. and Moore, E. (2011).Potential impacts of increasing coastal flooding in California due to sea-level rise. *Climatic Change*, 109 (supplement), S229–49.

Heckmann, M. (2014). Farmers, smelters and caravans: Two thousand years of land use and soil erosion in North Pare, NE Tanzania. *Catena*, 113, 187–201.

Heine, K., Niller H. P., Nuber, T., and Scheibe, R. (2005). Slope and valley sediments as evidence of deforestation and land-use in prehsitporic and historic Eastern Bavaria. *Zeitschrift für Geomorphologie Supplementband*, 139, 147–71.

Heitkamp, F., Sylvester, S. P., Kessler, M., Sylvester, M. D., and Jungkunst, H. F. (2014). Inaccessible Andean sites reveal human-induced weathering in grazed soils. *Progress in Physical Geography*, 38, 576–601.

Hernández, A. J., Lacasta, C., and Pastor, J. (2005). Effects of different management practices on soil conservation and soil water in a rainfed olive orchard. *Agricultural Water Management*, 77, 232–48.

Hernández-Calvento, L., Jackson, D. W. T., Medina, R., Hernández-Cordero, A. I., Cruz, N., and Requejo, S. (2014). Downwind effects on an arid dunefield from an evolving urbanised area. *Aeolian Research*, 15, 301–9.

Hesp, P. A. (2001). The Manawatu Dunefield: environmental change and human impacts. *New Zealand Geographer*, 57, 33–40.

Hesp P. A., Schmutz, P., Martinez, M. L., Driskell, L., Orgera, R., Renken, K., Revelo, N. A. R., and Orocio, O. A. J. (2010). The effect on coastal vegetation of trampling on a parabolic dune. *Aeolian Research*, 2, 105–11.

Hewitt, K. (2005).The Karakoram Anomaly? Glacier expansion and the "elevation effect," Karakoram Himalaya. *Mountain Research and Development*, 25, 332–40.

Hewitt, K. and Liu, J. (2010). Ice-dammed lakes and outburst floods, Karakoram Himalaya: historical perspectives on emerging threats. *Physical Geography*, 31, 528–51.

Higgitt, D. L. and Lee, E. M. (eds.) (2001). *Geomorphological Processes and Landscape Change: Britain in the Last 1000 Years*, Chichester: Wiley, pp. 269–87.

Hilfinger IV, M. F., Mullins, H. T., Burnett, A., and Kirby, M. E. (2001). A 2500 year sediment record from Fayetteville Green Lake, New York: evidence for anthropogenic impacts and historic isotope shift. *Journal of Paleolimnology*, 26, 293–305.

Hilmes, M. M. and Wohl, E. E. (1995). Changes in channel morphology associated with placer mining. *Physical Geography*, 16, 223–42.

Hilton, M. J. (2006). The loss of New Zealand's active dunes and the spread of marram grass (*Ammophila arenaria*). *New Zealand Geographer*, 62, 105–20.

Hinkel, J., Nicholls, R. J., Tol, R. S., Wang, Z. B., Hamilton, J. M., Boot, G., and Klein, R. J. (2013). A global analysis of erosion of sandy beaches and sea-level rise: an application of DIVA. *Global and Planetary Change*, 111, 150–8.

Hirabayashi, Y., Mahendran, R., Koirala, S., Konoshima, L., Yamazaki, D., Watanabe, S., and Kanae, S. (2013). Global flood risk under climate change. *Nature Climate Change*, 3, 816–21.

Hiscock, K. M., Lister, D. H., Boar, R. R., and Green, F. M. L. (2001). An integrated assessment of long-term changes in the hydrology of three lowland rivers in eastern England. *Journal of Environmental Management*, 61, 195–214.

Hodgen, M. T. (1939). Domesday water mills. *Antiquity*, 13, 261–79.

Hoegh-Guldberg, O. (1999). Climate change, coral bleaching and the future of the world's coral reefs. *Marine and Freshwater Research*, 50, 839–66.

(2001). Sizing the impact: coral reef ecosystems as early casualties of climate change. In *Fingerprints of Climate Change*, ed. G. R. Walther, C. A. Burga, and P. J. Edwards. New York: Kluwer/Plenum, pp 203–28.

(2011). Coral reef ecosystems and anthropogenic climate change. *Regional Environmental Change*, 11 (suppl.1), S215–27.

(2014). Coral reefs in the Anthropocene: persistence or the end of the line? *Geological Society, London, Special Publications*, 395, 167–83.

Hoeksema, R. J. (2007). Three stages in the history of land reclamation in the Netherlands. *Irrigation and Drainage*, 56(S1), S113–26.

Hoffmann, T., Erkens, G., Gerlach, R., Klostermann, J., and Lang, A. (2009). Trends and controls of Holocene floodplain sedimentation in the Rhine catchment. *Catena*, 77, 96–106.

Hoffmann, T., Thorndycraft, V. R., Brown, A. G., Coulthard, T. J., Damnati, B., Kale, V. S., and Walling, D. E. (2010). Human impact on fluvial regimes and sediment flux during the Holocene: review and future research agenda. *Global and Planetary Change*, 72, 87–98.

Hoffmann, T., Mudd, S. M., Van Oost, K., Verstraeten, G., Erkens, G., Lang, A., and Aalto, R. (2013). Short communication: humans and the missing C-sink: erosion and burial of soil carbon through time. *Earth Surface Dynamics*, 1, 45–52.

Hollis, G. E. (1975). The effects of urbanization on floods of different recurrence interval. *Water Resources Research*, 11, 431–5.

Holm, K., Bovis, M., and Jacob, M. (2004). The landslide response of alpine basins to post-Little Ice Age glacial thinning and retreat in southwestern British Columbia. *Geomorphology*, 57, 201–16.

Holmes, S. C. A. (1980). *Geology of the Country around Faversham*. London: IGS/HMSO.

Hoogland, T., Van Den Akker, J. J. H., and Brus, D. J. (2012). Modeling the subsidence of peat soils in the Dutch coastal area. *Geoderma*, 171, 92–7.

Hooijer, A., Page, S., Jauhiainen, J., Lee, W. A., Lu, X. X., Idris, A., and Anshari, G. (2012). Subsidence and carbon loss in drained tropical peatlands. *Biogeosciences*, 9, 1053–71.

Hooke, J. M. and Mant, J. (2002). Floodwater use and management strategies in valleys of southeast Spain. *Land Degradation & Development*, 13, 165–75.

Hooke, R. L. (1994). On the efficacy of humans as geomorphic agents. *GSA Today*, 4, 217, 224–5.

(1999). Spatial distribution of human geomorphic activity in the United States: comparison with rivers. *Earth Surface Processes and Landforms*, 24, 687–92.

(2000). On the history of humans as geomorphic agents. *Geology*, 28, 843–6.

Hooke, R. L., Martín-Duque, J. F., and Pedraza, J. (2012). Land transformation by humans: a review. *GSA Today*, 22(12), 4–10.

Hoomehr, S., Schwartz, J. S., and Yoder, D. C. (2015). Potential changes in rainfall erosivity under GCM climate change scenarios for the southern Appalachian region, USA. *Catena*, doi:10.1016/j.catena.2015.01.012.

Hope, G. (1999). Vegetation and fire response to late Holocene human occupation in island and mainland north west Tasmania. *Quaternary International*, 59, 47–60.

Hopkins, K. G., Morse, N. B., Bain, D. J., Bettez, N. D., Grimm, N. B., Morse, J. L., and Palta, M. M. (2015). Type and timing of stream flow changes in urbanizing watersheds in the Eastern US. *Elementa: Science of the Anthropocene*, 3(1), 000056.

Hopkins, R. L., Altier, B. M., Haselman, D., Merry, A. D., and White, J. J. (2013). Exploring the legacy effects of surface coal mining on stream chemistry. *Hydrobiologia*, 713, 87–95.

Horn, J. D., Joeckel, R. M., and Fielding, C. R. (2012). Progressive abandonment and planform changes of the central Platte River in Nebraska, central USA, over historical timeframes. *Geomorphology*, 139, 372–83.

Hornbach, M. J., DeShon, H. R., Ellsworth, W. L., Stump, B. W., Hayward, C., Frohlich, C., and Luetgert, J. H. (2015). Causal factors for seismicity near Azle, *Texas. Nature Communications*, 6, doi:10.1038/ncomms7728.

Horrocks M., Nichol, S. L., D-Costa D., Augustinius, P., Jacobi, T., Shane, P. A., and Middleton, A. (2007). A late quaternary record of natural change and human impact from Rangihoua Bay, Bay of Islands, northern New Zealand. *Journal of Coastal Research*, 23, 592–604.

Hoshino, S., Esteban, M., Mikami, T., Takagi, H., and Shibayama, T. (2015). Estimation of increase in storm surge damage due to climate change and sea level rise in the Greater Tokyo area. *Natural Hazards*, doi: 10.1007/s11069–015–1983–4.

Houston, S. L., Houston, W. N., Zapata, C. E., and Lawrence, C. (2001). Geotechnical engineering practice for collapsible soils. *Geotechnical and Geological Engineering*, 19, 333–55.

Howard, J. L. (2014). Proposal to add anthrostratigraphic and technostratigraphic units to the stratigraphic code for classification of anthropogenic Holocene deposits. *The Holocene*, 24, 1856–61.

Hoyos, C.D., Agudelo, P. A., Webster, P. J., and Curry, J.A. (2006). Deconvolution of the factors contributing to the increase in global hurricane intensity. *Science*, 312, 94–7.

Hu, A. and Deser, C. (2013). Uncertainty in future regional sea level rise due to internal climate variability. *Geophysical Research Letters*, 40, 2768–72.

Hu, D., Clift, P. D., Böning, P., Hannigan, R., Hillier, S., Blusztajn, J., and Fuller, D. Q. (2013). Holocene evolution in weathering and erosion patterns in the Pearl River delta. *Geochemistry, Geophysics, Geosystems*, 14, 2349–68.

Hu, R. (2006). Urban land subsidence in China. *IAEG2006 Paper* 786, Geological Society of London.

Hu, R. L., Yue, Z. Q., Wang, L. U., and Wang, S. J. (2004). Review on current status and challenging issues of land subsidence in China. *Engineering Geology*, 76, 65–77.

Huang, L., Liu, J., Shao, Q., and Liu, R. (2011). Changing inland lakes responding to climate warming in Northeastern Tibetan Plateau. *Climatic Change*, 109, 479–502.

Huang, R. and Chan, L. (2004). Human-induced landslides in China: mechanism study and its implications on slope management. *Chinese Journal of Rock Mechanics and Engineering*, 23, 2766–77.

Hübner, R., Herbert, R. J. H. and Astin, K. B. (2010). Cadmium release caused by the die-back of the saltmarsh cord grass Spartina anglica in Poole Harbour (UK). *Estuarine, Coastal and Shelf Science*, 87, 553–60.

Huckleberry, G. and Duff, A. I. (2008). Alluvial cycles, climate and Pueblan settlement shifts near Zuni Salt Lake, New Mexico, USA. *Geoarchaeology*, 23, 107–30.

Hudson, P. and Middelkoop, H. (eds.) (2015). *Geomorphic Approaches to Integrated Floodplain Management of Lowland Fluvial Systems in North America and Europe*. New York: Springer.

Hudson, P. F., Middelkoop, H., and Stouthamer, E. (2008). Flood management along the Lower Mississippi and Rhine Rivers (The Netherlands) and the continuum of geomorphic adjustment. *Geomorphology*, 101, 209–36.

Hudson-Edwards, K., Macklin, M., and Taylor, M. (1997). Historic metal mining inputs to Tees river sediment. *Science of the Total Environment*, 194, 437–45.

Huggel, C., Allen, S., Deline, P., Fischer, L., Noetzli, J., and Ravanel, L. (2012). Ice thawing, mountains falling—are alpine rock slope failures increasing? *Geology Today*, 28, 98–104.

Huggel, C., Clague, J. J., and Korup, O. (2012). Is climate change responsible for changing landslide activity in high mountains? *Earth Surface Processes and Landforms*, 37, 77–91.

Hughes, R. G. and Paramor, O. A. L. (2004). On the loss of saltmarshes in south-east England and methods for their restoration. *Journal of Applied Ecology*, 41, 440–8.

Hughes, R. J., Sullivan, M. E., and Yok, D. (1991). Human-induced erosion in a highlands catchment in Papua New Guinea: the prehistoric and contemporary records. *Zeitschrift für Geomorphologie, supplementband*, 83, 227–39.

Hughes, T. P., Bellwood, D. R., Baird, A. H., Brodie, J., Bruno, J. F., and Pandolfi, J. M. (2011). Shifting base-lines, declining coral cover, and the erosion of reef resilience: comment on Sweatman et al. (2011). *Coral Reefs*, 30, 653–60.

Hughes, T. P., Linares, C., Dakos, V., van de Leemput, I. A., and van Nes, E. H. (2013). Living dangerously on borrowed time during slow, unrecognized regime shifts. *Trends in Ecology & Evolution*, 28, 149–55.

Hultine, K. R. and Bush, S. E. (2011). Ecohydrological consequences of non-native riparian vegetation in the southwestern United States: a review from an ecophysiological perspective. *Water Resources Research*, 47, W07542, doi:10.1029/2010WR010317.

Humboldt, A. von and Bonpland, A. (1815). *Personal Narrative of Travels to the Equinoctial Regions of the New Continent, during the Years 1799–1804*. London: M. Carey.

Huntington, E. (1914). *The Climatic Factor as Illustrated in Arid America*. Carnegie Institution of Washington publication, 192.

Hupp, C. R. and Osterkamp, W. R. (1996). Riparian vegetation and fluvial geomorphic processes. *Geomorphology*, 14, 277–95.

Hupp, C. R., Noe, G. B., Schenk, E. R., and Benthem, A. J. (2013). Recent and historic sediment dynamics along Difficult Run, a suburban Virginia Piedmont stream. *Geomorphology*, 180, 156–69.

Hupy, J. P. and Koehler, T. (2012). Modern warfare as a significant form of zoogeomorphic disturbance upon the landscape. *Geomorphology*, 157, 169–82.

Hupy, J. P. and Schaetzl, R. J. (2006). Introducing "bombturbation," a singular type of soil disturbance and mixing. *Soil Science*, 171, 823–36.

Husen, S., Kissling, E., and von Deschwanden, A. (2012). Induced seismicity during the construction of the Gotthard Base Tunnel, Switzerland: hypocenter locations and source dimensions. *Journal of Seismology*, 16, 195–213.

Hussain, I., Abu-Rizaiza, O. S., Habib, M. A., and Ashfaq, M. (2008). Revitalizing a traditional dryland water supply system: the karezes in Afghanistan, Iran, Pakistan and the Kingdom of Saudi Arabia. *Water International*, 33, 333–49.

Hüttl, R. F. (1998). Ecology of post strip-mining landscapes in Lusatia, Germany. *Environmental Science & Policy*, 1, 129–35.

Huxman, T. E., Wilcox, B. P., Breshears, D. D., Scott, R. L., Snyder, K. A., Small, E. E., and Jackson, R. B. (2005). Ecohydrological implications of woody plant encroachment. *Ecology*, 86, 308–19.

Hyde, K. D., Wilcox, A. C., Jencso, K., and Woods, S. (2014). Effects of vegetation disturbance by fire on channel initiation thresholds. *Geomorphology*, 214, 84–96.

Ibáñez, C., Day, J. W., and Reyes, E. (2014). The response of deltas to sea-level rise: natural mechanisms and management options to adapt to high-end scenarios. *Ecological Engineering*, 65, 122–30.

Ilyés, Z. (2010) Military activities: warfare and defence. In *Anthropogenic Geomorphology: a Guide to Man-Made Landforms*, ed. J. Szabó, L. Dávid, and D. Loczy. Dordrecht: Springer, pp. 217–31.

Imaizumi, F., Sidle, R. C., and Kamei, R. (2008). Effects of forest harvesting on the occurrence of landslides and debris flows in steep terrain of central Japan. *Earth Surface Processes and Landforms*, 33, 827–40.

Immer Zeel, W. W., van Beek, L. P. H., and Bierkens, M. F. P. (2010). Climate change will affect the Asian water towers. *Science*, 328, 1382–5.

Improta, L., Valoroso, L., Piccinini, D., and Chiarabba, C. (2015). A detailed analysis of wastewater-induced seismicity in the Val d'Agri oil field (Italy). *Geophysical Research Letters*, 42, 2682–90.

Inbar, M. and Llerena, C. A. (2000). Erosion processes in high mountain agricultural terraces in Peru. *Mountain Research and Development*, 20, 72–9.

Inbar, M., Tamir, M. I., and Wittenberg, L. (1998). Runoff and erosion processes after a forest fire in Mount Carmel, a Mediterranean area. *Geomorphology*, 24, 17–33.

Indoitu, R., Kozhoridze, G., Batyrbaeva, M., Vitkovskaya, I., Orlovsky, N., Blumberg, D., and Orlovsky, L. (2015). Dust emission and environmental changes in the dried bottom of the Aral Sea. *Aeolian Research*, 17, 101–15.

Inkpen, R., Viles, H., Moses, C., and Baily, B. (2012a). Modelling the impact of changing atmospheric pollution levels on limestone erosion rates in central London, 1980–2010. *Atmospheric Environment*, 61, 476–81.

Inkpen, R. J., Viles, H. A., Moses, C., Baily, B., Collier, P., Trudgill, S. T., and Cooke, R. U. (2012b). Thirty years of erosion and declining atmospheric pollution at St Paul's Cathedral, London. *Atmospheric Environment*, 62, 521–9.

Innes, J. L. (1983). Lichenometric dating of debris-flow deposits in the Scottish Highlands. *Earth Surface Processes and Landforms*, 8, 579–88.

Innes, J., Blackford, J., and Simmons, I. (2010). Woodland disturbance and possible land-use regimes during the Late Mesolithic in the English uplands: pollen, charcoal and non-pollen palynomorph evidence from Bluewath Beck, North York Moors, UK. *Vegetation History and Archaeobotany*, 19, 439–52.

IPCC (Intergovernmental Panel on Climate Change) (2007). *Climate Change: The Physical Science Basis. Contribution of Working Group I to the Fourth Assessment Report of the*

Intergovernmental Panel on Climate Change, ed. S. Solomon et al., Cambridge and New York: Cambridge University Press.

(2013). *Climate Change 2013: the Physical Science Basis: Contribution of Working Group I to the Fifth Assessment Report of the Intergovernmental Panel on Climate Change*, ed. T. F. Stocker, D. Qin, G.-K. Plattner, M. Tignor, S. K. Allen, J. Boschung, A. Nauels, Y. Xia, V. Bex, and P. M. Midgley. Cambride and New York: Cambridge University Press.

Irabien, M. J., García-Artola, A., Cearreta, A., and Leorri, E. (2015). Chemostratigraphic and lithostratigraphic signatures of the Anthropocene in estuarine areas from the eastern Cantabrian coast (N. Spain). *Quaternary International*, 364, 196–205.

Ireland, A. W., Clifford, M. J., and Booth, R. K. (2014). Widespread dust deposition on North American peatlands coincident with European land-clearance. *Vegetation History and Archaeobotany*, 23, 693–700.

Irish, J. L. and 8 others (2010). Potential implications of global warming and barrier island degradation on future hurricane inundation, property damages, and population impacted. *Ocean and Coastal Management*, 53, 645–57.

Isebrands, J. G. and Richardson, J. (eds.) 2014. *Poplars and Willows: Trees for Society and the Environment*. Wallingford, U.K.: CABI.

Isermann, M., Diekmann, M., and Heemann, S. (2007). Effects of the expansion by Hippophaë rhamnoides on plant species richness in coastal dunes. *Applied Vegetation Science*, 10, 33–42.

Jabaloy-Sánchez, A., Lobo, F. J., Azor, A., Bárcenas, P., Fernández-Salas, L. M., del Río, V. D., and Pérez-Peña, J. V. (2010). Human-driven coastline changes in the Adra River deltaic system, southeast Spain. *Geomorphology*, 119, 9–22.

Jacks, G. V. and Whyte, R. O. (1939). *The Rape of the Earth: a World Survey of Soil Erosion*. London: Faber and Faber.

Jaeger, K. L. (2015). Reach-scale geomorphic differences between headwater streams draining mountaintop mined and unmined catchments. *Geomorphology*, 236, 25–33.

Jaeger, K. L. and Wohl, E. (2011). Channel response in a semiarid stream to removal of tamarisk and Russian olive. *Water Resources Research*, 47(2), W02536, doi: 10.1029/2009WR008741.

Jaffe, B. E., Smith, R. E., and Foxgrover, A. C. (2007). Anthropogenic influence on sedimentation and intertidal mudflat change in San Pablo Bay, California: 1856–1983. *Estuarine, Coastal and Shelf Science*, 73, 175–87.

Jalowska, A. M., Rodriguez, A. B., and McKee, B. A. (2015). Responses of the Roanoke Bayhead Delta to variations in sea level rise and sediment uupply during the Holocene and Anthropocene. *Anthropocene*, 9, 41–55.

James, L. A. (1989). Sustained storage and transport of hydraulic gold mining sediment in the Bear River, California. *Annals of the Association of American Geographers*, 79, 570–92.

(1991). Incision and morphologic evolution of an alluvial channel recovering from hydraulic mining sediment. *Geological Society of America Bulletin*, 103, 723–36.

(2004). Tailings fans and valley-spur cutoffs created by hydraulic mining. *Earth Surface Processes and Landforms*, 29, 869–82.

(2010). Secular sediment waves, channel bed waves, and legacy sediment. *Geography Compass*, 4, 576–98.

(2013). Legacy sediment: definitions and processes of episodically produced anthropogenic sediment. *Anthropocene*, 2, 16–26.

James, L. A. and Marcus, W. A. (2006). The human role in changing fluvial systems: retrospect, inventory and prospect. *Geomorphology*, 79, 152–71.

Jeffries, M. J. (2012). Ponds and the importance of their history: an audit of pond numbers, turnover and the relationship between the origins of ponds and their contemporary plant communities in south-east Northumberland, UK. *Hydrobiologia*, 689, 11–21.

Jennings, J. N. (1952). *The Origin of the Broads*. London: Royal Geographical Society.

Jensen, A. (2002). Artificial reefs of Europe: perspective and future. *ICES Journal of Marine Science: Journal du Conseil*, 59(suppl), S3-S13.

Jensen, K., Trepel, M., Merritt, D., and Rosenthal, G. (2006). Restoration ecology of river valleys. *Basic and Applied Ecology*, 7, 383–7.

Jeppesen, E., Brucet, S., Naselli-Flores, L., Papastergiadou, E., Stefanidis, K., Nõges, T., and Beklioğlu, M. (2015). Ecological impacts of global warming and water abstraction on lakes and reservoirs due to changes in water level and related changes in salinity. *Hydrobiologia*, 750, 201–27.

Jewell, M., Houser, C., and Trimble, S. (2014). Initiation and evolution of blowouts within Padre Island National Seashore, Texas. *Ocean & Coastal Management*, 95, 156–64.

Jiang, Y. Luo, Y., Zhao, Z., and Tao, S. (2009). Changes in wind speed over China during 1956–2004. *Theoretical and Applied Climatology*, doi: 10.1007/s00704-009-0152-7.

Jiang, Z., Lian, Y., and Qin, X. (2014). Rocky desertification in Southwest China: impacts, causes, and restoration. *Earth-Science Reviews*, 132, 1–12.

Jilbert, T., Reichard, G.-J., Aeschlimann, B., Günther, D., Boer, W., and de Lange, G. (2010). Climate-controlled multidecadal variability in North African dust transport to the Mediterranean. *Geology*, 38, 19–22.

Jin, H., Li, S., Cheng, G., Shaoling, W., and Li, X. (2000). Permafrost and climatic change in China. *Global and Planetary Change*, 26, 387–404.

Jin, R., Li, X., Che, T., Wu, L., and Mool, P. (2005). Glacier area changes in the Pumqu river basin, Tibetan Plateau, between the 1970s and 2001. *Journal of Glaciology*, 51, 607–10.

Johnson, A. I. (ed.) (1991). *Land Subsidence*. International Association of Hydrological Sciences Publication 200.

Johnson, K. S. (1997). Evaporite karst in the United States. *Carbonates and Evaporites*, 12, 2–14.

    (2005). Subsidence hazards due to evaporite dissolution in the United States. *Environmental Geology*, 48, 395–409.

Jomelli, V., Brunstein, D., Déqué, M., Vrac, M., and Grancher, D. (2009). Impacts of future climatic change (2070–2099) on the potential occurrence of debris flows: a case study in the Massif des Ecrins (French Alps). *Climatic Change*, 97, 171–91.

Jonah, F. E., Adjei-Boateng, D., Agbo, N. W., Mensah, E. A., and Edziyie, R. E. (2015). Assessment of sand and stone mining along the coastline of Cape Coast, Ghana. *Annals of GIS*, doi:10.1080/19475683.2015.1007894.

Jones, B. M., Arp, C. D., Jorgenson, M. T., Hinkel, K. M., Schmutz, J. A., and Flint, P. L. (2009). Increase in the rate and uniformity of coastline erosion in Arctic Alaska. *Geophysical Research Letters*, 36, doi: 10.1029/2008GL036205.

Jones, B. M., Grosse, G., Arp, C. D., Jones, M. C., Walter Anthony, K. M., and Romanovsky, V. E. (2011). Modern thermokarst lake dynamics in the continuous permafrost zone, northern Seward Peninsula, Alaska. *Journal of Geophysical Research: Biogeosciences* 116(G2), doi: 10.1029/2011JG001666.

Jones, B. M., Stoker, J. M., Gibbs, A. E., Grosse, G., Romanovsky, V. E., Douglas, T. A., and Richmond, B. M. (2013). Quantifying landscape change in an arctic coastal lowland using repeat airborne LiDAR. *Environmental Research Letters*, 8, 045025.

Jones, C. G., Lawton, J. H., and Shachak, M. (1997). Positive and negative effects of organisms as physical ecosystem engineers. *Ecology*, 78, 1946–57.

Jones, J. A., Swanson, F. J., Wemple, B. C., and Snyder, K. U. (2000). Effects of roads on hydrology, geomorphology, and disturbance patches in stream networks. *Conservation Biology*, 14, 76–85.

Jones, L. S., Rosenburg, M., Figueroa, M. D. M., McKee, K., Haravitch, B., and Hunter, J. (2010). Holocene valley-floor deposition and incision in a small drainage basin in western Colorado, USA. *Quaternary Research*, 74, 199–206.

Jones, M. D., Djamali, M., Holmes, J., Weeks, L., Leng, M. J., Lashkari, A., and Metcalfe, S. E. (2015). Human impact on the hydroenvironment of Lake Parishan, SW Iran, through the late Holocene. *The Holocene*, 25, 1651–61.

Jones, R., Benson-Evans, K., and Chambers, F. M. (1985). Human influence upon sedimentation in Llangorse Lake, Wales. *Earth Surface Processes and Landforms*, 10, 227–35.

Jordán, A. and Martínez-Zavala, L. (2008). Soil loss and runoff rates on unpaved forest roads in southern Spain after simulated rainfall. *Forest Ecology and Management*, 255, 913–9.

Jordan, H., Hamilton, K., Lawley, R., and Price, S. J. (2016). Anthropogenic contribution to the geological and geomorphological record: a case study from Great Yarmouth, Norfolk, UK. *Geomorphology*, 253, 534–46, doi:10.1016/j.geomorph.2014.07.008.

Jordán-López, A., Martínez-Zavala, L., and Bellinfante, N. (2009). Impact of different parts of unpaved forest roads on runoff and sediment yield in a Mediterranean area. *Science of the Total Environment*, 407, 937–44.

Jorgenson, M. T., Shur, Y. L., and Pullman, E. R. (2006). Abrupt increase in permafrost degradation in Arctic Alaska. *Geophysical Research Letters*, 33, L02503, doi: 10.1029/2005GL024960.

Joughlin, I. and Alley. R. B. (2011). Stability of the West Antarctic ice sheet in a warming world. *Nature Geoscience*, 4, 506–13.

Jouvet, G., Huss, M., Funk, M., and Blatter, H. (2011). Modelling the retreat of Grosser Aletschgletscher, Switzerland, in a changing climate. *Journal of Glaciology*, 57, 1033–45.

Joyce, E. B. (2010). Australia's geoheritage: history of study, a new inventory of geosites and applications to geotourism and geoparks. *Geoheritage*, 2(1–2), 39–56.

Julian, J. P., Wilgruber, N. A., de Beurs, K. M., Mayer, P. M., and Jawarneh, R. N. (2015). Long-term impacts of land cover changes on stream channel loss. *Science of the Total Environment*, 537, 399–410.

Julian, M. and Anthony, E. (1996). Aspects of landslide activity in the Mercantour Massif and the French Riviera, southeastern France. *Geomorphology*, 15, 275–89.

Kabakci, H., Chevalier, P. M., and Papendick, R. I. (1993). Impact of tillage and residue management on dryland spring wheat development. *Soil and Tillage Research*, 26, 127–37.

Kadiri, M., Ahmadian, R., Bockelmann-Evans, B., Rauen, W., and Falconer, R. (2012). A review of the potential water quality impacts of tidal renewable energy systems. *Renewable and Sustainable Energy Reviews*, 16, 329–41.

Kadomura H. (1983). Some aspects of large-scale land transformation due to urbanization and agricultural evelopment in recent Japan. *Advances in Space Research*, 2, 169–78.

Kairis, O., Karavitis, C., Salvati, L., Kounalaki, A., and Kosmas, K. (2015). Exploring the impact of overgrazing on soil erosion and land degradation in a dry Mediterranean agro-forest landscape (Crete, Greece). *Arid Land Research and Management*, 29, 360–74.

Kamh, G. M. E., Kallash, A., and Azzam, R. (2008). Factors controlling building susceptibility to earthquakes: 14-year recordings of Islamic archaeological sites in Old Cairo, Egypt: a case study. *Environmental Geology*, Special Issue, 56, 269–79, doi: 10.10007/s00254–007–1162–3.

Kanevskiy, M., Shur, Y., Strauss, J., Jorgenson, T., Fortier, D., Stephani, E., and Vasiliev, A. (2016). Patterns and rates of riverbank erosion involving ice-rich permafrost (yedoma) in northern Alaska. *Geomorphology*, 253, 370–84, doi: 10.1016/j.geomorph.2015.10.023.

Kang, S., Xu, Y., You, Q., Flügel, W. A., Pepin, N., and Yao, T. (2010). Review of climate and cryospheric change in the Tibetan Plateau. *Environmental Research Letters*, 5(1), 015101.

Karl, T. R., Melillo, J. M., and Peterson, T. C. (2009). *Global Climate Change Impacts in the United States*. Cambridge: Cambridge University Press.

Kasai, M., Brierley, G. J., Page, M. J., Marutani, T., and Trustrum, N. A. (2005). Impacts of land use change on patterns of sediment flux in Weraamaia catchment, New Zealand. *Catena*, 64, 27–60.

Kaser, G. (1999). A review of the modern fluctuations of tropical glaciers. *Global and Planetary Change*, 22, 93–103.

Kaspari, S., McKenzie Skiles, S., Delaney, I., Dixon, D., and Painter, T. H. (2015). Accelerated glacier melt on Snow Dome, Mount Olympus, Washington, USA, due to deposition of black carbon and mineral dust from wildfire. *Journal of Geophysical Research: Atmospheres*, 120, 2793–807.

Kassam, A., Derpsch, R., and Friedrich, T. (2014). Global achievements in soil and water conservation: the case of conservation agriculture. *International Soil and Water Conservation Research*, 2, 5–13.

Kates, R. W., Turner, B. L. I., and Clark, W. C. (1990). The great transformation. In *The Earth as Transformed by Human Action*, ed. B. L. Turner, W. C. Clark, R. W. Kates, J. F. Richards, J. T. Matthews, and W. B. Meyer. Cambridge: Cambridge University Press, pp. 1–17.

Kaushal, S. S., Likens, G. E., Utz, R. M., Pace, M. L., Grese, M., and Yepsen, M. (2013). Increased river alkalinization in the Eastern US. *Environmental Science & Technology*, 47, 10302–11.

Kearns, T. J., Wang, G., Bao, Y., Jiang, J., and Lee, D. (2015). Current land subsidence and groundwater level changes in the Houston Metropolitan Area (2005–2012). *Journal of Surveying Engineering*, 05015002.

Keatings, K., Tassie, G. J., Flower, R. J., Hassan, F. A., Hamdan, M. A. T., Hughes, M., and Arrowsmith, C. (2007). An examination of groundwater within the Hawara Pyramid, Egypt. *Geoarchaeology*, 22, 533–54.

Keenan, T. F., Hollinger, D. Y., Bohrer, G., Dragoni, D., Munger, J. W., Schmid, H. P., and Richardson, A. D. (2013). Increase in forest water-use efficiency as atmospheric carbon dioxide concentrations rise. *Nature*, 499, 324–7.

Kellerer-Pirklbauer, A. Lieb, G. K., Avian, M., and Carrivick, J. (2012). Climate change and rock fall events in high mountain areas: numerous and extensive rock falls in 2007 at Mittlerer Burgstall, Central Austria. *Geografiska Annaler: Series A, Physical Geography*, 94, 59–78.

Kench, P. S., Smithers, S. G., McLean, R. F., and Nichol, S. L. (2009). Holocene reef growth in the Maldives: evidence for a mid-Holocene sea-level highstand in the central Indian Ocean. *Geology*, 37, 455–8.

Kennett, D. J. and Beach, T. P. (2013). Archeological and environmental lessons for the Anthropocene from the Classic Maya collapse. *Anthropocene*, 4, 88–100.

Kennish, M. J. (2001). Coastal salt marsh systems in the U.S.: a review of anthropogenic impacts. *Journal of Coastal Research*, 17, 731–48.

Keranen, K. M., Savage, H. M., Abers, G. A., and Cochran, E. S. (2013). Potentially induced earthquakes in Oklahoma, USA: links between wastewater injection and the 2011 Mw 5.7 earthquake sequence. *Geology*, 41, 699–702.

Khanal, S., Anex, R. P., Anderson, C. J., Herzmann, D. E., and Jha, M. K. (2013). Implications of biofuel policy-driven land cover change for rainfall erosivity and soil erosion in the United States. *Global Change Biology: Bioenergy*, 5, 713–22.

Khazendar, A., Borstad, C. P., Scheuchl, B., Rignot, E., and Seroussi, H. (2015). The evolving instability of the remnant Larsen B Ice Shelf and its tributary glaciers. *Earth and Planetary Science Letters*, 419, 199–210.

Khromova, T., Nosenko, G., Kutuzov, S., Muraviev, A., and Chernova, L. (2014). Glacier area changes in Northern Eurasia. *Environmental Research Letters*, 9, 015003.

Khvorostyanov, D. V., Ciais, P., Krinner, G., and Zimov, S. A. (2008). Vulnerability of east Siberia's frozen carbon stores to future warming. *Geophysical Research Letters*, 35, L10703, doi: 10.1029/2008GL033639.

Kibler, K., Tullos, D., and Kondolf, M. (2011). Evolving expectations of dam removal outcomes: downstream geomorphic effects following removal of a small, gravel-filled dam. *Journal of the American Water Resources Association*, 47, 408–22.

Kiernan, K. (2015). Nature, severity and persistence of geomorphological damage caused by armed conflict. *Land Degradation and Development*, 26, 380–96.

Killick, D. (2015). Invention and innovation in African iron-smelting technologies. *Cambridge Archaeological Journal*, 25, 307–19.

Kim, K. D., Lee, S., Oh, H. J., Choi, J. K., and Won, J. S. (2006). Assessment of ground subsidence hazard near an abandoned underground coal mine using GIS. *Environmental Geology*, 50, 1183–91.

Kim, W. Y. (2013). Induced seismicity associated with fluid injection into a deep well in Youngstown, Ohio. *Journal of Geophysical Research: Solid Earth*, 118, 3506–18.

Kirby, M. X. (2004). Fishing down the coast: Historical expansion and collapse of oyster fisheries along continental margins. *Proceedings of the National Academy of Sciences of the United States of America*, 101, 13096–9.

Kirk, S. and Herbert, A. W. (2002). Assessing the impact of groundwater abstractions on river flows. *Geological Society, London, Special Publications*, 193, 211–33.

Kirkbride, M. P. and Warren, C. R. (1999). Tasman Glacier, New Zealand: 20th century thinning and predicted calving retreat. *Global and Planetary Change*, 22, 11–28.

Kirkels, F. M. S. A., Cammeraat, L. H., and Kuhn, N. J. (2014). The fate of soil organic carbon upon erosion, transport and deposition in agricultural landscapes—A review of different concepts. *Geomorphology*, 226, 94–105.

Kirpotin, S. N., Polishchuk, Y., and Bryksina, N. (2009). Abrupt changes of thermokarst lakes in Western Siberia: impacts of climatic warming on permafrost melting. *International Journal of Environmental Studies*, 66, 423–31.

Kirwan, M. L. and Megonigal, J. P. (2013). Tidal wetland stability in the face of human impacts and sea-level rise. *Nature*, 504, 53–60.

Kirwan, M. L., Murray, A. B., Donnelly, J. P., and Corbett, D. R. (2011). Rapid wetland expansion during European settlement and its implication for marsh survival under modern sediment delivery rates. *Geology*, 39, 507–10.

Kiss T., Sipos, G., Mauz, B., and Mezösi, G. (2012). Holocene aeolian sand mobilization, vegetation history and human impact on the stabilized sand dune area of the southern Nyírség, Hungary. *Quaternary Research*, 78, 492–501.

Knight, J. and Burningham, H. (2011). Sand dune morphodynamics and prehistoric human occupation in NW Ireland. *Geological Society of America Special Papers*, 476, 81–92.

Knight, J. and Harrison, S. (2014a). Limitations of uniformitarianism in the Anthropocene. *Anthropocene*, 5, 71–5.

(2014b). Mountain glacial and paraglacial environments under global climate change: lessons from the past, future directions and policy implications. *Geografiska Annaler: Series A, Physical Geography*, 96, 245–64.

Knight, M., Thomas, D. S. G., and Wiggs, G. F. S. (2004). Challenges of calculating dunefield mobility over the 21st century. *Geomorphology*, 59, 197–213.

Knighton, A. D. (1989). River adjustment to changes in sediment load: the effects of tin mining on the Ringarooma River, Tasmania, 1875–1984. *Earth Surface Processes and Landforms*, 14, 333–59.

Knippertz, P. and Stuut, J. -B. W. (eds.) (2014). *Mineral Dust*. Dordrecht: Springer.

Knox, J. C. (1977). Human impacts on Wisconsin stream channels. *Annals of the Association of American Geographers*, 67, 323–42.

(1987). Historical valley floor sedimentation in the Upper Mississippi Valley. *Annals of the Association of American Geographers*, 77, 224–44.

Knutson, T. R., Sirutis, J. J., Garner, S. T., Vecchi, G. A., and Held, I. M. (2008). Simulated reduction in Atlantic hurricane frequency under twenty-first-century warming conditions. *Nature Geoscience*, 1, 359–64.

Koluvek, P. K., Tanji, K. K., and Trout, T. J. (1993). Overview of soil erosion from irrigation. *Journal of Irrigation and Drainage Engineering*, 119, 929–46.

Komar, P. D. and Allan, J. C. (2008). Increasing hurricane-generated wave heights along the U.S. East Coast and their climate controls. *Journal of Coastal Research*, 24, 479–88.

Komori, J. (2008). Recent expansions of glacial lakes in the Bhutan Himalayas. *Quaternary International*, 184, 177–86.

Kondolf, G. M. (1997). Profile: hungry water: effects of dams and gravel mining on river channels. *Environmental Management*, 21, 533–51.

(2013). Sustainable sediment management in reservoirs and regulated rivers: experiences from five continents, *Earth's Future*, 2, 256–80 doi: 10.1002/2013EF000184.

Kondolf, G. M. and Podolak, K. (2014). Space and time scales in human-landscape systems. *Environmental Management*, 53, 76–87.

Kondolf, G. M., Anderson, S., Lave, R., Pagano, L., Merenlender, A., and Bernhardt, E. S. (2007). Two decades of river restoration in California: what can we learn? *Restoration Ecology*, 15, 516–23.

Korpak, J. (2007). The influence of river training on mountain channel changes (Polish Carpathian Mountains). *Geomorphology*, 92, 166–81.

Kortekaas, S., Bagdanaviciute, I., Gyssels, P., Huerta, J. M. A., and Héquette, A. (2010). Assessment of the effects of marine aggregate extraction on the coastline: an example from the German Baltic Sea Coast. *Journal of Coastal Research*, S1, 51, 205–14.

Kossoff, D., Dubbin, W. E., Alfredsson, M., Edwards, S. J., Macklin, M. G., and Hudson Edwards, K. A. (2014). Mine tailings dams: characteristics, failure, environmental impacts, and remediation. *Applied Geochemistry*, 51, 229–45.

Koulouri, M. and Giourga, C. (2007). Land abandonment and slope gradient as key factors of soil erosion in Mediterranean terraced lands. *Catena*, 69, 274–81.

Kozlov, M. V. and Zvereva, E. L. (2007). Industrial barrens: extreme habitats created by non-ferrous metallurgy. *Review of Environmental Science and Bio-Technology*, 6, 231–59.

Kraaijpoel, D. and Dost, B. (2013). Implications of salt-related propagation and mode conversion effects on the analysis of induced seismicity. *Journal of Seismology*, 17, 95–107.

Krabill, W., Frederick, E., Manizade, S., Martin, C., Sonntag, J., Swift, R., Thomas, R., Wright, W., and Yungel, J. (1999). Rapid thinning of parts of the southern Greenland Ice Sheet. *Science*, 283, 1522–4.

Krauss, K. W., Lovelock, C. E., McKee, K. L., López-Hoffman, L., Ewe, S. M., and Sousa, W. P. (2008). Environmental drivers in mangrove establishment and early development: a review. *Aquatic Botany*, 89, 105–27.

Krauss, K. W., McKee, K. L., Lovelock, C. E., Cahoon, D. R., Saintilan, N., Reef, R., and Chen, L. (2014). How mangrove forests adjust to rising sea level. *New Phytologist*, 202, 19–34.

Kravtsova, V. I. and Tarasenko, T. V. (2010). Space monitoring of Aral Sea degradation. *Water Resources*, 37, 285–96.

Kroeker, K. J., Kordas, R. L., Crim, R., Hendriks, I. E., Ramajo, L., Singh, G. S., and Gattuso, J. P. (2013). Impacts of ocean acidification on marine organisms: quantifying sensitivities and interaction with warming. *Global Change Biology*, 19, 1884–96.

Kroes, D. E., and Hupp, C. R. (2010). The effect of channelization on floodplain sediment deposition and subsidence along the Pocomoke River, Maryland. *Journal of the American Water Resources Association*, 46, 686–99.

Kueny, J. A. and Day, M. J. (2002). Designation of protected karstlands in Central America: a regional assessment. *Journal of Cave and Karst Studies*, 64, 165–74.

Kuenzer, C. and Stracher, G. B. (2012). Geomorphology of coal seam fires. *Geomorphology*, 138, 209–22.

Kuenzer, C., Zhang, J., Tetzlaff, A., Van Dijk, P., Voigt, S., Mehl, H., and Wagner, W. (2007). Uncontrolled coal fires and their environmental impacts: investigating two arid mining regions in north-central China. *Applied Geography*, 27, 42–62.

Kull, C. A. (2000). Deforestation, erosion, and fire: degradation myths in the environmental history of Madagascar. *Environment and History*, 6, 423–50.

Kumar, A. (2009). Reclaimed islands and new offshore townships in the Arabian Gulf: potential natural hazards. *Current Science*, 96, 480–5.

Kumar, M., Goossens, E., and Goossens, R. (1993). Assessment of sand dune change detection in Rajasthan (Thar) Desert, India. *International Journal of Remote Sensing*, 14, 1689–703.

Kumar, S. V. and Bhagavanulu, D. V. S. (2008). Effect of deforestation on landslides in Nilgiris district—a case study. *Journal of the Indian Society of Remote Sensing*, 36, 105–8.

Kummu, M., Penny, D., Sarkkula, J., and Koponen, J. (2008). Sediment: curse or blessing for Tonle Sap Lake? *Ambio*, 37, 158–63.

Küster, M., Fülling, A., Kaiser, K., and Ulrich, J. (2014). Aeolian sands and buried soils in the Mecklenburg Lake District, NE Germany: Holocene land-use history and pedo-geomorphic response. *Geomorphology* 211, 64–76.

Kwong, Y. T. J. and Tau, T. Y. (1994). Northward migration of permafrost along the Mackenzie Highway and climatic warming. *Climatic Change*, 26, 399–419.

Labadz, J. C., Burt, T. P., and Potter, A. W. L. (1991). Sediment yield and delivery in the blanket peat moorlands of the southern Pennines. *Earth Surface Processes and Landforms*, 16, 255–71.

Labat, D., Goddéris, Y., Probst, J. L., and Guyot, J. L. (2004). Evidence for global runoff increase related to climate warming. *Advances in Water Resources*, 27, 631–42.

Labrière, N., Locatelli, B., Laumonier, Y., Freycon, V., and Bernoux, M. (2015). Soil erosion in the humid tropics: a systematic quantitative review. *Agriculture, Ecosystems & Environment*, 203, 127–39.

Lach, J. and Wyżga, B. (2002). Channel incision and flow increase of the upper Wisłoka River, southern Poland, subsequent to the reafforestation of its catchment. *Earth Surface Processes and Landforms*, 27, 445–62.

Lahaye, C., Guérin, G., Boëda, E., Fontugne, M., Hatté, C., Frouin, M., and Da Costa, A. (2015). New insights into a late-Pleistocene human occupation in America: the Vale da Pedra Furada complete chronological study. *Quaternary Geochronology* 30, 445–51, doi: 10.1016/j.quageo.2015.03.009.

Lai, S., Loke, L. H., Hilton, M. J., Bouma, T. J., and Todd, P. A. (2015). The effects of urbanisation on coastal habitats and the potential for ecological engineering: a Singapore case study. *Ocean & Coastal Management*, 103, 78–85.

Laity, J. E. (2003). Aeolian destabilization along the Mojave River, Mojave Desert, California: linkages among fluvial, groundwater, and aeolian systems. *Physical Geography*, 24, 196–221.

Lambert, J. H., Jennings, J. N., Smith, C. T., Green, C., and Hutchinson, J. N. (1970). *The Making of the Broads: a Reconsideration of their Origin in the Light of New Evidence*. London: Royal Geographical Society.

Lammers, J. M., van Soelen, E. E., Donders, T. H., Wagner-Cremer, F., Damsté, J. S., and Reichart, G. J. (2013). Natural environmental changes versus human impact in a Florida estuary (Rookery Bay, USA). *Estuaries and Coasts*, 36, 149–57.

Lamoureux, S. F. and Lafrenière, M. J. (2009). Fluvial impact of extensive active layer detachments, Cape Bounty, Melville Island, Canada. *Arctic, Antarctic, and Alpine Research*, 41, 59–68.

Lane, P. (2009). Environmental narratives and the history of soil erosion in Kondoa District, Tanzania: an archaeological perspective. *International Journal of African Historical Studies*, 42, 457–83.

Lane, P. N. and Sheridan, G. J. (2002). Impact of an unsealed forest road stream crossing: water quality and sediment sources. *Hydrological Processes*, 16, 2599–612.

Lang, A. (2003). Phases of soil erosion-derived colluviation in the loess hills of South Germany. *Catena*, 51, 209–21.

Langedal, M. (1997). The influence of a large anthropogenic sediment source on the fluvial geomorphology of the Knabeåna-Kvina rivers, Norway. *Geomorphology*, 19, 117–32.

Langdon, J. (1991). Water–mills and windmills in the west midlands, 1086–1500. *The Economic History Review*, 44, 424–44.

Lantuit, H. and Pollard, W. H. (2008). Fifty years of coastal erosion and retrogressive thaw slump activity on Herschel Island, southern Beaufort Sea, Yukon Territory, Canada. *Geomorphology*, 95, 84–102.

Larsen, I. J., MacDonald, L. H., Brown, E., Rough, D., Welsh, M. J., Pietraszek, J. H., and Schaffrath, K. (2009). Causes of post-fire runoff and erosion: water repellency, cover, or soil sealing? *Soil Science Society of America Journal*, 73, 1393–407.

Laruelle, G. G., Roubeix, V., Sferratore, A., Brodherr, B., Ciuffa, D., Conley, D. J., and Van Cappellen, P. (2009). Anthropogenic perturbations of the silicon cycle at the global scale: Key role of the land-ocean transition. *Global Biogeochemical Cycles*, 23, GB4031, doi: 10.1029/2008GB003267.

Lasanta, T., García-Ruiz, J. M., Pérez-Rontomé, C., and Sancho-Marcén (2000). Runoff and sediment yield in a semi-ard environment: the effect of land management after farmland abandonment. *Catena*, 38, 265–78.

Laumets, L., Kalm, V., Poska, A., Kele, S., Lasberg, K., and Amon, L. (2014). Palaeoclimate inferred from δ18O and palaeobotanical indicators in freshwater tufa of Lake Äntu Sinijärv, Estonia. *Journal of Paleolimnology*, 51, 99–111.

Lavee, H., Poesen, J., and Yair, A. (1997). Evidence of high efficiency water-harvesting by ancient farmers in the Negev Desert, Israel. *Journal of Arid Environments*, 35, 341–8.

Lavee, H., Kutiel, P., Segev, M., and Benyamini, Y. (1995). Effect of surface roughness on runoff and erosion in a Mediterranean ecosystem: the role of fire. *Geomorphology*, 11, 227–34.

Lawson, D. E. (1986). Response of permafrost terrain to disturbance: a synthesis of observations from northern Alaska, USA. *Arctic and Alpine Research*, 18, 1–17.

Le Maitre, D.C., Versfield, D. B., and Chapman, R.A. (2000). The impact of invading alien plants on surface water resources in South Africa: a preliminary assessment. *Water SA*, 26, 397–408.

Lecce, S. A. and Pavlowsky, R. T. (2014). Floodplain storage of sediment contaminated by mercury and copper from historic gold mining at Gold Hill, North Carolina, USA. *Geomorphology*, 206, 122–32.

Lecce, S., Pavlowsky, R., and Schlomer, G. (2008). Mercury contamination of active channel sediment and floodplain deposits from historic gold mining at Gold Hill, North Carolina, USA. *Environmental Geology*, 55, 113–21.

Leclercq, N., Gattuso, J.-P., and Jaubert, J. (2000). CO_2 partial pressure controls the calcification rate of a coral community. *Global Change Biology*, 6, 329–34.

Lee, J. A. and Gill, T. E. (2015). Multiple causes of wind erosion in the Dust Bowl. *Aeolian Research*, 19, 15–36.

Lei, X., Ma, S., Chen, W., Pang, C., Zeng, J., and Jiang, B. (2013). A detailed view of the injection-induced seismicity in a natural gas reservoir in Zigong, southwestern Sichuan Basin, China. *Journal of Geophysical Research: Solid Earth*, 118, 4296–311.

Lei, Y., Yang, K., Wang, B., Sheng, Y., Bird, B. W., Zhang, G., and Tian, L. (2014). Response of inland lake dynamics over the Tibetan Plateau to climate change. *Climatic Change*, 125, 281–90.

Lenzi, M. A. (2002). Stream bed stabilization using boulder check dams that mimic step-pool morphology features in Northern Italy. *Geomorphology*, 45, 243–60.

Leopold, L. B. (1951). Rainfall frequency: an aspect of climatic variation. *Transactions of the American Geophysics Union*, 32, 347–57.

Lespez, L. (2003). Geomorphic responses to long-term land use changes in Eastern Macedonia (Greece). *Catena*, 51, 181–208.

Lesschen, J. P., Cammeraat, L. H., and Nieman, T. (2008). Erosion and terrace failure due to agricultural land abandonment in a semi-arid environment. *Earth Surface Processes and Landforms*, 33, 1574–84.

Lesschen, J. P., Kok, K., Verburg, P. H., and Cammeraat, L. H. (2007). Identification of vulnerable areas for gully erosion under different scenarios of land abandonment in Southeast Spain. *Catena*, 71, 110–21.

Lestrelin, G., Vigiak, O., Pelletreau, A., Keohavong, B., and Valentin, C. (2012, May). Challenging established narratives on soil erosion and shifting cultivation in Laos. *Natural Resources Forum*, 36, 63–75.

Letey, J. (2001). Cases and consequences of fire-induced soil water repellency. *Hydrological Processes*, 15, 2867–75.

Levermann, A., Clark, P. U., Marzeion, B., Milne, G. A., Pollard, D., Radic, V., and Robinson, A. (2013). The multimillennial sea-level commitment of global warming. *Proceedings of the National Academy of Sciences*, 110, 13745–50.

Lewis, K. C., Zyvoloski, G. A., Travis, B., Wilson, C., and Rowland, J. (2012). Drainage subsidence associated with Arctic permafrost degradation. *Journal of Geophysical Research: Earth Surface* 117(F4), F04019, doi: 10.1029/2011JF002284.

Lewis, R. W., Makurat, A., and Pao, W. K. (2003). Fully coupled modeling of seabed subsidence and reservoir compaction of North Sea oil fields. *Hydrogeology Journal*, 11, 142–61.

Lewis, S. L. and Maslin, M. A. (2015). Defining the Anthropocene. *Nature*, 519, 171–80.

Leyva, J. C., Martínez, J. F., and Roa, M. G. (2007). Analysis of the adoption of soil conservation practices in olive groves: the case of mountainous areas in southern Spain. *Spanish Journal of Agricultural Research*, 5, 249–58.

Li, C., Tang, X., and Ma, T. (2006). Land subsidence caused by groundwater exploitation in the Hangzhou-Jiaxing-Huzhou Plain, China. *Hydrogeology Journal*, 14, 1652–65.

Li, D. D., Lerman, A., and Mackenzie, F. T. (2011). Human perturbations on the global biogeochemical cycles of coupled Si–C and responses of terrestrial processes and the coastal ocean. *Applied Geochemistry*, 26, S289–91.

Li, J., Okin, G. S., Tatarko, J., Webb, N. P. and Herrick, J. E. (2014). Consistency of wind erosion assessments across land use and land cover types: a critical analysis. *Aeolian Research*, 15, 253–60.

Li, X. and 9 others (2008). Cryospheric change in China. *Global and Planetary Change*, 62, 210–8.

Li, Y., Cui, J., Zhangh, T., Okuro, T., and Drake, S. (2009). Effectiveness of sand-fixing measures on desert land restoration in Kerqin Sandy Land, northern China. *Ecological Engineering*, 35, 118–27.

Liébault, F. and Piégay, H. (2002). Causes of 20th century channel narrowing in mountain and piedmont rivers of southeastern France. *Earth Surface Processes and Landforms*, 27, 425–44.

Lightfoot, D. R. (1996a). Moroccan khettara: traditional irrigation and progressive desiccation. *Geoforum*, 27, 261–73.

(1996b). Syrian qanat Romani: history, ecology, abandonment. *Journal of Arid Environments*, 33, 321–6.

(2000). The origin and diffusion of qanats in Arabia: new evidence from the northern and southern peninsula. *Geographical Journal*, 166, 215–26.

Lightfoot, K. G. and Cuthrell, R. Q. (2015). Anthropogenic burning and the Anthropocene in late Holocene California. *The Holocene*, 25, 1581–7.

Limondin-Lozouet, N., Preece, R. C., and Antoine, P. (2013). The Holocene tufa at Daours (Somme Valley, northern France): Malacological succession and palaeohydrological implications. *Boreas*, 42, 650–63.

Lin, C., Yang, K., Qin, J., and Fu, R. (2013). Observed coherent trends of surface and upper-air wind speed over China since 1960. *Journal of Climate*, 26, 2891–903.

Lin, N., Emanuel, K., Oppenheimer, M., and Vanmarcke, E. (2012). Physically based assessment of hurricane surge threat under climate change. *Nature Climate Change*, 2, 462–7.

Lin, Y. and Wei, X. (2008). The impact of large-scale forest harvesting on hydrology in the Willow watershed of Central British Columbia. *Journal of Hydrology*, 359, 141–49.

Lindberg, T. T., Bernhardt, E. S., Bier, R., Helton, A. M., Merola, R. B., Vengosh, A., and Di Giulio, R. T. (2011). Cumulative impacts of mountaintop mining on an Appalachian watershed. *Proceedings of the National Academy of Sciences*, 108, 20929–34.

Ling, F. and Zhang, T. (2003). Impact of the timing and duration of seasonal snow cover on the active layer and permafrost in the Alaskan Arctic. *Permafrost and Periglacial Processes*, 14, 141–50.

List, J. H., Sallenger, A. H., Hansen, M. E., and Jaffe, B. E. (1997). Accelerated sea level rise and rapid coastal erosion: testing a causal relationship for the Louisiana barrier islands. *Marine Geology*, 140, 437–65.

Liu, C. H., Pan, Y. W., Liao, J. J., Huang, C. T., and Ouyang, S. (2004). Characterization of land subsidence in the Choshui River alluvial fan, Taiwan. *Environmental Geology*, 45, 1154–66.

Liu, C. W., Lin, W. S., Shang, C., and Liu, S. H. (2001). The effect of clay dehydration on land subsidence in the Yun-Lin coastal area, Taiwan. *Environmental Geology*, 40, 518–27.

Liu, S. and Wang, T. (2014). Aeolian processes and landscape change under human disturbances on the Sonid grassland of inner Mongolian Plateau, northern China. *Environmental Earth Sciences*, 71, 2399–407.

Liu, S., Xu, L., and Talwani, P. (2011). Reservoir-induced seismicity in the Danjiangkou Reservoir: a quantitative analysis. *Geophysical Journal International*, 185, 514–28.

Liu, T., Kinouchi, T., and Ledezma, F. (2013). Characterization of recent glacier decline in the Cordillera Real by LANDSAT, ALOS, and ASTER data. *Remote Sensing of Environment*, 137, 158–72.

Liu, W., Chen, W., and Peng, C. (2014). Assessing the effectiveness of green infrastructures on urban flooding reduction: a community scale study. *Ecological Modelling*, 291, 6–14.

Lixin, Y., Fang, Z., He, X., Shijie, C., Wei, W., and Qiang, Y. (2011). Land subsidence in Tianjin, China. *Environmental Earth Sciences*, 62, 1151–61.

Llenos, A. L. and Michael, A. J. (2013). Modeling earthquake rate changes in Oklahoma and Arkansas: possible signatures of induced seismicity. *Bulletin of the Seismological Society of America*, 103, 2850–61.

Lóczy, D. and Sütö, L. (2011). Human agency and geomorphology. In *The SAGE Handbook of Geomorphology*, ed. K. J. Gregory and A. S. Goudie. Sage: London, pp. 260–78.

Logsdon, S. D. (2013). Depth dependence of chisel plow tillage erosion. *Soil and Tillage Research*, 128, 119–24.

Lokier, S. W. (2013). Coastal sabkha preservation in the Arabian Gulf. *Geoheritage*, 5, 11–22.

Londoño, A. C. (2008). Pattern and rate of erosion inferred from Inca agricultural terraces in arid southern Peru. *Geomorphology*, 99, 13–25.

Lopez-Bermudez, F., Romero-Díaz, A., Martínez-Fernandez, J., and Martínez-Fernandez, J. (1998). Vegetation and soil erosion under a semi-arid Mediterranean climate: a case study from Murcia (Spain). *Geomorphology*, 24, 51–8.

López-Moreno, J. I., Beguería, S., and García-Ruiz, J. M. (2006). Trends in high flows in the central Spanish Pyrenees: response to climatic factors or to land-use change? *Hydrological Sciences Journal*, 51, 1039–50.

López-Moreno, J. I., Fontaneda, S., Bazo, J., Revuelto, J., Azorin-Molina, C., Valero-Garcés, B., and Alejo-Cochachín, J. (2014). Recent glacier retreat and climate trends in Cordillera Huaytapallana, Peru. *Global and Planetary Change*, 112, 1–11.

Lorimer, J. (2012). Multinatural geographies for the Anthropocene. *Progress in Human Geography*, 36, 593–612.

Loughran, R. J., Campbell, B. L., Elliott, G. L., and Shelly, D. J. (1990). Determination of the rate of sheet erosion on grazing land using caesium-137. *Applied Geography*, 10, 125–33.

Lowdermilk, W. C. (1934). Acceleration of erosion above geologic norms. *Eos, Transactions American Geophysical Union*, 15, 505–9.

   (1935). Man-made deserts. *Pacific Affairs*, 8, 409–19.

Lowe, A. (1986). Bronze Age burial mounds on Bahrain. *Iraq*, 48, 73–84.

Lowenthal, D. (2016). Origins of Anthropocene awareness. *The Anthropocene Review*, 3, 52–63, doi: 10.1177/2053019615609953

Lu, H. and 12 others (2009). Earliest domestication of common millet (*Panicum miliaceum*) in East Asia extended to 10,000 years ago. *Proceedings of the National Academy of Sciences* 106, 7367–72.

Lu, Z., Streets, D. G., Zhang, Q., Wang, S., Carmichael, G. R., Cheng, Y. F., Wei, C., Chin, M., Diehl, T., and Tan, Q. (2010). Sulfur dioxide emissions in China and sulphur trends in East Asia since 2000. *Atmospheric Chemistry and Physics*, 10, 6311–31.

Lubke, R. A. (2013). Restoration of dune ecosystems following mining in Madagascar and Namibia: contrasting restoration approaches adopted in regions of high and low population density. In *Restoration of Coastal Dunes*, ed. M. L.

Martínez, J. B. Gallego-Fernández, and P. A. Hesp. Berlin and Heidelberg: Springer, pp 199–215.

Lubos, C. C. M., Dreibrodt, S., Nelle, O., Klamm, M., Friederich, S., Meller, H., and Bork, H. R. (2011). A multi-layered prehistoric settlement structure (tell?) at Niederröblingen, Germany and its implications. *Journal of Archaeological Science*, 38, 1101–10.

Luce, C. H. (2002). Hydrological processes and pathways affected by forest roads: what do we still need to learn? *Hydrological Processes*, 16, 2901–4.

Luce, C. H. and Black, T. A. (1999). Sediment production from forest roads in western Oregon. *Water Resources Research*, 35, 2561–70.

Lucha, P., Cardona, F., Gutiérrez, F., and Guerrero, J. (2008). Natural and human-induced dissolution and subsidence processes in the salt outcrop of the Cardona Diapir (NE Spain). *Environmental Geology*, 53, 1023–35.

Luo, H. R., Smith, L. M., Allen, B. L., and Haukos, D. A. (1997). Effects of sedimentation on playa wetland volume. *Ecological Applications*, 7, 247–52.

Luzón, M. A., Pérez, A., Borrego, A. G., Mayayo, M. J., and Soria, A. R. (2011). Interrelated continental sedimentary environments in the central Iberian Range (Spain): facies characterization and main palaeoenvironmental changes during the Holocene. *Sedimentary Geology*, 239, 87–103.

Lwenya, C. and Yongo, E. (2010). Human aspects of siltation of Lake Baringo: causes, impacts and interventions. *Aquatic Ecosystem Health & Management*, 13, 437–41.

Lyell, C. (1835). *Principles of Geology* (4th edn). London: Murray

   (1875). *Principles of Geology* (12th edn). *London: Murray*

Lynas, M. (2011). *The God species*. London: Fourth Estate.

Lyons, R., Tooth, S., and Duller, G. A. (2013). Chronology and controls of donga (gully) formation in the upper Blood River catchment, KwaZulu-Natal, South Africa: Evidence for a climatic driver of erosion. *The Holocene*, 23, 1875–87.

Ma, R., Duan, H., Hu, C., Feng, X., Li, A., Ju, W., and Yang, G. (2010). A half-century of changes in China's lakes: Global warming or human influence? *Geophysical Research Letters*, 37, L24106, DOI: 10.1029/2010GL045514.

Macdonald, K. C. (1997). More forgotten tells of Mali: an archaeologist's journey from here to Timbuktu. *Archaeology International*, 1, 40–2.

Macklin, M. G. and Lewin, J. (1986). Terraced fills of Pleistocene and Holocene age in the Rheidol Valley, Wales. *Journal of Quaternary Science*, 1, 21–34.

Macklin, M. G. and Woodward, J. C. (2009). River systems and environmental change. In *The*

Physical Geography of the Mediterranean, ed. J. C. Woodward, Oxford: Oxford University Press. pp. 319–52.

Macklin, M. G., Jones, A. F. and Lewin, J. (2010). River response to rapid Holocene environmental change: evidence and explanation in British catchments. *Quaternary Science Reviews,* 29, 1555–76.

Macklin, M. G., Lewin, J., and Jones, A. F. (2013). River entrenchment and terrace formation in the UK Holocene. *Quaternary Science Reviews,* 76, 194–206.

(2014). Anthropogenic alluvium: an evidence-based meta-analysis for the UK Holocene. *Anthropocene,* 6, 26–38.

Macklin, M. G., Passmore, D. G., Stevenson, A. C., Colwey, A. C., Edwards, D. N., and O'Brien, C. F. (1991). Holocene alluviation and land-use change on Callaly Moor, Northumberland, England. *Journal of Quaternary Science,* 6, 225–32.

Maghsoudi, M., Simpson, I. A., Kourampas, N., and Nashli, H. F. (2014). Archaeological sediments from settlement mounds of the Sagzabad Cluster, central Iran: human-induced deposition on an arid alluvial plain. *Quaternary International,* 324, 67–83.

Maguregui, M., Sarmiento, A., Martinez-Arkarazo, I., Angulo, M., Castro, K., Arana, G., and Madariaga, J. M. (2008). Analytical diagnosis methodology to evaluate nitrate impact on historical building materials. *Analytical and Bioanalytical Chemistry,* 391, 1361–1370.

Mahowald, N., Albani, S., Kok, J. F., Engelstaeder, S., Scanza, R., Ward, D. S., and Flanner, M. G. (2014). The size distribution of desert dust aerosols and its impact on the Earth system. *Aeolian Research,* 15, 53–71.

Mahowald, N. M. and 19 others. (2010). Observed 20th century desert dust variability: impact on climate and biogeochemistry. *Atmospheric Chemistry and Physics,* 10, 10875–93.

Maina, J., de Moel, H., Zinke, J., Madin, J., McClanahan, T., and Vermaat, J. E. (2013). Human deforestation outweighs future climate change impacts of sedimentation on coral reefs. *Nature Communications,* 4, article number: 1986, doi:, doi: 10.1038/ncomms2986.

Majer, E. L., Baria, R., Start, M., Oates, S., Bommer, J., Smith, B., and Asanuma, H. (2007). Induced seismicity associated with Enhanced Geothermal Systems. *Geothermics,* 36, 185–222.

Majewski, M. (2014). Human impact on Subatlantic slopewash processes and landform development at Lake Jasień (northern Poland). *Quaternary International,* 324, 56–66.

Major, J. and 10 others. (2008). Initial fluvial response to the removal of Oregon's Marmot Dam. *Eos,* 27, 241–3.

Maley, J., Giresse, P., Doumenge, C., and Favier, C. (2012). Comment on "Intensifying weathering and land use in Iron Age Central Africa." *Science,* 337, 1040.

Malm, W. C., Schichtel, B. A., Ames, R. B., and Gebhart, K. A. (2002). A 10-year spatial and temporal trend of sulfate across the United States. *Journal of Geophysical Research,* 107 (D22), article no. 4627.

Mancini, F., Stecchi, F., Zanni, M., and Gabbianelli, G. (2009). Monitoring ground subsidence induced by salt mining in the city of Tuzla (Bosnia and Herzegovina). *Environmental Geology,* 58, 381–9.

Mann, D. H. and Meltzer, D. J. (2007). Millennial-scale dynamics of valley fills over the past 12000 ^{14}C yr in northeastern New Mexico. *Bulletin of the Geological Society of America,* 119, 1433–48.

Mann, K. C., Peck, J. A., and Peck, M. C. (2013). Assessing dam pool sediment for understanding past, present and future watershed dynamics: an example from the Cuyahoga River, Ohio. *Anthropocene,* 2, 76–88.

Mao, D., Wang, Z., Li, L., Song, K., and Jia, M. (2014). Quantitative assessment of human-induced impacts on marshes in Northeast China from 2000 to 2011. *Ecological Engineering,* 68, 97–104.

Marden, M., Arnold, G. Seymour, A., and Hambling, R. (2012). History and distribution of steepland gullies in response to land use change, East Coast Region, North Island, New Zealand. *Geomorphology,* 153–154, 81–90.

Marfai, M. A. and King, L. (2007). Monitoring land subsidence in Semarang, Indonesia. *Environmental Geology,* 53, 651–9.

Mark, A. F. and McSweeney, G. D. (1990). Patterns of impoverishment in natural communities: case studies in forest ecosystems – New Zealand. In *The Earth in Transition: Patterns and Processes of Biotic Impoverishment,* ed. G. M. Woodwell. Cambridge: Cambridge University Press, pp. 151–76.

Marker, M. E. (1967). The Dee estuary: its progressive silting and salt marsh development. *Transactions of the Institute of British Geographers,* 41, 65–71.

Marren, P. M., Grove, J. R., Webb, J. A., and Stewardson, M. J. (2014). The potential for dams to impact lowland meandering river floodplain geomorphology. *The Scientific World Journal,* 309673, http://dx.doi.org/10.1155/2014/309673 (accessed, 12th November, 2015).

Mars, J. C. and Houseknecht, D. W. (2007). Quantitative remote sensing study indicates doubling of coastal erosion rate in past 50 yr along a segment of the Arctic coast of Alaska. *Geology,* 35, 583–6.

Marsh, B. and Kealhofer, L. (2014). Scales of impact: Settlement history and landscape change in the Gordion Region, central Anatolia. *The Holocene,* 24, 689–701.

Marsh, G. P. (1864). *Man and Nature*. New York: Scribner.

Marshall, E., Weinberg, M., Wunder, S., and Kaphengst, T. (2011). Environmental dimensions of bioenergy development. *EuroChoices*, 10, 43–8.

Marshall, E. J. P., Wade, P. M., and Clare, P. (1978). Land drainage channels in England and Wales. *Geographical Journal*, 144, 254–63.

Marston, R. A. and Dolan, L. S. (1999). Effectiveness of sediment control structures relative to spatial patterns of upland soil loss in an arid watershed, Wyoming. *Geomorphology*, 31, 313–23.

Marston, R. A., Bravard, J. P., and Green, T. (2003). Impacts of reforestation and gravel mining on the Malnant River, Haute-Savoie, French Alps. *Geomorphology*, 55, 65–74.

Marszałek, M., Alexandrowicz, Z., and Rzepa, G. (2014). Composition of weathering crusts on sandstones from natural outcrops and architectonic elements in an urban environment. *Environmental Science and Pollution Research*, 21, 14023–36.

Martin, M. A., Levermann, A., and Winkelmann, R. (2015). Comparing ice discharge through West Antarctic Gateways: Weddell vs. Amundsen Sea warming. *The Cryosphere Discussions*, 9, 1705–33.

Martin, Y. E., Johnson, E. A., Gallaway, J. M., and Chaikina, O. (2011). Negligible soil erosion in a burned mountain watershed, Canadian Rockies: field and modelling investigations considering the role of duff. *Earth Surface Processes and Landforms*, 36, 2097–113.

Martínez, M. L., Gallego-Fernández, J. B., and Hesp, P. A. (2013a). *Restoration of Coastal Dunes*. Berlin and Heidelberg: Springer.

Martínez M. L., Hesp, P. A., and Gallego-Fernández, J. B. (2013b). Coastal dune restoration: trends and perspectives. In *Restoration of Coastal Dunes*, ed. M. L. Martínez, J. B. Gallego-Fernández, and P. A. Hesp. Berlin and Heidelberg: Springer, pp. 323–39.

Martínez-Casasnovas, J. A. and Ramos, M. C. (2006). The cost of soil erosion in vineyard fields in the Penedès–Anoia Region (NE Spain). *Catena*, 68, 194–9.

(2009). Soil alteration due to erosion, ploughing and levelling of vineyards in north east Spain. *Soil Use and Management*, 25, 183–92.

Martínez-Casasnovas, J. A., Ramos, M. C., and García-Hernández, D. (2009). Effects of land-use changes in vegetation cover and sidewall erosion in a gully head of the Penedès region (northeast Spain). *Earth Surface Processes and Landforms*, 34, 1927–37.

Martínez-Casasnovas, J. A., Ramos, M. C., and Ribes-Dasi, M. (2005). On-site effects of concentrated flow erosion in vineyard fields: some economic implications. *Catena*, 60, 129–46.

Martinez Raya, A., Duran Zuazo, V. H. and Francia Martinez, J. R. (2006). Soil erosion and runoff response to plant-cover strips on semiarid slopes (SE Spain). *Land Degradation and Development*, 17, 1–11.

Martín-Vide, J. P., Ferrer-Boix, C., and Ollero, A. (2010). Incision due to gravel mining: modeling a case study from the Gállego River, Spain. *Geomorphology*, 117, 261–71.

Marx, S. K., McGowan, H. A., Kamber, B. S., Knight, J. M., Denholm, J., and Zawadzki, A. (2014). Unprecedented wind erosion and perturbation of surface geochemistry marks the Anthropocene in Australia. *Journal of Geophysical Research: Earth Surface* 119, 45–61.

Maselli, V. and Trincardi, F. (2013). Man made deltas. *Scientific Reports*, 3, doi:10.1038/srep01926.

Massey, S. W. (1999). The effects of ozone and NOx on the deterioration of calcareous stone. *Science of the Total Environment*, 227, 109–21.

Massonnet, D., Holzer, T., and Vadon, H. (1997). Land subsidence caused by the East Mesa geothermal field, California, observed using SAR interferometry. *Geophysical Research Letters*, 24, 901–4.

Matless, D. (2014). *Nature of Landscape: Cultural Geography of the Norfolk Broads*. Wiley Blackwell: Chichester and Oxford.

Mattheus, C. R., Rodriguez, A. B., McKee, B. A., and Curring, C. A. (2010). Impact of land-use change and hard structures on the evolution of fringing marsh shorelines. *Estuarine, Coastal and Shelf Science*, 88, 365–76.

Mattsson, T., Kortelainen, P., Räike, A., Lepistö, A., and Thomas, D. N. (2015). Spatial and temporal variability of organic C and N concentrations and export from 30 boreal rivers induced by land use and climate. *Science of the Total Environment*, 508, 145–54.

May, V. (2015). Coastal cliff conservation and management: the Dorset and East Devon Coast World Heritage Site. *Journal of Coastal Conservation*, 19, 821–9, doi:10.1007/s11852-014-0358-4.

May, V. J. and Hansom, J. D. (2003). *Coastal Geomorphoogy of Great Britain*. Peterborough: Joint Nature Conservation Committee.

Mazzoldi, A., Rinaldi, A. P., Borgia, A., and Rutqvist, J. (2012). Induced seismicity within geological carbon sequestration projects: maximum earthquake magnitude and leakage potential from undetected faults. *International Journal of Greenhouse Gas Control*, 10, 434–42.

McCabe, S., Smith, B. J., McAlister, J. J., Gomez-Heras, M., McAllister, D., Warke, P. A., and Basheer, P. A. M. (2013). Changing climate, changing process: implications for salt transportation and weathering within building sandstones in the UK. *Environmental Earth Sciences*, 69, 1225–35.

McCarney-Castle, K., Voulgaris, G., and Kettner, A. J. (2010). Analysis of fluvial suspended sediment load contribution through Anthropocene history to the South Atlantic Bight coastal zone, U.S.A. *Journal of Geology*, 118, 399–416.

McConnell, J. R., Aristarain, A. J., Banta, J. R., Edwards, P. R., and Simões, J. C. (2007). 20th-century doubling in dust archived in an Antarctic Peninsula ice core parallels climate change and desertification in South America. *Proceedings National Academy of Sciences*, 104, 5743–8.

McCorriston, J. and Oches, E. (2001). Two early Holocene check dams from Southern Arabia. *Antiquity*, 75, 675–6.

McCulloch, M., Fallon, S., Wyndham, T., Hendy, E., Lynch, J., and Barnes, D. (2003). Coral record of increased sediment flux to the inner Great Barrier Reef since European settlement. *Nature*, 421, 727–30.

McCulloch, M., Falter, J., Trotter, J., and Montagna, P. (2012). Coral resilience to ocean acidification and global warming through pH up-regulation. *Nature Climate Change*, 2, 623–7.

McDonald, A., Lane, S. N., Haycock, N. E., and Chalk, E. A. (2004). Rivers of dreams: on the gulf between theoretical and practical aspects of an upland river restoration. *Transactions of the Institute of British Geographers*, 29, 257–81.

McFadden, L., Spencer, T., and Nicholls, R. J. (2007). Broad-scale modelling of coastal wetlands: what is required? *Hydrobiologia*, 577, 5–15.

McFadden, L. D. and McAuliffe, J. R. (1997). Lithologically influenced geomorphic responses to Holocene climatic changes in the Southern Colorado Plateau, Arizona: a soil-geomorphic and ecologic perspective. *Geomorphology*, 19, 303–32.

McGarr, A. (2014). Maximum magnitude earthquakes induced by fluid injection. *Journal of Geophysical Research: Solid Earth*, 119, 1008–19.

McGarr, A., Bekins, B., Burkardt, N., Dewey, J., Earle, P., Ellsworth, W., and Sheehan, A. (2015). Coping with earthquakes induced by fluid injection. *Science*, 347, 830–1.

McGlone, M. S. and Wilmshurst, J. M. (1999). Dating initial Maori environmental impact in New Zealand. *Quaternary International*, 59, 5–16.

McMillan, B. R., Pfeiffer, K. A., and Kaufman, D. W. (2011). Vegetation responses to an animal-generated disturbance (bison wallows) in tallgrass prairie. *The American Midland Naturalist*, 165, 60–73.

McNeill, J. R. (2000). *Something New Under the Sun: an Environmental History of the Twentieth-Century World*. New York: W.W. Norton & Company.

McNeill, J. R. (2003). Resource exploitation and over-exploitation: a look at the 20th century. In *Exploitation and Overexploitation in Societies Past and Present*, ed. T. S. Benzing and B. Herrmann. Münster: LIT Verlag, pp. 51–60.

McWethy, D. B., Whitlock, C., Wilmshurst, J. M., McGlone, M. S., and Li, X. (2009). Rapid deforestation of South Island, New Zealand, by early Polynesian fires. *The Holocene*, 19, 883–97.

McWethy, D. B., Wilmshurst, J. M., Whitlock, C., Wood, J. R., and McGlone, M. S. (2014) A high-resolution chronology of rapid forest transitions following Polynesian arrival in New Zealand. *PLoS ONE* 9(11), e111328. doi:10.1371/journal.pone.0111328.

Meade, R. H. (1991). Reservoirs and earthquakes. *Engineering Geology*, 30, 245–62.

Meade, R. H. and Moody, J. A. (2010). Causes for the decline of suspended-sediment discharge in the Mississippi River system, 1940–2007. *Hydrological Processes*, 24, 35–49.

Meade, R. H. and Parker, R. S. (1985). Sediment in rivers in the United States. *United States Geological Survey Water Supply Paper*, 2276, 49–60.

Meade, R. H. and Trimble, S. W. (1974). Changes in sediment loads in rivers of the Atlantic drainage of the United States since 1900. *Publication of the International Association of Hydrological Science*, 113, 99–104.

Meadows, P. S. and Meadows, A. (1999). *The Indus River: Biodiversity, Resources, Humankind*. Karachi: Oxford University Press.

Megahan, W. F., Wilson, M. and Monsen, S. B. (2001). Sediment production from granitic cut-slopes on forest roads in Idaho, USA. *Earth Surface Processes and Landforms*, 26, 153–63.

Meijer, A. D., Heitman, J. L., White, J. G., and Austin, R. E. (2013). Measuring erosion in long-term tillage plots using ground-based lidar. *Soil and Tillage Research*, 126, 1–10.

Mekonnen, M. M., Hoekstra, A. Y., and Becht, R. (2012). Mitigating the water footprint of export cut flowers from the Lake Naivasha Basin, Kenya. *Water Resources Management*, 26, 3725–42.

Mendonça, A., Fortes, C. J., Capitão, R., Neves, M. G., Antunes do Carmo, J. S., and Moura, T. (2011). Hydrodynamics around an artificial surfing reef at Leirosa, Portugal. *Journal of Waterway, Port, Coastal, and Ocean Engineering*, 138, 226–35.

Mendoza-González, G., Martínez, M. L., Rojas-Soto, O. R., Vázquez, G., and Gallego-Fernández, J. B. (2013). Ecological niche modeling of coastal dune plants and future potential distribution in response to climate change and sea level rise. *Global Change Biology*, 19, 2524–35.

Menze, B. H. and Ur, J. A. (2012). Mapping patterns of long-term settlement in Northern Mesopotamia at a large scale. *Proceedings of the National Academy of Sciences*, 109, E778–87.

Metzidakis, I., Martinez-Vilela, A., Castro Nieto, G., and Basso, B. (2008). Intensive olive orchards on sloping land: good water and pest management are essential. *Journal of Environmental Management*, 89, 120–8.

Meulen, F. van der and Salman, A. H. P. M. (1996). Management of Mediterranean coastal dunes. *Ocean and Coastal Management*, 30, 177–95.

Meulen, F. van der and Jungerius, P. D. (1989). Landscape development in Dutch coastal dunes: the breakdown and restoration of geomorphological and geohydrological processes. *Proceedings of*

the Royal Society of Edinburgh. Section
B. Biological Sciences, 96, 219–29.

Meunier, J. D., Braun, J. J., Riotte, J., Kumar, C., and
Sekhar, M. (2011). Importance of weathering
and human perturbations on the riverine transport of Si. *Applied Geochemistry*, 26, S360–2.

Meusberger, K. and Alewell, C. (2008). Impacts of
anthropogenic and environmental factors on the
occurrence of shallow landslides in an alpine
catchment (Urseren Valley, Switzerland). *Natural
Hazards and Earth System Sciences*, 8, 509–20.

Meybeck, M. (2003). Global analysis of river systems:
from Earth system controls to Anthropocene syn-
dromes. *Philosophical Transactions of the Royal
Society B: Biological Sciences*, 358, 1935–55.

Meybeck, M. and Vörösmarty, C. (2005). Fluvial
filtering of land-to-ocean fluxes: from natural
Holocene variations to Anthropocene. *Comptes
Rendus Geoscience*, 337, 107–23.

Meyer, A. and Martınez-Casasnovas, J. A. (1999).
Prediction of existing gully erosion in vineyard
parcels of the NE Spain: a logistic modelling
approach. *Soil and Tillage Research*, 50,
319–31.

Meyfroidt, P. and Lambin, E. (2009). Geographic and
historical patterns of reforestation. *Bulletin des
Séances Academie Royale des Sciences d'outre-
mer*, 55, 477–502.

Micheli, F., Halpern, B. S., Walbridge, S., Ciriaco, S.,
Ferretti, F., Fraschetti, S., and Rosenberg, A. A.
(2013). Cumulative human impacts on mediterranean and black sea marine ecosystems: assessing current pressures and opportunities. *PloS
one*, 8(12), e79889.

Micheletti, N., Lane, S. N., and Chandler, J. H. (2015).
Application of archival aerial photogrammetry
to quantify climate forcing of Alpine landscapes.
The Photogrammetric Record, doi: 10.1111/
phor.12099.

Middleton, N. J. (1990). Wind erosion and dust storm
prevention. In *Desert Reclamation*, ed. A. S.
Goudie. Wiley, Chichester: Wiley, pp. 87–108.

Mietton, M., Cordier, S., Frechen, M., Dubar, M.,
Beiner, M., and Andrianaivoarivony, R. (2014).
New insights into the age and formation of the
Ankarokaroka lavaka and its associated sandy
cover (NW Madagascar, Ankarafantsika natural
reserve). *Earth Surface Processes and Landforms*, 39, 1467–77.

Migoń, P. (Ed.). (2010). *Geomorphological Landscapes of the World*. Springer.

(2014). The significance of landforms–the contribution of geomorphology to the World Heritage
Programme of UNESCO. *Earth Surface Processes and Landforms*, 39, 836–843.

Miller, A. J. (2011). Identifying landslide activity as a
function of economic development: a case study
of increased landslide frequency surrounding
Dominical, Costa Rica. *Environment, Development and Sustainability*, 13, 901–21.

Miller, A. J. and Zégre, N. P. (2014). Mountaintop
removal mining and catchment hydrology.
Water, 6, 472–99.

Miller, J. D., Kim, H., Kjeldsen, T. R., Packman, J.,
Grebby, S., and Dearden, R. (2014). Assessing
the impact of urbanization on storm runoff in a
peri-urban catchment using historical change in
impervious cover. *Journal of Hydrology*, 515,
59–70.

Miller L. and Douglas, B. C. (2004). Mass and volume
contribution to twentieth-century global sea
level rise. *Nature*, 428, 406–9.

Miller, R. L. and Tegen, I. (1998). Climate response to
soil dust aerosols. *Journal of Climate*, 11,
3247–67.

Miller Rosen, A. (1997). The geoarchaeology of Holocene environments and land use at Kazane Höyük,
SE Turkey. *Geoarchaeology*, 12, 395–416.

Milliman, J. D., Qin, Y. S., Ren, M. E., and Yoshiki
Saita (1987). Man's influence on erosion and
transport of sediment by Asian rivers: the
Yellow River (Huanghe) example. *Journal of
Geology*, 95, 751–62.

Millspaugh, S. H. and Whitlock, C. (1995). A 750-year
fire history based on lake sediment records in
central Yellowstone National Park, USA. *The
Holocene*, 5, 283–92.

Min, S.-K., Zhang, X., Zwiers, F. W., and Hegerl, G.
C. (2011). Human contribution to more-intense
precipitation extremes. *Nature*, 470, 378–81.

Miyasaka, T., Okuro, T., Miyamori, E., Zhao, X., and
Takeuchi, K. (2014). Effects of different restoration measures and sand dune topography on
short-and long-term vegetation restoration in
northeast China. *Journal of Arid Environments*
111, 1–6.

Moffet, C. A., Pierson, F. B., Robichaud, P. R.,
Spaeth, K. E., and Hardegree, S. P. (2007).
Modeling soil erosion on steep sagebrush rangeland before and after prescribed fire. *Catena*, 71,
218–28.

Mokhtari, D., Karami, F., and Bayati Khatibi, M.
(2011). An investigation on Parsian alluvial
fan's tufa in northwest of Iran and its implications for the Holocene tufa decline. *Geography
and Environmental Planning*, 22, 1–22.

Mölg, N. (2014). Hasty retreat of glaciers in northern
Patagonia from 1985–2011. *Journal of Glaciology*, 60, 1033–43.

Molg, T., Georges, C., and Kaser, G. (2003). The
contribution of increased incoming shortwave
radiation to the retreat of the Rwenzori Glaciers,
East Africa, during the 20th Century. *International Journal of Climatology*, 23, 291–303.

Moncel, M.-H. (2010). Oldest human expansions in
Eurasia: favouring and limiting factors. *Quaternary International*, 223–224, 1–9.

Monteith, D. T., Stoddard, J. L., Evans, C. D., de Wit,
H. A., Forsius, M., Høgåsen, T., and Vesely, J.
(2007). Dissolved organic carbon trends

resulting from changes in atmospheric deposition chemistry. *Nature*, 450, 537–40.

Montgomery, D. R. (1997). What's best on the banks? *Nature*, 388, 328–9.

(2007). *Dirt: The Erosion of Civilizations*. Berkeley: University of California Press.

Moody, J. A. and Martin, D. A. (2001). Initial hydrologic and geomorphic response following a wildfire in the Colorado Front Range. *Earth Surface Processes and Landforms*, 26, 1049–70.

Moody, J. A., Shakesby, R. A., Robichaud, P. R., Cannon, S. H., and Martin, D. A. (2013). Current research issues related to post-wildfire runoff and erosion processes. *Earth-Science Reviews*, 122, 10–37.

Moore, P. D. (1987). Man and mire: a long and wet relationship. *Transactions of the Botanical Society of Edinburgh*, 45, 77–95.

(2002). The future of cool temperate bogs. *Environmental Conservation*, 29, 3–20.

Moosdorf, N., Renforth, P., and Hartmann, J. (2014). Carbon dioxide efficiency of terrestrial enhanced weathering. *Environmental Science & Technology*, 48, 4809–16.

Mora, J. W. and Burdick, D. M. (2013). The impact of man-made earthen barriers on the physical structure of New England tidal marshes (USA). *Wetlands Ecology and Management*, 21, 387–98.

Morales C. (ed.) (1979). *Saharan Dust*, Chichester: John Wiley.

Morgan, J. A. and 14 others (2004). Water relations in grassland and desert ecosystems exposed to elevated atmospheric CO_2. *Oecologia*, 140, 11–25.

Morgan, R. P. C. (2005). *Soil Erosion and Conservation* (3rd edn). Oxford: Blackwell Publishing.

Moritz, M. A., Parisien, M. A., Batllori, E., Krawchuk, M. A., Van Dorn, J., Ganz, D. J., and Hayhoe, K. (2012). Climate change and disruptions to global fire activity. *Ecosphere*, 3(6), art49.

Morris, S. E. and Moses, T. A. (1987). Forest fire and the natural soil erosion regime in the Colorado Front Range. *Annals of the Association of American Geographers*, 77, 245–54.

Mortimore, M. (1989). *Adapting to Drought: Farmers, Famines, and Desertification in West Africa*. Cambridge: Cambridge University Press.

Morton, R. A., Bernier, J. C., and Barras, J. A. (2006). Evidence of regional subsidence and associated wetland loss induced by hydrocarbon production, Gulf Coast region, USA. *Environmental Geology*, 50, 261–74.

Morton, R. A., Bernier, J. C., Barras, J. A., and Ferina, N. F. (2005). Historical subsidence and wetland loss in the Mississippi Delta Plain. *Gulf Coast Association of Geological Societies, Transactions*, 55, 555–71.

Mortsch, L., Hengeveld, H., Lister, M., Wenger, L., Lofgren, B., Quinn, F., and Slivitzky, M. (2000). Climate change impacts on the hydrology of the Great Lakes-St. Lawrence system. *Canadian Water Resources Journal*, 25, 153–79.

Morzaria-Luna, H. N., Turk-Boyer, P., Rosemartin, A., and Camacho-Ibar, V. F. (2014). Vulnerability to climate change of hypersaline salt marshes in the Northern Gulf of California. *Ocean & Coastal Management*, 93, 37–50.

Mossa, J. and McLean, M. (1997). Channel planform and land cover changes on a mined river floodplain: Amite River, Louisiana, USA. *Applied Geography*, 17, 43–54.

Motyka, R. J., O'Neal, S., Connor, C. L., and Echelmeyer, K. A. (2002). Twentieth century thinning of Mendenhall Glacier, Alaska, and its relationship to climate, lake calving and glacier run-off. *Global and Planetary Change*, 35, 93–112.

Moulin, C. and Chiapello, I. (2006). Impact of human-induced desertification on the intensification of Sahel dust emission and export over the last decades. *Geophysical Research Letters*, 33, doi: 10.1029/2006GL025923.

Mousavi, M. E., Irish, J. L., Frey, A. E., Olivera, F., and Edge, B. L. (2011). Global warming and hurricanes: the potential impact of hurricane intensification and sea level rise on coastal flooding. *Climatic Change*, 104, 575–97.

Mugagga, F., Kakembo, V., and Buyinza, M. (2012). Land use changes on the slopes of Mount Elgon and the implications for the occurrence of landslides. *Catena*, 90, 39–46.

Muhly, J. D. (1997). Artifacts of the Neolithic, Bronze and Iron Ages. In *The Oxford Encyclopaedia of Archaeology in the Near East*, ed. E. M. Myers. New York: Oxford University Press, vol. 4, pp. 5–15.

Mulitza, S. and 10 others. (2010). Increase in African dust flux at the onset of commercial agriculture in the Sahel region. *Nature*, 466, 226–8.

Mulla, D. J. and Sekely, A. C. (2009). Historical trends affecting accumulation of sediment and phosphorus in Lake Pepin, upper Mississippi River, USA. *Journal of Paleolimnology*, 41, 589–602.

Mullan, D. (2013). Soil erosion under the impacts of future climate change: assessing the statistical significance of future changes and the potential on-site and off-site problems. *Catena*, 109, 234–46.

Müller-Nedebock, D. and Chaplot, V. (2015). Soil carbon losses by sheet erosion: a potentially critical contribution to the global carbon cycle. *Earth Surface Processes and Landforms*, 40, 1803–13.

Munson, S., Belnap, J., and Okin, G. S. (2011). Responses of wind erosion to climate-induced vegetation changes on the Colorado Plateau. *Proceedings of the National Academy of Sciences*, 108, 3854–9.

Murray, N. J., Clemens, R. S., Phinn, S. R., Possingham, H. P., and Fuller, R. A. (2014). Tracking the rapid loss of tidal wetlands in the Yellow

Sea. *Frontiers in Ecology and the Environment*, 12, 267–72.

Murray-Rust, D. H. (1972). Soil erosion and reservoir sedimentation in a grazing area west of Arusha, northern Tanzania. *Geografiska Annaler*, 54A, 325–43.

Murria, J. (1991). Subsidence due to oil production in Western Venezuela: engineering problems and solutions. *IAHS Publication*, 200, 129–39.

Muturi, G. M., Mohren, G. M. J., and Kimani, J. N. (2009). Prediction of Prosopis species invasion in Kenya using geographical information system techniques. *African Journal of Ecology*, 48, 628–36.

Myer, L. R. and Daley, T. M. (2011). Elements of a best practices approach to induced seismicity in geologic storage. *Energy Procedia*, 4, 3707–13.

Nadeu, E., Gobin, A., Fiener, P., Wesemael, B., and Oost, K. (2015). Modelling the impact of agricultural management on soil carbon stocks at the regional scale: the role of lateral fluxes. *Global Change Biology*, doi: 10.1111/gcb.12889.

Nagel, N. B. (2001). Compaction and subsidence issues within the petroleum industry: from Wilmington to Ekofisk and beyond. *Physics and Chemistry of the Earth, Part A: Solid Earth and Geodesy*, 26, 3–14.

Naik, P. K. and Jay, D. A. (2011). Distinguishing human and climate influences on the Columbia River: changes in mean flow and sediment transport. *Journal of Hydrology*, 404, 259–77.

Nakamura, F., Sudo, T., Kameyama, S., and Jitsu, M. (1997). Influences of channelization on discharge of suspended sediment and wetland vegetation in Kushiro Marsh, northern Japan. *Geomorphology*, 18, 279–89.

Nakamura, K., Tockner, K., and Amano, K. (2006). River and wetland restoration: lessons from Japan. *BioScience*, 56, 419–29.

Nakano, T. and Matsuda, I. (1976). A note on land subsidence in Japan. *Geographical Reports of Tokyo Metropolitan University*, 11, 147–62.

Natural England. (2009). Assessing impacts of wind farm development on blanket peatland in England. Project Report and Guidance, Final Report, http://publications.naturalengland.org.uk/publication/43010 (Accessed June 4, 2015).

Navarro-Pons, M., Muñoz-Perez, J. J., Román-Sierra, J., Tejedor, B., Rodriguez, I., and Gomez-Pina, G. (2007). Morphological evolution in the migrating dune of Valdevaqueros (SW Spain) during an eleven-year period. In *International Conference on Management and Restoration of Coastal Dunes*, Santander, Spain, pp. 80–5.

Navratil, O., Breil, P., Schmitt, L., Grosprêtre, L., and Albert, M. B. (2013). Hydrogeomorphic adjustments of stream channels disturbed by urban runoff (Yzeron River basin, France). *Journal of Hydrology*, 485, 24–36.

Nawaz, M. F., Bourrie, G., and Trolard, F. (2013). Soil compaction impact and modelling. *A review*.

Agronomy for Sustainable Development, 33, 291–309.

Nearing, M. A. (2001). Potential changes in rainfall erosivity in the US with climate change during the 21st century. *Journal of Soil and Water Conservation*, 56, 229–32.

Nearing, W. and 11 others (2005). Modeling response of soil erosion and runoff to changes in precipitation and cover. *Catena*, 61, 131–54.

Neave, M., Rayburg, S., and Swan, A. (2009). River channel change following dam removal in an ephemeral stream. *Australian Geographer*, 40, 235–46.

Neff, J. C. and 9 others. (2008). Increasing eolian dust deposition in the western United States linked to human activity. *Nature Geoscience*, 1, 189–95.

Nelson, F. E. and Anisimov, O. A. (1993). Permafrost zonation in Russia under anthropogenic climate change. *Permafrost and Periglacial Processes*, 4, 137–48.

Neris, J., Tejedor, M., Fuentes, J., and Jiménez, C. (2013). Infiltration, runoff and soil loss in Andisols affected by forest fire (Canary Islands, Spain). *Hydrological Processes*, 27, 2814–24.

Nesje, A., Lie, O., and Dahl, S. O. (2000). Is the North Atlantic Oscillation reflected in glacier mass balance records?' *Journal of Quaternary Science*, 15, 587–601.

Nesje, A., Bakke, J., Dahl, S. O., Lie, O., and Matthews, J. A. (2008). Norwegian mountain glaciers in the past, present and future. *Global and Planetary Change*, 60, 10–27.

New, M., Todd, M., Hulme, M., and Jones, P. (2001). Precipitation measurements and trends in the twentieth century. *International Journal of Climatology*, 21, 1899–922.

Newton, J. G. (1976). Induced and natural sinkholes in Alabama: continuing problem along highway corridors. In *Subsidence over Mines and Caverns*, ed. F. R. Zwanig. Washington, DC: National Academy of Sciences, pp. 9–16.

Nichol, S. L., Augustinus, P. L., Gregory, M. R., Creese, R., and Horrocks, M. (2000). Geomorphic and sedimentary evidence of human impact on the New Zealand landscape. *Physical Geography*, 21, 109–32.

Nicholls, R. J., Hoozemans, F. M. J., and Marchand, M. (1999). Increasing flood risk and wetland losses due to global sea level rise: regional and global analyses. *Global Environmental Change*, 9, S69–87.

Nicholls, R. J., Marinova, N., Lowe, J. A., Brown, S., Vellinga, P., de Gusmão, D., Hinkel, J., and Tol, R. S. J. (2011). Sea-level rise and its possible impacts given a "beyond 4°C world" in the twenty-first century. *Philosophical Transactions of the Royal Society, A*, 369, 161–81.

Nichols, R. A. and Ketcheson, G. L. (2013). A two-decade watershed approach to stream restoration log jam design and stream recovery monitoring:

Finney Creek, Washington. *Journal of the American Water Resources Association*, 49, 1367–84.

Nick, F. M., Vieli, A., Andersen, M. L., Joughin, I., Payne, A., Edwards, T. L., and van de Wal, R. S. (2013). Future sea-level rise from Greenland/'s main outlet glaciers in a warming climate. *Nature*, 497, 235–8.

Nicod, J. (1986). Facteurs physico-chimiques de l'accumulation des formations travertineuses. *Mediterranée*, 10, 161–4.

Nicol, A., Gerstenberger, M., Bromley, C., Carne, R., Chardot, L., Ellis, S., and Viskovic, P. (2013). Induced seismicity; observations, risks and mitigation measures at CO_2 Storage Sites. *Energy Procedia*, 37, 4749–56.

Nicolay, A., Raab, A., Raab, T., Rösler, H., Bönisch, E., and Murray A. S. (2014). Evidence of (pre-) historic to modern landscape and land use history near Jänschwalde (Brandenburg, Germany). *Zeitschrift für Geomorphologie* 58, Suppl. 2, 7–31.

Nieuwenhuis, H. S. and Schokking, F. (1997). Land subsidence in drained peat areas of the Province of Friesland, The Netherlands. *Quarterly Journal of Engineering Geology and Hydrogeology*, 30, 37–48.

Nijssen, B., O'Donnell, G. M., Hamlet, A. F. and Lettenmaier, D. P. (2001). Hydrologic sensitivity of global rivers to climate change. *Climatic Change*, 50, 143–75.

Nilsson, C., Reidy, C. A., Dynesius, M., and Revenga, C. (2005). Fragmentation and flow regulation of the world's large river systems. *Science*, 308, 405–8.

Nir, D. (1983). *Man, a Geomorphological Agent: an Introduction to Anthropic Geomorphology*. Jerusalem: Keter.

Nitto, D. D., Neukermans, G., Koedam, N., Defever, H., Pattyn, F., Kairo, J. G., and Dahdouh-Guebas, F. (2014). Mangroves facing climate change: landward migration potential in response to projected scenarios of sea level rise. *Biogeosciences*, 11, 857–71.

Nolte, S., Esselink, P., Bakker, J. P., and Smit, C. (2014). Effects of livestock species and stocking density on accretion rates in grazed salt marshes. *Estuarine, Coastal and Shelf Science*, 185, 41–7.

Nordstrom, K. F. (1994). Beaches and dunes of human-altered coasts. *Progress in Physical Geography*, 18, 497–516.

(2000). *Beaches and Dunes of Developed Coasts*. Cambridge University Press: Cambridge.

(2014). Living with shore protection structures: a review. *Estuarine, Coastal and Shelf Science*, 150, 11–23.

Nordstrom, K. F., and Hotta, S. (2004). Wind erosion from cropland in the USA: a review of problems, solutions and prospects. *Geoderma*, 121, 157–67.

Nordstrom, K. F., Jackson, N. L., Freestone A. L., Korotky, K. H., and Puleo, J. A. (2012). Effects of beach raking and sand fences on dune dimensions and morphology. *Geomorphology*, 179, 106–15.

Nordstrom K. F., Lampe, R., and Vandemark, L. M. (2000). Reestablishing naturally functioning dunes on developed coasts. *Environmental Management* 25, 37–51.

Noss, R. F. (2011). Between the devil and the deep blue sea: Florida's unenviable position with respect to sea level rise. *Climatic Change*, 107, 1–16.

Notebaert, B. and Verstraeten, G. (2010). Sensitivity of West and Central European river systems to environmental changes during the Holocene: a review. *Earth-Science Reviews*, 103, 163–82.

Novara, A., Gristina, L., Saladino, S. S., Santoro, A., and Cerda, A. (2011). Soil erosion assessment on tillage and alternative soil managements in a Sicilian vineyard. *Soil and Tillage Research*, 117, 140–7.

Nyssen, J. and Vermeersch, D. (2010). Slope aspect affects geomorphic dynamics of coal mining spoil heaps in Belgium. *Geomorphology*, 123, 109–21.

Nyssen, J., Debever, M., Poesen, J., and Deckers, J. (2014). Lynchets in eastern Belgium—a geomorphic feature resulting from non-mechanised crop farming. *Catena*, 121, 164–75.

Nyssen, J., Haile, M., Moeyersons, J., Poesen, J., and Deckers, J. (2000). Soil and water conservation in Tigray (Northern Ethiopia): the traditional dagat technique and its integration with introduced techniques. *Land Degradation & Development*, 11, 199–208.

Nyssen, J., Poesen, J., Moeyersons, J., Luyten, E., Veyret-Picot, M., Deckers, J., and Govers, G. (2002). Impact of road building on gully erosion risk: a case study from the northern Ethiopian highlands. *Earth Surface Processes and Landforms*, 27, 1267–83.

Nyssen, J., Veyret-Picot, M., Poesen, J., Moeyersons, J., Haile, M., Deckers, J., and Govers, G. (2004). The effectiveness of loose rock check dams for gully control in Tigray, northern Ethiopia. *Soil Use and Management*, 20, 55–64.

Olthof, I., Fraser, R. H., and Schmitt, C. (2015). Landsat-based mapping of thermokarst lake dynamics on the Tuktoyaktuk Coastal Plain, Northwest Territories, Canada since 1985. *Remote Sensing of Environment*, 168, 194–204.

O'Neal, M. R., Nearing, M. A., Vining, R. C., Southworth, J., and Pfeifer, R. A. (2005). Climate change impacts on soil erosion in Midwest United States with changes in crop management. *Catena*, 61, 165–84.

O'Sullivan, P. E., Coard, M. A., and Pickering, D. A. (1982). The use of laminated lake sediments in the estimation and calibration of erosion rates. *Publication of the International Association of Hydrological Science*, 137, 385–96.

O'Hara, S. L., Street-Perrott, F. A., and Burt, T. P. (1993). Accelerated soil erosion around a

Mexican highland lake caused by prehispanic agriculture. *Nature*, 362, 48–51.

Oliver, F. W. (1946). Dust-storms in Egypt as noted in Maryut: a supplement. *Geographical Journal*, 108, 221–6.

Oliveira, P. T. S., Nearing, M. A., and Wendland, E. (2015). Orders of magnitude increase in soil erosion associated with land use change from native to cultivated vegetation in a Brazilian savannah environment. *Earth Surface Processes and Landforms*, 40, 1524–33.

Olivier, S., Blaser, C., Brütsch, S., Frolova, N., Gäggeler, H. W., Henderson, K. A., Palmer, A. S., Papina, T., and Schwikowski, M. (2006). Temporal variations of mineral dust, biogenic tracers, and anthropogenic species during the past two centuries from Belukha ice core, Siberian Altai. *Journal of Geophysical Research: Atmospheres* 111, D5, doi: 10.1029/2005JD005830.

Olley, J. M. and Wasson, R. J. (2003). Changes in the flux of sediment in the Upper Murrumbidgee catchment, Southeastern Australia, since European settlement. *Hydrological Processes*, 17, 3307–20.

Olson, K. R., Gennadiyev, A. N., Zhidkin, A. P., Markelov, M. V., Golosov, V. N., and Lang, J. M. (2013). Use of magnetic tracer and radiocesium methods to determine past cropland soil erosion amounts and rates. *Catena*, 104, 103–10.

Oppenheimer, M. (1998). Global warming and the stability of the West Antarctic ice sheet. *Nature*, 393, 325–32.

Ore, G. and Bruins, H. J. (2012). Design features of ancient agricultural terrace walls in the Negev Desert: human-made geodiversity. *Land Degradation & Development*, 23, 409–18.

Orme, A. R. (2002). Human imprints on the primeval landscape. In *The Physical Geography of North America*, ed. A. R. Orme. New York: Oxford University Press, pp. 459–81.

Ormerod, S. J. (2004). A golden age of river restoration science? *Aquatic Conservation: Marine and Freshwater Ecosystems*, 14, 543–9.

Orr, J. C, Pantoja, S., and Pörtner, H.-O. (2005a). Introduction to special section: the ocean in a high-CO2 world. *Journal of Geophysical Research*, 110 (C), doi: 10 1029/2005 JC 003036.

Orr, J. C., and 26 others. (2005b). Anthropogenic ocean acidification over the twenty-first century and its impact on organisms. *Nature*, 437, 681–686.

Orts, W. J., Roa-Espinosa, A., Sojka, R. E., Glenn, G. M., Imam, S. H., Erlacher, K., and Pedersen, J. S. (2007). Use of synthetic polymers and biopolymers for soil stabilization in agricultural, construction and military applications. *Journal of Materials in Civil Engineering*, 19, 58–66.

Orwig, D. A. (2002). Ecosystem to regional impacts of introduced pests and pathogens: historical context, questions and issues. *Journal of Biogeography*, 29, 1471–4.

Osterkamp, T. E. and Romanovsky, V. E. (1999). Evidence for warming and thawing of discontinuous permafrost in Alaska. *Permafrost and Periglacial Processes*, 10, 17–37.

Osterkamp, T. E., Jorgenson, M. T., Schuur, E. A. G., Shur, Y. L., Kanevskiy, M. Z., Vogel, J. G., and Tumskoy, V. E. (2009). Physical and ecological changes associated with warming permafrost and thermokarst in interior Alaska. *Permafrost and Periglacial Processes*, 20, 235–6.

Out, W. A. and Verhoeven, K. (2014). Late Mesolithic and Early Neolithic human impact at Dutch wetland sites: the case study of Hardinxveld-Giessendam De Bruin. *Vegetation History and Archaeobotany*, 23, 41–56.

Overeem, I., Anderson, R. S., Wobus, C. W., Clow, G. D., Urban, F. E., and Matell, N. (2011). Sea ice loss enhances wave action at the Arctic coast. *Geophysical Research Letters*, 38, doi: 10.1029/2011GL048681.

Owen, L. A., Kamp, U., Khattak, G. A., Harp, E. L., Keefer, D. K., and Bauer, M. A. (2008). Landslides triggered by the 8 October 2005 Kashmir earthquake. *Geomorphology*, 94, 1–9.

Ozer, P. (2003). Fifty years of African mineral dust production. *Bulletin Scientifique Academie Royale Sciences d'Outre-Mer*, 49, 371–96.

Özkan, H., Willcox, G., Graner, A., Salamini, F., and Kilian, B. (2011). Geographic distribution and domestication of wild emmer wheat (Triticum dicoccoides). *Genetic Resources and Crop Evolution*, 58, 11–53.

Padonou, E. A., Assogbadjo, A. E., Bachmann, Y., and Sinsin, B. (2013). How far bowalization affects phytodiversity, life forms and plant morphology in Sub-humid tropic in West Africa. *African Journal of Ecology*, 51, 255–62.

Padonou, E. A., Fandohan, B., Bachmann, Y., and Sinsin, B. (2014). How farmers perceive and cope with 'bowalization': a case study from West Africa. *Land Use Policy*, 36, 461–7.

Page, M. J. and Trustrum, N. A. (1997). A late Holocene lake sediment record of the erosion response to land use change in a steepland catchment, New Zealand. *Zeitschrift für Geomorphologie*, 41, 369–92.

Painter, T. H., Flanner, M. G., Kaser, G., Marzeion, B., VanCuren, R. A., and Abdalati, W. (2013). End of the Little Ice Age in the Alps forced by industrial black carbon. *Proceedings of the National Academy of Sciences*, 110, 15216–21.

Palmer, M., Allan, J. D., Meyer, J., and Bernhardt, E. S. (2007). River restoration in the twenty-first century: data and experiential knowledge to inform future efforts. *Restoration Ecology*, 15, 472–81.

Palmer, M. A., Bernhardt, E. S., Allan, J. D., Lake, P. S., Alexander, G., Brooks, S., and Sudduth, E. (2005). Standards for ecologically successful river restoration. *Journal of Applied Ecology*, 42, 208–17.

Palumbi, S. R., Barshis, D. J., Traylor-Knowles, N. and Bay, R. A. (2014). Mechanisms of reef coral resistance to future climate change. *Science*, 344, 895–8.

Pandey, D. N. (2000). Sacred water and sanctified vegetation: tanks and trees in India. In *Conference of the International Association for the Study of Common Property (IASCP), in the Panel "Constituting the Riparian Commons," Bloomington, Indiana, USA*, 31, 21 p.

Pando, L., Pulgar, J. A., and Gutiérrez-Claverol, M. (2013). A case of man-induced ground subsidence and building settlement related to karstified gypsum (Oviedo, NW Spain). *Environmental Earth Sciences*, 68, 507–19.

Pandolfi, J. M., and 10 others. (2005). Are U.S.coral reefs on the slippery slope to slime? *Science*, 307, 1725–6.

Pandolfi, J. M., Connolly, S. R., Marshall, D. J., and Cohen, A. L. (2011). Projecting coral reef futures under global warming and ocean acidification. *Science*, 333, 418–22.

Pardini, G., Gispert, M., and Dunjó, G. (2004). Relative influence of wildfire on soil properties and erosion processes in different Mediterranean environments in NE Spain. *Science of the Total Environment*, 328, 237–46.

Parfitt, S. A., and 15 others (2010). Early Pleistocene human occupance at the edge of the boreal zone in northwest Europe. *Nature*, 466, 229–33.

Parizek, B. R. and Alley, R. B. (2004). Implications of increased Greenland surface melt under global-warming scenarios: ice-sheet simulations. *Quaternary Science Reviews*, 23, 1013–27.

Park, R. A., Armentano, T. V., and Cloonan, C. L. (1986). Predicting the effects of sea level rise on coastal wetlands. In *Effects of Changes in Stratospheric Ozone and Global Climate, Vol. 4, Sea Level Rise*, ed. J. G. Titus. Washington, DC: UNEP/USEPA, pp. 129–52.

Parkinson, R. W., Harlem, P. W., and Meeder, J. F. (2014). Managing the Anthropocene marine transgression to the year 2100 and beyond in the State of Florida, USA. *Climatic Change*, 128, 85–98.

Paroissien, J. B., Lagacherie, P., and Le Bissonnais, Y. (2010). A regional-scale study of multi-decennial erosion of vineyard fields using vine-stock unearthing–burying measurements. *Catena*, 82, 159–68.

Parry, L. E., Holden, J., and Chapman, P. J. (2014). Restoration of blanket peatlands. *Journal of Environmental Management*, 133, 193–205.

Parshall, T. and Foster, D. R. (2002). Fire on the New England landscape: regional and temporal variation, cultural and environmental controls. *Journal of Biogeography*, 29, 1305–17.

Parson, E. A., Carter, L., Anderson, P., Wang, B., and Weller, G. (2001). Potential consequences of climate variability and change for Alaska. In *Climate Change Impacts on the United States: the Potential Consequences of Climate Variability and Change*, ed. National Assessment Synthesis Team. Cambridge: Cambridge University Press, pp. 283–312.

Parsons, A. J., Abrahams, A. D., and Wainwright, J. (1996). Responses of interrill runoff and erosion rates to vegetation change in southern Arizona. *Geomorphology*, 14, 311–7.

Pasternack, G. B., Brush, G. S., and Hilgartner, W. B. (2001). Impact of historic land-use change on sediment delivery to a Chesapeake Bay subestuarine delta. *Earth Surface Processes and Landforms*, 26, 409–27.

Pausas, J. G., Llovet, J., Rodrigo, A., and Vallejo, R. (2009). Are wildfires a disaster in the Mediterranean basin? – A review. *International Journal of Wildland Fire*, 17, 713–23.

Pearson, A. J., Snyder, N. P., and Collins, M. J. (2011). Rates and processes of channel response to dam removal with a sand-filled impoundment. *Water Resources Research* 47, W08504, doi:10.1029/2010WR009733.

Pechony, O. and Shindell, D. T. (2010). Driving forces of global wildfires over the past millennium and the forthcoming century. *Proceedings of the National Academy of Sciences*, 107, 19167–70.

Peck, J. A. and Kasper, N. R. (2013). Multiyear assessment of the sedimentological impacts of the removal of the Munroe Falls Dam on the middle Cuyahoga River, Ohio. *Reviews in Engineering Geology*, 21, 81–92.

Pederson, G. T., Gray, S. T., Woodhouse, C. A., Betancourt, J. L., Fagre, D. B., Littell, J. S., and Graumlich, L. J. (2011). The unusual nature of recent snowpack declines in the North American Cordillera. *Science*, 333, 332–5.

Pelejero, C., Calvo, E., and Hoegh-Guldberg, O. (2010). Paleo-perspectives on ocean acidification. *Trends in Ecology and Evolution*, 25, 332–44.

Pellatt, M. G. and Gedalof, Z. E. (2014). Environmental change in Garry oak (Quercus garryana) ecosystems: the evolution of an eco-cultural landscape. *Biodiversity and Conservation*, 23, 2053–67, doi: 10.1007/s10531-014-0703-9.

Pelletier, J. D., Brad Murray, A., Pierce, J. L., Bierman, P. R., Breshears, D. D., Crosby, B. T., and Yager, E. M. (2015). Forecasting the response of Earth's surface to future climatic and land-use changes: A review of methods and research needs. *Earth's Future*, 3, 220–51, doi: 10.1002/2014EF000290.

Pellicciotti, F., Carenzo, M., Bordoy, R., and Stoffel, M. (2014). Changes in glaciers in the Swiss Alps and impact on basin hydrology: current state of the art and future research. *Science of the Total Environment*, 493, 1152–70.

Peng, H., Ma, W., Mu, Y. H., Jin, L., and Yuan, K. (2015). Degradation characteristics of permafrost under the effect of climate warming and engineering disturbance along the Qinghai–Tibet Highway. *Natural Hazards*, 75, 2589–605.

Pennington, W. (1981). Records of a lake's life in time: the sediments. *Hydrobiologia*, 79, 197–219.

Pereira, H. C. (1973). *Land Use and Water Resources in Temperate and Tropical Climates.* Cambridge: Cambridge University Press.

Perevolotsky, A. and Seligman, N. A. G. (1998). Role of grazing in Mediterranean rangeland ecosystems. *Bioscience*, 48, 1007–17.

Perkol-Finkel, S., Shashar, N., and Benayahu, Y. (2006). Can artificial reefs mimic natural reef communities? The roles of structural features and age. *Marine Environmental Research*, 61, 121–35.

Perroy, R. L., Bookhagen, B., Chadwick, O. A., and Howarth, J. T. (2012). Holocene and anthropocene landscape change: arroyo formation on Santa Cruz Island, California. *Annals of the Association of American Geographers*, 102, 1229–50.

Perry, C. T., Murphy, G. N., Kench, P. S., Smithers, S. G., Edinger, E. N., Steneck, R. S., and Mumby, P. J. (2013). Caribbean-wide decline in carbonate production threatens coral reef growth. *Nature Communications*, 4, 1402.

Perski, Z., Hanssen, R., Wojcik, A., and Wojciechowski, T. (2009). InSAR analyses of terrain deformation near the Wieliczka Salt Mine, Poland. *Engineering Geology*, 106, 58–67.

Peters, F. (1999). Bronze Age barrows: factors influencing their survival and destruction. *Oxford Journal of Archaeology*, 18, 255–64.

Pethick, J. (1993). Shoreline adjustments and coastal management: physical and biological processes under accelerated sea level rise. *Geographical Journal*, 159, 162–8.

Pethick, J. and Orford, J. D. (2013). Rapid rise in effective sea-level in southwest Bangladesh: its causes and contemporary rates. *Global and Planetary Change*, 111, 237–45.

Petit, F., Poinsart, D., and Bravard, J. P. (1996). Channel incision, gravel mining and bedload transport in the Rhône river upstream of Lyon, France ("Canal de Miribel"). *Catena*, 26, 209–26.

Petley, D. N., Hearn, G. J., Hart, A., Rosser, N., Dunning, S. A., Oven, K., and Mitchell, W. A. (2007). Trends in landslide occurrence in Nepal. *Natural Hazards*, 43, 23–44.

Péwé, T. L. (ed.) (1981). Desert dust: origin, characteristics and effect on man. *Geological Society of America Special Paper* 186.

Phien-wej, N., Giao, P. H., and Nutalaya, P. (2006). Land subsidence in Bangkok, Thailand. *Engineering Geology*, 82, 187–201.

Phillips, J. D. (1997). Humans as geological agents and the question of scale. *American Journal of Science*, 297, 98–115.

Pianalto, F. S. and Yool, S. R. (2013). Monitoring fugitive dust emission sources arising from construction: a remote-sensing approach. *GIScience & Remote Sensing*, 50, 251–70.

Piccarreta, M., Caldara, M., Capolongo, D., and Boenzi, F. (2011). Holocene geomorphic activity related to climatic change and human impact in Basilicata, Southern Italy. *Geomorphology*, 128, 137–47.

Pierce, J. L., Meyer, G. A., and Jull, A. T. (2004). Fire-induced erosion and millennial-scale climate change in northern ponderosa pine forests. *Nature*, 432, 87–90.

Pierson, F. B., Robichaud, P. R., and Spaeth, K. E. (2001). Spatial and temporal effects of wildfire on the hydrology of a steep rangeland watershed. *Hydrological Processes*, 15, 2905–16.

Pierson, F. B., Moffet, C. A., Williams, C. J., Hardegree, S. P., and Clark, P. E. (2009). Prescribed-fire effects on rill and interrill runoff and erosion in a mountainous sagebrush landscape. *Earth Surface Processes and Landforms*, 34, 193–203.

Pierson, F. B., Robichaud, P. R., Moffet, C. A., Spaeth, K. E., Hardegree, S. P., Clark, P. E., and Williams, C. J. (2008). Fire effects on rangeland hydrology and erosion in a steep sagebrush-dominated landscape. *Hydrological Processes*, 22, 2916–29.

Pierson, F. B., Williams, C. J., Hardegree, S. P., Weltz, M. A., Stone, J. J., and Clark, P. E. (2011). Fire, plant invasions, and erosion events on western rangelands. *Rangeland Ecology & Management*, 64, 439–49.

Piggott, S. (1935). A note on the relative chronology of the English long barrows. *Proceedings of the Prehistoric Society (New Series)*, 1, 115–26.

Pimentel, D. (1976). Land degradation: effects on food and energy resources. *Science*, 194, 149–55.

Pimentel, D., Harvey, C., Resosudarmo, P., Sinclair, K., Kurz, D., McNair, M., Crist, S., Shpritz, L., Fitton, L., Saffouri, R., and Blair, R. (1995). Environmental and economic costs of soil erosion and conservation benefits. *Science*, 267, 1117–22.

Piperno, D. R., McMichael, C., and Bush, M. B. (2015). Amazonia and the Anthropocene: what was the spatial extent and intensity of human landscape modification in the Amazon Basin at the end of prehistory? *The Holocene*, 25, 1588–97.

Pirazzoli, P. A. (1996). *Sea-level Changes—The Last 20,000 Years.* Chichester: Wiley.

Pizzuto, J. and O'Neal, M. (2009). Increased mid-twentieth century riverbank erosion rates related to the demise of mill dams, South River, Virginia. *Geology*, 37, 19–22.

Plug, L. J., Walls, C., and Scott, B. M. (2008). Tundra lake changes from 1978 to 2001 on the Tuktoyaktuk Peninsula, western Canadian Arctic. *Geophysical Research Letters*, 35, doi: 10.1029/2007GL032303.

Poeppl, R. E., Keesstra, S. D., and Hein, T. (2015). The geomorphic legacy of small dams—An Austrian study. *Anthropocene*, 10, 43–55.

Poesen, J. W., Torri, D., and Bunte, K. (1994). Effects of rock fragments on soil erosion by water and different spatial scales: a review. *Catena*, 23, 141–66.

Poesen, J. W. E., Nachtergaele, J., Verstraeten, G., and Valentin, C. (2003). Gully erosion and environmental change: importance and research needs. *Catena*, 50, 91–133.

Poirier, C., Chaumillon, E., and Arnaud, F. (2011). Siltation of river-influenced coastal environments: respective impact of late Holocene land use and high-frequency climate changes. *Marine Geology*, 290, 51–62.

Polidoro, B. A. and 20 others. (2010). The loss of species: mangrove extinction risk and geographic areas of global concern. *PLoS One*, 5, e10095.

Pollard, E. and Miller, A. (1968). Wind erosion of the East Anglian Fens. *Weather*, 23, 414–7.

Pollen-Bankhead, N., Simon, A., Jaeger, K., and Wohl, E. (2009). Destabilization of streambanks by removal of invasive species in Canyon de Chelly National Monument, Arizona. *Geomorphology*, 103, 363–74.

Pomeroy, J., Fang, X., and Ellis, C. (2012). Sensitivity of snowmelt hydrology in Marmot Creek, Alberta, to forest cover disturbance. *Hydrological Processes*, 26, 1891–904.

Pongratz, J., Reick, C., Raddatz, T., and Claussen, M. (2008). A reconstruction of global agricultural areas and land cover for the last millennium. *Global Biogeochemical Cycles*, 22(3), GB 3018, doi: 10.1029/2007GB003153.

Pope, G. A. and Rubenstein, R. (1999). Anthroweathering: theoretical framework and case study for human-impacted weathering. *Geoarchaeology*, 14, 247–64.

Potemkina, T. G. and Potemkin, V. L. (2014). Sediment load of the main rivers of Lake Baikal in a changing environment (east Siberia, Russia). *Quaternary International*, 380–381, 342–9.

Pranzini, E., Wetzel, L., and Williams, A. T. (2015). Aspects of coastal erosion and protection in Europe. *Journal of Coastal Conservation*, 19, 445–59, doi: 10.1007/s11852-015-0399-3

Pratt, W. E. and Johnson, D. W. (1926). Local subsidence of the Goose Creek oil field. *Journal of Geology*, 34, 577–90.

Price, K. (2011). Effects of watershed topography, soils, land use, and climate on baseflow hydrology in humid regions: A review. *Progress in Physical Geography*, 35, 465–92.

Price, S. J., Ford, J. R., Cooper, A. H., and Neal, C. (2011). Humans as major geological and geomorphological agents in the Anthropocene: the significance of artificial ground in Great Britain. *Philosophical Transactions of the Royal Society, A*, 369, 1056–84.

Prince, H. C. (1962). Pits and ponds in Norfolk. *Erdkunde*, 16, 10–31.

(1964). The origin of pits and depressions in Norfolk. *Geography*, 49, 15–32.

Pringle, A. W. (1996). History, geomorphological problems and effects of dredging in Cleveland Bay, Queensland. *Australian Geographical Studies*, 34, 58–80.

Pritchard, H. D. and Vaughan, D. G. (2007). Widespread acceleration of tidewater glaciers on the Antarctic Peninsula. *Journal of Geophysical Research*, 112: F03S29, doi:10.1029/2006JF000597.

Prosser, C. D., Burek, C. V., Evans, D. H., Gordon, J. E., Kirkbride, V. B., Rennie, A. F., and Walmsley, C. A. (2010). Conserving geodiversity sites in a changing climate: management challenges and responses. *Geoheritage*, 2, 123–36.

Prosser, I. P. and Williams, L. (1998). The effect of wildfire on runoff and erosion in native Eucalyptus forest. *Hydrological Processes*, 12, 251–65.

Provoost, S., Jones, M. L. M. and Edmondson. (2011). Changes in landscape and vegetation of coastal dunes in northwest Europe: a review. *Journal of Coastal Conservation*, 15, 207–26.

Purvis, K. G., Gramling, J. M., and Murren, C. J. (2015). Assessment of beach cccess paths on dune vegetation: diversity, abundance, and cover. *Journal of Coastal Research*, 31, 1222–8.

Pye, K. and Blott, S. J. (2014). The geomorphology of UK estuaries: the role of geological controls, antecedent conditions and human activities. *Estuarine, Coastal and Shelf Science*, 150B, 196–214.

Pye, K., Blott, S. J., and Howe, M. A. (2014). Coastal dune stabilization in Wales and requirements for rejuvenation. *Journal of Coastal Conservation*, 18, 27–54.

Pye, K. and Neal, A. (1994). Coastal dune erosion at Formby Point, north Merseyside, England: causes and mechanisms. *Marine Geology*, 119, 39–56.

Pye, K. and Tsoar, H. (1990). *Aeolian Sand and Sand Dunes*. London: Unwin Hyman.

Quataert, E., Storlazzi, C., Rooijen, A., Cheriton, O., and Dongeren, A. (2015). The influence of coral reefs and climate change on wave-driven flooding of tropical coastlines. *Geophysical Research Letters*, 42, 6407–15. doi: 10.1002/2015GL064861.

Qiu, J. (2012). Evidence mounts for dam-quake link. *Science*, 336, 291.

Qiu, X. and Fenton, C. (2015). Factors controlling the occurrence of reservoir-induced seismicity. *Engineering Geology for Society and Territory*, 6, 567–70.

Quinton, J. N., Govers, G., Van Oost, K., and Bardgett, R. D. (2010). The impact of agricultural soil erosion on biogeochemical cycling. *Nature Geoscience*, 3, 311–4.

Quiquerez, A., Brenot, J., Garcia, J. P., and Petit, C. (2008). Soil degradation caused by a high-intensity rainfall event: implications for medium-term soil sustainability in Burgundian vineyards. *Catena*, 73, 89–97.

Quisthoudt, K., Adams, J., Rajkaran, A., Dahdouh-Guebas, F., Koedam, N., and Randin, C. F.

(2013). Disentangling the effects of global climate and regional land-use change on the current and future distribution of mangroves in South Africa. *Biodiversity and Conservation*, 22, 1369–90.

Raabe, E. A. and Stumpf, R. P. (2015). Expansion of tidal marsh in response to sea-level rise: Gulf Coast of Florida, USA. *Estuaries and Coasts*, doi: 10.1007/s12237-015-9974-y.

Raclot, D., Le Bissonnais, Y., Louchart, X., Andrieux, P., Moussa, R., and Voltz, M. (2009). Soil tillage and scale effects on erosion from fields to catchment in a Mediterranean vineyard area. *Agriculture, Ecosystems and Environment*, 134, 201–10.

Raczky, P. (2015). Settlements in South-east Europe. In *The Oxford Handbook of Neolithic Europe*, ed. C. Fowler, J. Harding, D. Hofmann. Oxford: Oxford University Press. pp. 235–53.

Radic, V. and Hock, R. (2011). Regionally differentiated contribution of mountain glaciers and ice caps to future sea-level rise. *Nature Geoscience*, 4, 91–4.

Radić, V., Bliss, A., Beedlow, A. C., Hock, R., Miles, E., and Cogley, J. G. (2014). Regional and global projections of twenty-first century glacier mass changes in response to climate scenarios from global climate models. *Climate Dynamics*, 42, 37–58.

Radivojević, M., Rehren, T., Pernicka, E., Šljivar, D., Brauns, M., and Borić, D. (2010). On the origins of extractive metallurgy: new evidence from Europe. *Journal of Archaeological Science*, 37, 2775–87.

Radley, J. (1962). Peat erosion on the high moors of Derbyshire and west Yorkshire. *East Midlands Geographer*, 3, 40–50.

Rădoane, M., Obreja, F., Cristea, I., and Mihăilă, D. (2013). Changes in the channel-bed level of the eastern Carpathian rivers: climatic vs. human control over the last 50 years. *Geomorphology*, 193, 91–111.

Raharimahefa, T. and Kusky, T. M. (2010). Environmental monitoring of Bombetoka Bay and the Betsiboka Estuary, Madagascar, using multi-temporal satellite data. *Journal of Earth Science*, 21, 210–26.

Rahm, D. (2011). Regulating hydraulic fracturing in shale gas plays: the case of Texas. *Energy policy*, 39, 2974–81.

Rahman, M. (1981). Ecology of Karez irrigation: a case of Pakistan. *GeoJournal*, 5, 7–15.

Rahmstorf, S. (2007). A semi-empirical approach to projecting future sea-level rise. *Science*, 315, 368–70.

Räike, A., Kortelainen, P., Mattsson, T., and Thomas, D. N. (2012). 36 year trends in dissolved organic carbon export from Finnish rivers to the Baltic Sea. *Science of the Total Environment*, 435, 188–201.

Raji, B. A., Utovbisere, E. O. and Momodu, A. B. (2004). Impact of sand dune stabilization structures on soil and yield of millet in the semi-arid region of NW Nigeria. *Environmental Monitoring and Assessment*, 99, 181–96.

Ramankutty, N., Heller, E., and Rhemtulla, J. (2010). Prevailing myths about agricultural abandonment and forest regrowth in the United States. *Annals of the Association of American Geographers*, 100, 502–12.

Ramos, M. C. and Martínez-Casasnovas, J. A. (2007). Soil loss and soil water content affected by land levelling in Penedès vineyards, NE Spain. *Catena*, 71, 210–7.

Ramos-Scharrón, C. E. and MacDonald, L. H. (2005). Measurement and prediction of sediment production from unpaved roads, St John, US Virgin Islands. *Earth Surface Processes and Landforms*, 30, 1283–304.

Ran, L., Lu, X. X., and Xin, Z. (2014). Erosion-induced massive organic carbon burial and carbon emission in the Yellow River basin, China. *Biogeosciences*, 11, 945–59.

Ran, L., Lu, X. X., Xin, Z., and Yang, X. (2013). Cumulative sediment trapping by reservoirs in large river basins: a case study of the Yellow River basin. *Global and Planetary Change*, 100, 308–19.

Ranwell, D. S. (1964). *Spartina* salt marshes in southern England, II: rate and seasonal pattern of sediment accretion. *Journal of Ecology*, 52, 79–94.

Ranwell, D. S. and Boar, R. (1986). *Coast Dune Management Guide*. Abbots Ripton: Institute of Terrestrial Ecology.

Rao, K. N., Subraelu, P., Naga Kumar, K., Ch., V., Demudu, G., Hema Malini, B., and Rajawat, A. S. (2010). Impacts of sediment retention by dams on delta shoreline recession: evidences from the Krishna and Godavari deltas, India. *Earth Surface Processes and Landforms*, 35, 817–27.

Rapp, A., Murray-Rust, D. H., Christiansson, C., and Berry, L. (1972). Soil erosion and sedimentation in four catchments near Dodoma, Tanzania. *Geografiska Annaler*, 54A, 255–318.

Rasmussen, P. and Bradshaw, E. G. (2005). Mid-to late-Holocene land-use change and lake development at Dallund S0, Denmark: study aims, natural and cultural setting, chronology and soil erosion history. *The Holocene*, 15, 1105–15.

Ravanel, L. and Deline, P. (2011). Climate influence on rockfalls in high-Alpine steep rockwalls: the north side of the Aiguilles de Chamonix (Mont Blanc massif) since the end of the 'Little Ice Age'. *The Holocene*, 21, 357–65.

(2015). Rockfall hazard in the Mont Blanc Massif increased by the current atmospheric warming. In *Engineering Geology for Society and Territory-Volume 1*, ed. G. Lollino, A. Manconi, J. Clague, W. Shan, and M. Chiarle. Switzerland: Springer International Publishing, pp. 425–8.

Ravi, S., D'Odorico, P., Wang, J., White, C. S., Okin, G. S., Macko, S. A., and Collins, S. L. (2009).

Post-fire resource redistribution in desert grasslands: a possible negative feedback on land degradation. *Ecosystems*, 12, 434–44.

Ravi, S., D'Odorico, P., Breshears, D. D., Field, J. P., Goudie, A. S., Huxman, T. E., Li, J., Okin, G. S., Swap, R. J., Thomas, A. D., and Van Pelt, S. (2011). Aeolian processes and the biosphere. *Review of Geophysics*, 49, RG3001, doi: 10.1029/2010RG000328.

Raymond, P. A. and Oh, N. H. (2009). Long term changes of chemical weathering products in rivers heavily impacted from acid mine drainage: Insights on the impact of coal mining on regional and global carbon and sulfur budgets. *Earth and Planetary Science Letters*, 284, 50–6.

Reed, A. J., Mann, M. E., Emanuel, K. A., Lin, N., Horton, B. P., Kemp, A. C., and Donnelly, J. P. (2015). Increased threat of tropical cyclones and coastal flooding to New York City during the anthropogenic era. *Proceedings of the National Academy of Sciences*, 201513127.

Reed, D. J. (1990).The impact of sea level rise on coastal salt marshes. *Progress in Physical Geography*, 14, 465–81.

  (1995). The response of coastal marshes to sea level rise: survival or submergence? *Earth Surface Processes and Landforms*, 20, 39–48.

  (2002). Sea-level rise and coastal marsh sustainability: geological and ecological factors in the Mississippi delta plain. *Geomorphology*, 48, 233–43.

Rees, H. G. and Collins, D. N. (2006). Regional differences in response of flow in glacier-fed Himalayan rivers to climatic warming. *Hydrological Processes*, 20, 2157–69.

Reid, D. and Church, M. (2015). Geomorphic and ecological consequences of riprap placement in river systems. *Journal of the American Water Resources Association*, 51, 1043–59.

Reid, L. M. and Dunne, T. (1984). Sediment production from forest road surfaces. *Water Resources Research*, 20, 1753–61.

Remini, B., Achour, B., and Albergel, J. (2011). Timimoun's foggara (Algeria): an heritage in danger. *Arabian Journal of Geosciences*, 4, 495–506.

Remondo, J., Soto, J., González-Díez, A., Díaz de Terán, J. R., and Cendrero, A. (2005). Human impact on geomorphic processes and hazards in mountain areas in northern Spain. *Geomorphology*, 66, 69–84.

Rempel, L. L. and Church, M. (2009). Physical and ecological response to disturbance by gravel mining in a large alluvial river. *Canadian Journal of Fisheries and Aquatic Sciences*, 66, 52–71.

Renaud, F. G., Syvitski, J. P., Sebesvari, Z., Werners, S. E., Kremer, H., Kuenzer, C., and Friedrich, J. (2013). Tipping from the Holocene to the Anthropocene: how threatened are major world deltas? *Current Opinion in Environmental Sustainability*, 5, 644–54.

Rendle, E. J. and Esteves, L. (2011). Developing protocols for assessing the performance of artificial surfing reefs – a new breed of coastal engineering. In *Littoral 2010 – Adapting to Global Change at the Coast: Leadership, Innovation, and Investment* (p. 12008). EDP Sciences.

Rendle, E. J. and Rodwell, L. D. (2014). Artificial surf reefs: A preliminary assessment of the potential to enhance a coastal economy. *Marine Policy*, 45, 349–58.

Restrepo, A. and Juan, D. (2012). Assessing the effect of sea-level change and human activities on a major delta on the Pacific coast of northern South America: the Patía River. *Geomorphology*, 151, 207–23.

Reusser, L., Bierman, P., and Rood, D. (2015). Quantifying human impacts on rates of erosion and sediment transport at a landscape scale. *Geology*, 43, 171–4.

Revell, D. L., Battalio, R., Spear, B., Ruggiero, P., and Vandever, J. (2011). A methodology for predicting future coastal hazards due to sea-level rise on the California Coast. *Climatic Change*, 109, (supplement), S251–76.

Reynard, E. (2007). Geomorphosites and geodiversity: a new domain of research. *Géomorphologie*, 3, 181–188.

Rhind P. and Jones, R. (2009). A framework for the management of sand dune systems in Wales. *Journal of Coastal Conservation*, 13, 15–23.

Ricaurte, L. F., Boesch, S., Jokela, J., and Tockner, K. (2012). The distribution and environmental state of vegetated islands within human-impacted European rivers. *Freshwater Biology*, 57, 2539–49.

Ricci, M., Bertini, A., Capezzuoli, E., Horvatinčić, N., Andrews, J. E., Fauquette, S., and Fedi, M. (2014). Palynological investigation of a Late Quaternary calcareous tufa and travertine deposit: a case study of Bagnoli in the Valdelsa Basin (Tuscany, central Italy). *Review of Palaeobotany and Palynology*, 218, 184–97.

Richards, J. F. (1991). Land transformation. In *The Earth as Transformed by Human Action*, ed. B. L. Turner, W. C. Clark, R. W. Kates, J. F. Richards, J. T. Matthews, and W. B. Meyer. Cambridge: Cambridge University Press, pp. 163–78.

Richardson, D. M., Holmes, P. M., Esler, K. J., Galatowitsch, S. M., Stromberg, J. C., Kirkman, S. P., and Hobbs, R. J. (2007). Riparian vegetation: degradation, alien plant invasions, and restoration prospects. *Diversity and Distributions*, 13, 126–39.

Richardson, J. A. (1976). Pit heap into pasture. In *Reclamation*, ed. J. Lenihan and W. W. Fletcher. Glasgow: Blackie, pp. 60–93.

Richardson, J. M., Fuller, I. C., Holt, K. A., Litchfield, N. J., and Macklin, M. G. (2014). Rapid post-settlement floodplain accumulation in Northland, New Zealand. *Catena*, 113, 292–305.

Richardson, S. J. and Smith, J. (1977). Peat wastage in the East Anglian Fens. *Journal of Soil Science*, 28, 485–9.

Rick, T. C., Sillett, T. S., Ghalambor, C. K., Hofman, C. A., Ralls, K., Anderson, R. S., and Morrison, S. A. (2014). Ecological Change on California's Channel Islands from the Pleistocene to the Anthropocene. *BioScience*, doi: 10.1093/biosci/biu094.

Rickson, R. J. (2006). Controlling sediment at source: an evaluation of erosion control geotextiles. *Earth Surface Processes and Landforms*, 31, 550–60.

Ridley, D. A., Heald, C. L., and Prospero, J. M. (2014). What controls the recent changes in African mineral dust aerosol across the Atlantic? *Atmospheric Chemistry and Physics Discussions*, 14, 3583–627.

Ries, J. B., Andres, K., Wirtz, S., Tumbrink, J., Wilms, T., Peter, K. D., Burczyk, M., Butzen, V., and Seeger, M. (2014). Sheep and goat erosion–experimental geomorphology a as an approach for the quantification of underestimated processes. *Zeitschrift für Geomorphologie, Supplement*, 58, 23–45.

Rignot, E. and Kanagaratnam, P. (2006). Changes in the velocity structure of the Greenland Ice Sheet. *Science*, 311, 986–90.

Rignot, E. and Thomas, R. H. (2002). Mass balance of polar ice sheets. *Science*, 297, 1502–6.

Rignot, E., Velicogna, I., van der Broejke, M. R. Monaghan, A., and Lenaerts, J. (2011). Acceleration of the contribution of the Greenland and Antarctica ice sheets to sea level rise. *Geophysical Research Letters*, 38, L05503, doi: 10.1029/2011GL046583.

Riksen, M., Spaan, W., Arrué, J. L., and López, M. V. (2003). What to do about wind erosion. In *Wind Erosion on Agricultural Land in Europe*, ed. A. Warren. Luxembourg: European Commission, pp. 39–52.

Rinaldi, M., Wyżga, B., and Surian, N. (2005). Sediment mining in alluvial channels: physical effects and management perspectives. *River Research and Applications*, 21, 805–28.

Riordan, B., Verbyla, D., and McGuire, A. D. (2006). Shrinking ponds in subarctic Alaska based on 1950–2002 remotely sensed images. *Journal of Geophysical Research: Biogeosciences*, 111 (G4), doi: 10.1029/2005JG000150.

Rippon, S. (2000). *The Transformation of Coastal Wetlands: Exploitation and Management of Marshland Landscapes in North West Europe during the Roman and Medieval Periods*. Oxford: Oxford University Press.

(2006). *The Somerset Wetlands: an Ever Changing Environment*. Taunton: Somerset Archaeological & Natural History Society, pp. 47–56.

Risk, M. J. (2014). Assessing the effects of sediments and nutrients on coral reefs. *Current Opinion in Environmental Sustainability*, 7, 108–17.

Ritchie, W. and Gimingham, C. H. (1989). Restoration of coastal dunes breached by pipeline landfalls in north-east Scotland. *Proceedings of the Royal Society of Edinburgh. Section B. Biological Sciences*, 96, 231–45.

Rivas, V., Rix, K., Frances, E., Cendrero, A., and Brunsden, D. (1997). Geomorphological indicators for environmental impact assessment: consumable and non-consumable geomorphological resources. *Geomorphology*, 18, 169–82.

Rivas, V., Cendrero, A., Hurtado, M., Cabral, M., Giménez, J., Forte, L., and Becker, A. (2006). Geomorphic consequences of urban development and mining activities; an analysis of study areas in Spain and Argentina. *Geomorphology*, 73, 185–206.

Roberts, B. W., Thornton, C. P., and Pigott, V. C. (2009). Development of metallurgy in Eurasia. *Antiquity*, 83, 1012–22.

Roberts, R. G. and 10 others. (2001). New ages for the least Australian megafauna: continent-wide extinction about 46,000 years ago. *Science*, 292, 1888–92.

Robichaud, P. R. (2000). Fire effects on infiltration rates after prescribed fire in Northern Rocky Mountain forests, USA. *Journal of Hydrology*, 231, 220–9.

Robinson, D. N. (1968). Soil erosion by wind in Lincolnshire, March, 1968. *East Midland Geographer*, 4, 351–62.

Robinson, M. (1990). *Impact of Improved Land Drainage on River Flows*. Institute of Hydrology, Wallingford, Report 113.

Robinson, M. A. and Lambrick, G. H. (1984). Holocene alluviation and hydrology in the Upper Thames Basin. *Nature*, 308, 809–14.

Robroek, B. J., Smart, R. P., and Holden, J. (2010). Sensitivity of blanket peat vegetation and hydrochemistry to local disturbances. *Science of the Total Environment*, 408, 5028–34.

Rockström, J. (2015). In *The Observer*, November 15, 2015, p. 35.

Rockström, J. and 28 others. (2009). Planetary boundaries: exploring the safe operating space for humanity. *Ecology and Society*, 14, http://www.ecologyandsociety.org/vol14/iss2/art32/.

Rodgers, M., O'Connor, M., Robinson, M., Muller, M., Poole, R., and Xiao, L. (2011). Suspended solid yield from forest harvesting on upland blanket peat. *Hydrological Processes*, 25, 207–16.

Rodriguez, A. B., Fodrie, F. J., Ridge, J. T., Lindquist, N. L., Theuerkauf, E. J., Coleman, S. E., and Kenworthy, M. D. (2014). Oyster reefs can outpace sea-level rise. *Nature Climate Change*, 4, 493–7.

Rodway-Dyer, S. J. and Walling, D. E. (2010). The use of ^{137}Cs to establish longer-term soil erosion rates on footpaths in the UK. *Journal of Environmental Management*, 91, 1952–62.

Rogge, W. F., Medeiros, P. M., and Simoneit, B. R. (2006). Organic marker compounds for surface soil and fugitive dust from open lot dairies and cattle feedlots. *Atmospheric Environment* 40, 27–49.

Rojstaczer, S. and Deverel, S. J. (1995). Land subsidence in drained histosols and highly organic mineral soils of California. *Soil Science Society of America Journal*, 59, 1162–7.

Romero-Diaz, A., Belmonte-Serrato, F., and Ruiz-Sinoga, J. D. (2010). The geomorphic impact of afforestations on soil erosion in southeast Spain. *Land Degradation and Development*, 21, 188–95.

Romero Díaz, A., Marin Sanleandro, P., Sanchez Soriano, A., Belmonte Serrato, F., and Faulkner, H. (2007). The causes of piping in a set of abandoned agricultural terraces in southeast Spain. *Catena*, 69, 282–93.

Romps, D. M., Seeley, J. T., Vollaro, D., and Molinari, J. (2014). Projected increase in lightning strikes in the United States due to global warming. *Science*, 346, 851–4.

Roosevelt, A. C. (2014). The Amazon and the Anthropocene: 13,000 years of human influence in a tropical rainforest. *Anthropocene*, 4, 69–87.

Roosevelt, C. H. (2006). Tumulus survey and museum research in Lydia, western Turkey: Determining Lydian-and Persian-period settlement patterns. *Journal of Field Archaeology*, 31, 61–76.

Rose, N. L. (2015). Spheroidal carbonaceous fly-ash particles provide a globally synchronous stratigraphic marker for the Anthropocene. *Environmental Science & Technology*, 49, 4155–62.

Rose, N. L., Morley, D., Appleby, P. G., Battarbee, R. W., Alliksaar, T., Guilizzoni, P., and Punning, J. M. (2011). Sediment accumulation rates in European lakes since AD 1850: trends, reference conditions and exceedence. *Journal of Paleolimnology*, 45, 447–68.

Rosen, A. M., Lee, J., Li, M., Wright, J., Wright, H. T., and Fang, H. (2015). The Anthropocene and the landscape of Confucius: a historical ecology of landscape changes in northern and eastern China during the middle to late-Holocene. *The Holocene*, 25, 1640–50.

Rosenzweig, C. and Hillel, D. (1993). The dust bowl of the 1930s: analogy of greenhouse effect in the Great Plains? *Journal of Environmental Quality*, 22, 9–22.

Rosepiler, M. J. and Reilinger, R. (1977). Land subsidence due to water withdrawal in the vicinity of Pecos, Texas. *Engineering Geology*, 11, 295–304.

Roskin, J., Katra, I., and Blumberg, D. G. (2013). Late Holocene dune mobilizations in the northwestern Negev dunefield, Israel: a response to combined anthropogenic activity and short-term intensified windiness. *Quaternary International*, 303, 10–23.

Rott, H., Müller, F., Nagler, T., and Floricioiu, D. (2010). The imbalance of glaciers after disintegration of Larsen B ice shelf, Antarctic Peninsula. *The Cryosphere Discussions*, 4, 1607–33.

Rovira, A., Batalla, R. J., and Sala, M. (2005). Response of a river sediment budget after historical gravel mining (The lower Tordera, NE Spain). *River Research and Applications*, 21, 829–47.

Rowland, J. C., Jones, C. E., Altmann, G., Bryan, R., Crosby, B. T., Hinzman, L. D., and Geernaert, G. L. (2010). Arctic landscapes in transition: responses to thawing permafrost. *Eos, Transactions American Geophysical Union*, 91(26), 229–30.

Rowntree, K. (1991). An assessment of the potential impact of alien invasive vegetation on the geomorphology of river channels in South Africa. *Southern African Journal of Aquatic Science*, 17, 28–43.

Rowntree, K., Duma, M., Kakembo, V., and Thornes, J. (2004). Debunking the myth of overgrazing and soil erosion. *Land Degradation & Development*, 15, 203–14.

Royal Society (2005). *Ocean Acidification Due to Increased Atmospheric Carbon Dioxide*. London: Royal Society.

Rubinstein, J. L. and Mahani, A. B. (2015). Myths and facts on wastewater injection, hydraulic fracturing, enhanced oil recovery, and induced seismicity. *Seismological Research Letters*, 86, 1060–7.

Ruddiman, W., Vavrus, S., Kutzbach, J., and He, F. (2014). Does pre-industrial warming double the anthropogenic total? *The Anthropocene Review*, 1, 147–53.

Ruddiman, W. F. (2003). The anthropogenic greenhouse era began thousands of years ago. *Climatic Change*, 61, 261–93.

(2013). The Anthropocene. *Annual Review of Earth and Planetary Science* 41, 45–68.

(2014). *Earth Transformed*. New York: W.H. Freeman and Company.

Ruddiman, W. F., Ellis, E. C., Kaplan, J. O., and Fuller D. Q. (2015). Defining the epoch we live in. *Science*, 348, 38–9.

Ruecker, G., Schad, P., Alcubilla, M. M., and Ferrer, C. (1998). Natural regeneration of degraded soils and site changes on abandoned agricultural terraces in Mediterranean Spain. *Land Degradation & Development*, 9, 179–88.

Ruffing, C. M., Daniels, M. D., and Dwire, K. A. (2015). Disturbance legacies of historic tie-drives persistently alter geomorphology and large wood characteristics in headwater streams, southeast Wyoming. *Geomorphology*, 231, 1–14.

Rugel, K., Jackson, C. R., Romeis, J. J., Golladay, S. W., Hicks, D. W., and Dowd, J. F. (2012). Effects of irrigation withdrawals on streamflows in a karst environment: lower Flint River Basin, Georgia, USA. *Hydrological Processes*, 26, 523–34.

Ruiz-Colmenero, M., Bienes, R., and Marques, M. J. (2011). Soil and water conservation dilemmas associated with the use of green cover in steep vineyards. *Soil and Tillage Research*, 117, 211–23.

Ruiz-Colmenero, M., Bienes, R., Eldridge, D. J., and Marques, M. J. (2013). Vegetation cover reduces erosion and enhances soil organic carbon in a vineyard in the central Spain. *Catena*, 104, 153–60.

Rull, V. (2013). A futurist perspective on the Anthropocene. *The Holocene*, 23, 1198–201.

Rumschlag, J. H. and Peck, J. A. (2007). Short-term sediment and morphologic response of the Middle Cuyahoga River to the removal of the Munroe Falls Dam, Summit County, Ohio. *Journal of Great Lakes Research*, 33, 142–53.

Rutherford, I. (2000). Some human impacts on Australian stream channel morphology. In *River Management: the Australian Experience*, ed. S. Brizga and B. L. Linlayson. Chichester: Wiley, pp. 11–47.

Rutqvist, J., Rinaldi, A. P., Cappa, F., and Moridis, G. J. (2013). Modeling of fault reactivation and induced seismicity during hydraulic fracturing of shale-gas reservoirs. *Journal of Petroleum Science and Engineering*, 107, 31–44.

Rutz, K. (2012). Artificial Islands versus natural reefs: the environmental cost of development in Dubai. *International Journal of Islamic Architecture*, 1, 243–67.

Ruz, M.-H., Anthony, E. A. and Faucon, L. (2005). Coastal dune evolution on a shoreline subject to strong human pressure: the Dunkirk area, northern France. *Dunes and Estuaries*, 19, 441–9.

Ryken, N., Vanmaercke, M., Wanyama, J., Isabirye, M., Vanonckelen, S., Deckers, J., and Poesen, J. (2015). Impact of papyrus wetland encroachment on spatial and temporal variabilities of stream flow and sediment export from wet tropical catchments. *Science of the Total Environment*, 511, 756–66.

Saez, J. L., Corona, C., Stoffel, M., and Berger, F. (2013). Climate change increases frequency of shallow spring landslides in the French Alps. *Geology*, 41, 619–22.

Saiko, T. A. and Zonn, I. S. (2000). Irrigation expansion and dynamics of desertification in the Circum-Aral region of Central Asia. *Applied Geography*, 20, 349–67.

Saintilan, N., Wilson, N. C., Rogers, K., Rajkaran, A. and Krauss, K. W. (2014). Mangrove expansion and salt marsh decline at mangrove poleward limits. *Global Change Biology*, 20, 147–57.

Sakals, M. E., Innes, J. L., Wilford, D. J., Sidle, R. C., and Grant, G. E. (2006). The role of forests in reducing hydrogeomorphic hazards. *Forest Snow Landscape Research*, 80, 11–22.

Sale, P. F. (2013). The futures of coral reefs. In *The Balance of Nature and Human Impact*, ed.

K. Rhode. Cambridge: Cambridge University Press, pp. 325–34.

Sallenger Jr., A. H., Doran, K. S., and Howd, P. A. (2012). Hotspot of accelerated sea-level rise on the Atlantic coast of North America. *Nature Climate Change*, 2, 884–8.

Salman, A. B., Howari, F. M., El-Sankary, M. M., Wali, A. M., and Saleh, M. M. (2010). Environmental impact and natural hazards on Kharga Oasis monumental sites, Western Desert of Egypt. *Journal of African Earth Sciences*, 58, 341–53.

Salmoral, G., Willaarts, B. A., Troch, P. A., and Garrido, A. (2015). Drivers influencing streamflow changes in the Upper Turia basin, Spain. *Science of the Total Environment*, 503, 258–268.

Sandor, J. A., Gersper, P. L., and Hawley, J. W. (1990). Prehistoric agricultural terraces and soils in the Mimbres area, New Mexico. *World Archaeology*, 22, 70–86.

Sannel, A. B. K. and Kuhry, P. (2011). Warming-induced destabilization of peat plateau/thermokarst lake complexes. *Journal of Geophysical Research: Biogeosciences* 116(G3), doi: 10.1029/2010JG001635.

Santer, B. D., Wigley, T. M. L., Gleckler, P. J., Bonfils, C., Wehner, M. F., AchutaRao, K., Barnett, T. P., Boyle, J. S., Brüggemann, W., Fiorino, M., and Gillett, N. (2006). Forced and unforced ocean temperature changes in Atlantic and Pacific tropical cyclogenesis regions. *Proceedings of the National Academy of Sciences*, 103, 13905–10.

Sarkar, A. (2013), Tractor Production and Sales in India, 1989–2009. *Review of Agrarian Studies*, 3, 1, available at www.ras.org.in/tractor_production_and_sales_in_india_1989_2009 (Accessed 7th June, 2015).

Sattler, K., Keiler, M., Zischg, A., and Schrott, L. (2011). On the connection between debris flow activity and permafrost degradation: a case study from the Schnalstal, South Tyrolean Alps, Italy. *Permafrost and Periglacial Processes*, 22, 254–65.

Sauer, C. O. (1938). Destructive exploitation in modern colonial expansion. Proceedings International Geographical Congress, Amsterdam, Vol. III, section IIIC, 494–9.

Saunders M. A. and Lee, A. S. (2008). Large contribution of sea surface warming to recent increase in Atlantic hurricane activity. *Nature*, 451, 557–60.

Saussure, H. B. de (1796). *Voyages dans les Alpes*. Paris: Fauche.

Saye, S. E. and Pye, K. (2007). Implications of sea level rise for coastal dune habitat conservation in Wales, UK. *Journal of Coastal Conservation* 11, 31–52.

Sazhin, A. N. (1988). Regional aspects of dust storms in steppe regions of the east European and West Siberian plains. *Soviet Geography*, 29, 935–46.

Schaefer, K., Lantuit, H., Romanovsky, V. E., Schuur, E. A., and Witt, R. (2014). The impact of the permafrost carbon feedback on global climate. *Environmental Research Letters*, 9, 085003.

Schelker, J., Kuglerova, L., Eklöf, K., Bishop, K., and Laudon, H. (2013). Hydrological effects of clear-cutting in a boreal forest–Snowpack dynamics, snowmelt and streamflow responses. *Journal of Hydrology*, 484, 105–14.

Scherler, D., Bookhagen, B., and Strecker, M.R. (2011). Spatially variable response of Himalayan glaciers to climate change affected by debris cover. *Nature Geoscience*, 4, 156–9.

Schiavon, N., Chiavari, G., and Fabbri, D. (2004). Soiling of limestone in an urban environment characterized by heavy vehicular exhaust emissions. *Environmental Geology*, 46, 448–55.

Schiefer, E., Petticrew, E. L., Immell, R., Hassan, M. A., and Sonderegger, D. L. (2013). Land use and climate change impacts on lake sedimentation rates in western Canada. *Anthropocene*, 3, 61–71.

Schilling, K.E., Chan, K.-S., Liu, H., and Zhang, Y.-K. (2010). Quantifying the effect of land use land cover change on increasing discharge in the Upper Mississippi River. *Journal of Hydrology*, 387, 343–5.

Schmitt, A., Dotterweich, M., Schmidtchen, G., and Bork, H. R. (2003). Vineyards, hopgardens and recent afforestation: effects of late Holocene land use change on soil erosion in northern Bavaria, Germany. *Catena*, 51, 241–54.

Schmittbuhl, J., Lengliné, O., Cornet, F., Cuenot, N., and Genter, A. (2014). Induced seismicity in EGS reservoir: the creep route. *Geothermal Energy*, 2, 1–13.

Schoellhamer, D. H., Wright, S. A., and Drexler, J. Z. (2013). Adjustment of the San Francisco estuary and watershed to decreasing sediment supply in the 20th century. *Marine Geology*, 345, 63–71.

Schothorst, C. J. (1977). Subsidence of low moor peat soils in the western Netherlands. *Geoderma*, 17, 265–91.

Schottler, S. P., Ulrich, J., Belmont, P., Moore, R., Lauer, J., Engstrom, D. R., and Almendinger, J. E. (2014). Twentieth century agricultural drainage creates more erosive rivers. *Hydrological Processes*, 28, 1951–61.

Schumm, S. A., Harvey, M. D., and Watson, C. C. (1984). *Incised Channels: Morphology, Dynamics and Control*. Littleton, Colorado: Water Resources Publications.

Schuur, E. A. G., McGuire, A. D., Schädel, C., Grosse, G., Harden, J. W., Hayes, D. J., and Vonk, J. E. (2015). Climate change and the permafrost carbon feedback. *Nature*, 520, 171–9.

Schuster, R. L. (1979). Reservoir-induced landslides. *Bulletin of the International Association of Engineering Geology*, 20, 8–15.

Schwarz, H. E., Emel, J., Dickens, W. J., Rogers, P., and Thompson, J. (1991). Water quality and flows. In *The Earth as Transformed by Human Action*, ed.

B. L. Turner, W. C. Clark, R. W. Kates, J. F. Richards, J. T. Matthews, and W. B. Meyer. Cambridge: Cambridge University Press, pp. 253–70.

Scyphers, S. B., Powers, S. P., Heck Jr, K. L., and Byron, D. (2011). Oyster reefs as natural breakwaters mitigate shoreline loss and facilitate fisheries. *PloS one*, 6(8), e22396.

Sear, D., Newson, M., Hill, C., Old, J., and Branson, J. (2009). A method for applying fluvial geomorphology in support of catchment-scale river restoration planning. *Aquatic Conservation: Marine and Freshwater Ecosystems*, 19, 506–19.

Sefelnasr, A. and Sherif, M. (2014). Impacts of seawater rise on seawater intrusion in the Nile delta aquifer, Egypt. *Groundwater*, 52, 264–76.

Segura-Beltrán, F. and Sanchis-Ibor, C. (2013). Assessment of channel changes in a Mediterranean ephemeral stream since the early twentieth century. The Rambla de Cervera, eastern Spain. *Geomorphology*, 201, 199–214.

Seifan, N. (2009). Long-term effects of anthropogenic activities on semi-arid sand dunes. *Journal of Arid Environments*, 73, 332–7.

Senter, J. (2003). Live dunes and ghost forests: Stability and change in the history of North Carolina's maritime forests. *North Carolina Historical Review*, 80, 334–71.

Seyoum, W. M., Milewski, A. M., and Durham, M. C. (2015). Understanding the relative impacts of natural processes and human activities on the hydrology of the Central Rift Valley lakes, East Africa. *Hydrological Processes*, 29, 4312–24, doi: 10.1002/hyp.10490

Shakesby, R. A. (2011). Post-wildfire soil erosion in the Mediterranean: review and future research directions. *Earth-Science Reviews*, 105, 71–100.

Shakesby, R. A., Doerr, S. H., and Walsh, R. P. D. (2000). The erosional impact of soil hydrophobicity: current problems and future research directions. *Journal of Hydrology*, 231/2, 178–91.

Shakesby, R. A., Wallbrink, P. J., Doerr, S. H., English, P. M., Chafre, C. J., Humphreys, G. S., Blake, W. H. and Tomkins, K. M. (2007). Distinctiveness of wildfire effects on soil erosion in south-east Australian eucalyptus forests assessed in a global context. *Forest Ecology and Management*, 238, 347–64.

Shaler, N. S. (1912). *Man and the Earth*. New York: Duffield.

Shams, A. (2014). A rediscovered-new "Qanat" system in the high mountains of Sinai Peninsula, with Levantine reflections. *Journal of Arid Environments*, 110, 69–74.

Shan, W., Hu, Z., Guo, Y., Zhang, C., Wang, C., Jiang, H., and Xiao, J. (2015). The impact of climate change on landslides in southeastern of high-latitude permafrost regions of China. *Frontiers in Earth Science*, 3, doi: 10.3389/feart.2015.00007.

Shankman, D. and Smith, L. J. (2004). Stream channelization and swamp formation in the U.S. coastal plain. *Physical Geography*, 25, 22–38.

Shao, Y., Wyrwoll, K. H., Chappell, A., Huang, J., Lin, Z., McTainsh, G. H., and Yoon, S. (2011). Dust cycle: an emerging core theme in Earth system science. *Aeolian Research*, 2, 181–204.

Sheffield, A. T., Healy, T. R., and McGlone, M. S. (1995). Infilling rates of a steepland catchment estuary, Whangamata, New Zealand. *Journal of Coastal Research*, 11, 1294–308.

Sheng, J. and Wilson, J. P. (2009). Watershed urbanization and changing flood behaviour across the Los Angeles metropolitan region. *Natural Hazards*, 48, 41–57.

Shepard, C. C., Agostini, V. N., Gilmer, B., Allem, T., Stone, J., Brooks W., and Beck, M. W. (2012). Assessing future risk: quantifying the effects of sea level rise on storm surge risk for the southern shores of Long Island, New York. *Natural Hazards*, 60, 727–45.

Sheppard, C. R. C. (2003). Predicted recurrences of mass coral morality in the Indian Ocean. *Nature*, 425, 294–7.

Sheppard, C. R. C., Davy, S. K., and Pilling, G. M. (2009). *The Biology of Coral Reefs*. Oxford: Oxford University Press.

Sheridan, G. J. and Noske, P. J. (2007). A quantitative study of sediment delivery and stream pollution from different forest road types. *Hydrological Processes*, 21, 387–98.

Sheridan, G. J., Noske, P. J., Whipp, R. K., and Wijesinghe, N. (2006). The effect of truck traffic and road water content on sediment delivery from unpaved forest roads. *Hydrological Processes*, 20, 1683–99.

Sherlock, R. L. (1922). *Man as a Geological Agent*. London: Witherby.

Sherratt, A. (1983). The secondary exploitation of animals in the Old World. *World Archaeology*, 15, 90–104.

Shi, W., Wang, M., and Guo, W. (2014). Long-term hydrological changes of the Aral Sea observed by satellites. *Journal of Geophysical Research: Oceans*, 119, 3313–26.

Shiklomanov, I. A. (1985). Large scale water transfers. In *Facets of hydrology II*, ed. J. C. Rodda. Chichester: Wiley, pp. 345–87.

Shlemon, R. J. (1995). Groundwater rise and hydrocollapse: technical and political implications of Special Geologic Report Zones' in Riverside County, California, USA. *International Association of Hydrological Sciences, Publication*, 234, 481–6.

Shomurodov, H. F., Rakhimova, T. T., Saribaeva, S. U., Rakhimova, N. K., Esov, R. A., and Adilov, B. A. (2013). Perspective plant species for stabilization of sand dunes on the exposed Aral Sea bed. *Journal of Earth Science and Engineering*, 3, 655–62.

Shriner, D. S. and Street, R. B. (1998). North America. In *The Regional Impacts of Climate Change*, ed. R. T. Watson, M.C. Zinyowera and R.H.

Moss. Cambridge: Cambridge University Press, pp. 273–8.

Shuttleworth, E. L., Evans, M. G., Hutchinson, S. M., and Rothwell, J. J. (2015). Peatland restoration: controls on sediment production and reductions in carbon and pollutant export. *Earth Surface Processes and Landforms*, 40, 459–72.

Siakeu, J., Oguchi, T., Aoki, T., Esaki, Y., and Jarvie, H. P. (2004). Change in riverine suspended sediment concentration in central Japan in response to late 20th century human activities. *Catena*, 55, 231–54.

Sidle, R. C. and Dhakal, A. S. (2002). Potential effect of environmental change on landslide hazards in forest environments. In *Environmental Change and Geomorphic Hazards in Forests*, ed. R. C. Sidle. Wallingford: CABI, pp. 123–65.

Sidle, R. C. and T. P. Burt. (2009). Temperate forests and rangelands. In *Geomorphology and Global Environmental Change*, ed. O. Slaymaker, T. Spencer, and C. Embleton-Hamman. Cambridge: Cambridge University Press, pp. 321–43.

Sidle, R. C., Furuichi, T., and Kono, Y. (2011). Unprecedented rates of landslide and surface erosion along a newly constructed road in Yunnan, China. *Natural Hazards*, 57, 313–26.

Sidle, R. C., Ghestem, M., and Stokes, A. (2014). Epic landslide erosion from mountain roads in Yunnan, China–challenges for sustainable development. *Natural Hazards and Earth System Science*, 14, 3093–104.

Sidle, R. C., Sasaki, S., Otsuki, M., Noguchi, S., and Rahim Nik, A. (2004). Sediment pathways in a tropical forest: effects of logging roads and skid trails. *Hydrological Processes*, 18, 703–20.

Sidle, R. C., Kamil, I., Sharma, A., and Yamashita, S. (2000). Stream response to subsidence from underground coal mining in central Utah. *Environmental Geology*, 39, 279–91.

Siebert, S., Burke, J., Faures, J. M., Frenken, K., Hoogeven, J., Döll, P., and Portman, F. T. (2010). Groundwater use for irrigation – a global inventory. *Hydrology and Earth System Sciences Discussions*, 7, 3977–4021.

Silva, R., Martínez, M. L., Hesp, P. A., Catalan, P., Osorio, A. F., Martell, R., and Govaere, G. (2014). Present and future challenges of coastal erosion in Latin America. *Journal of Coastal Research*, 71(sp1), 1–16.

Silveira, L. and Alonso, J. (2009). Runoff modifications due to the conversion of natural grasslands to forests in a large basin in Uruguay. *Hydrological Processes*, 23, 320–9.

Silverberg, R. (2013). *Mound Builders*. Athens, Ohio: Ohio University Press.

Simas, T., Nunes, J. P., and Ferreira, J. G. (2001). Effects of global climate change on coastal salt marshes. *Ecological Modelling*, 139, 1–15.

Simpson, J. M., Darrow, M. M., Huang, S. L., Daanen, R. P., and Hubbard, T. D. (2015). Investigating movement and characteristics of a frozen debris lobe, South-central Brooks Range, Alaska. *Environmental & Engineering Geoscience*, 1078–7275, doi: 10.2113/EEG-1728.

Sinclair, W. C. (1982). *Sinkhole Development Resulting from Ground-Water Withdrawal in the Tampa Area, Florida*. US Geological Survey, Water Resources Division.

Siriwardena, L., Finlayson, B. L. and McMahon, T. A. (2006). The impact of land use change on catchment hydrology in large catchments: the Comet River, Central Queensland, Australia. *Journal of Hydrology*, 326, 199–214.

Six, D., Reynaud, L., and Letreguilly, A. (2001). Bilans de masse des glaciers alpines et scandinaves, leurs relations avec l'oscillation due climat de l'Atlantique nord. *Comptes Rendus Academie des Sciences, Sciences de la Terre et des Planètes*, 333, 693–8.

Skeat, A. J., East, T. J., and Corbett, L. K. (2012). Impact of feral water buffalo. In *Landscape and Vegetation Ecology of the Kakadu Region, Northern Australia*, ed. C. M. Finlayson and I. von Oetzen. Dordrecht: Springer, pp. 155–77.

Slater, L. J., Singer, M. B. and Kirchner, J. W. (2015). Hydrologic versus geomorphic drivers of trends in flood hazard. *Geophysical Research Letters*, 42, 370–6.

Slaymaker, O., Spencer, T., and Embleton-Hamann, C. (eds.) (2009). *Geomorphology and Global Environmental Change*. Cambridge: Cambridge University Press.

Sloss, C. R., Jones, B. G., Brooke, B. P., Heijnis, H. and Murray-Wallace, C. V. (2011). Contrasting sedimentation rates in Lake Illawarra and St George's Basin, two large barrier estuaries on the southeast coast of Australia. *Journal of Paleolimnology*, 46, 561–77.

Smil, V. (2011). Harvesting the biosphere: the human impact. *Population and Development Review*, 37, 613–36.

(2015). It's too soon to call this the Anthropocene. *Spectrum, IEEE*, 52, 28.

Smith, B. D. and Zeder, M. A. (2013). The onset of the Anthropocene. *Anthropocene*, 4, 8–13.

Smith, B. J., McCabe, S., McAllister, D., Adamson, C., Viles, H. A., and Curran, J. M. (2011). A commentary on climate change, stone decay dynamics and the 'greening'of natural stone buildings: new perspectives on 'deep wetting'. *Environmental Earth Sciences*, 63, 1691–700.

Smith, D. M., Zalasiewcz, J. A., Williams, M., Wilkinson, I. P., Redding, M., and Begg, C. (2010). Holocene drainage systems of the English Fenland: roddons and their environmental significance. *Proceedings of the Geologists' Association*, 121, 256–69.

Smith, H. G. and Dragovich, D. (2008). Post-fire hillslope erosion response in a sub-alpine environment, south-eastern Australia. *Catena*, 73, 274–85.

Smith, H. G., Sheridan, G. J., Lane, P. N., and Bren, L. J. (2011). Wildfire and salvage harvesting effects on runoff generation and sediment exports from radiata pine and eucalypt forest catchments, south-eastern Australia. *Forest Ecology and Management*, 261, 570–81.

Smith, L. M., Haukos, D. A., McMurry, S. T., LaGrange, T., and Willis, D. (2011). Ecosystem services provided by playas in the High Plains: potential influences of USDA conservation programs. *Ecological Applications*, 21(sp1), S82–92.

Sobota, I. and Nowak, M. (2014). Changes in the dynamics and thermal regime of the permafrost and active layer of the high Arctic coastal area in North-West Spitsbergen, Svalbard. *Geografiska Annaler: Series A, Physical Geography*, 96, 227–40.

Sofia, G., Prosdocimi, M., Dalla Fontana, G., and Tarolli, P. (2014). Modification of artificial drainage networks during the past half-century: Evidence and effects in a reclamation area in the Veneto floodplain (Italy). *Anthropocene*, 6, 46–82.

Sokolov, N. A. (1884). *Dunes, Their Formation, Development and Internal Structure*. St.Petersburg: St. Petersburg University (in Russian).

(1894) *Die Dünen, Bildung, Entwicklung und Ihrer Bau*, Berlin: Springer.

Solé-Benet, A., Lázaro, R., Domingo, F., Cantón, Y., and Puigdefábregas, J. (2010). Why most agricultural terraces in steep slopes in semiarid SE Spain remain well preserved since their abandonment 50 years go? *Pirineos*, 165, 215–35.

Sorg, A., Kääb, A., Roesch, A., Bigler, C., and Stoffel, M. (2015). Contrasting responses of Central Asian rock glaciers to global warming. *Scientific Reports*, 5, doi:10.1038/srep08228.

Soulsby, C., Birkel, C., and Tetzlaff, D. (2014). Assessing urbanization impacts on catchment transit times. *Geophysical Research Letters*, 41, 442–8.

Sousa, A., García-Barrón, L., García-Murillo, P., Vetter, M., and Morales, J. (2015). The use of changes in small coastal Atlantic brooks in southwestern Europe as indicators of anthropogenic and climatic impacts over the last 400 years. *Journal of Paleolimnology*, 53, 73–88.

Souter, D. W. and Linden, O. (2000). The health and future of coral reef systems. *Ocean and Coastal Management*, 43, 657–88.

Spalding, M. D. and Brown, B. E. (2015). Warm-water coral reefs and climate change. *Science*, 350, 769–71.

Spate, O. H. K. and Learmonth, A. T. A. (1967). *India and Pakistan*. London: Methuen.

Sperna Weiland, F. C., van Beek, L. P. H., Kwadijk, K. C. J., and Blerkens, M. F. P. (2011). Global patterns of change in discharge regimes for 2100. *Hydrology and Earth System Sciences Discussions*, 8, 10973–1014.

Spigel, K. M. and Robichaud, P. R. (2007). First-year post-fire erosion rates in Bitterroot National

Forest, Montana. *Hydrological Processes*, 21, 998–1005.

Stabile, T. A., Giocoli, A., Perrone, A., Piscitelli, S., and Lapenna, V. (2014). Fluid injection induced seismicity reveals a NE dipping fault in the southeastern sector of the High Agri Valley (southern Italy). *Geophysical Research Letters*, 41, 5847–54.

Stabile, T. A., Giocoli, A., Lapenna, V., Perrone, A., Piscitelli, S., and Telesca, L. (2014). Evidence of low-magnitude continued reservoir-induced seismicity associated with the Pertusillo Artificial Lake (Southern Italy). *Bulletin of the Seismological Society of America*, 104, 1820–8, doi: 10.1785/0120130333.

Stallins, J. A. (2006). Geomorphology and ecology: unifying themes for complex systems in biogeomorphology. *Geomorphology*, 77, 207–16.

Stancheva, M., Ratas, U., Orviku, K., Palazov, A., Rivis, R., Kont, A., and Stanchev, H. (2011). Sand dune destruction due to increased human impacts along the Bulgarian Black Sea and Estonian Baltic Sea coasts. *Journal of Coastal Research*, S1, 64, 324–8.

Stanley, D. J. (1996). Nile delta: extreme case of sediment entrapment on a delta plain and consequent coastal land loss. *Marine Geology*, 129, 189–95.

Steadman, D. W., Stafford, T. W., Donahue, D. J., and Jull, A. J. T. (1991). Chronology of Holocene vertebrate extinction in the Galápagos Islands. *Quaternary Research*, 36, 126–33.

Steffen, W. (2010). Observed trends in Earth System behaviour. *Interdisciplinary Reviews, Climate Change*, 1, 428–49.

Steffen W. and 10 others. (2004). *Global change and the Earth System*. Berlin: Springer

Steffen, W., Crutzen, P. J., and McNeill, J. R. (2007). The Anthropocene: are humans now overwhelming the great forces of nature? *Ambio*, 36, 614–21.

Steffen, W., Grinevald, J., Crutzen, P., and McNeill, J. (2011). The Anthropocene: conceptual and historical perspectives. *Philosophical Transactions of the Royal Society*, 369A, 842–67.

Steffen, W., Broadgate, W., Deutsch, L., Gaffney, O., and Ludwig, C. (2015). The trajectory of the Anthropocene: the Great Acceleration. *The Anthropocene Review*, 2, 81–98.

Stephens, J. C. (1956). Subsidence of organic soils in the Florida Everglades. *Soil Science Society of America Journal*, 20, 77–80.

Stephens, J. C., Allen, L. H., and Chen, E. (1984). Organic soil subsidence. *Reviews in Engineering Geology*, 6, 107–22.

Sterk, G. (2003). Causes, consequences and control of wind erosion in Sahelian Africa: a review. *Land Degradation and Development*, 14, 95–108.

Sterling, S. M., Ducharne, A., and Polcher, J. (2013). The impact of global land-cover change on the terrestrial water cycle. *Nature Climate Change*, 3, 385–90.

Stevenson, A. C., Jones, V. J., and Battarbee, R. W. (1990).The cause of peat erosion: a palaeolimnological approach. *New Phytologist*, 114, 727–35.

Stewart, I. T. (2009). Changes in snowpack and snowmelt runoff for key mountain regions. *Hydrological Processes*, 23, 78–94.

Stinchcomb, G. E., Stewart R.M., Messner, T.C., Nordt, L. C., Driese, S. G., and Allen, P. M., (2013). Using event stratigraphy to map the Anthropocene – an example from the historic coal mining region in eastern Pennsylvania, USA. *Anthropocene*, 2, 42–50.

Stive, M. J., de Schipper, M. A., Luijendijk, A. P., Aarninkhof, S. G., van Gelder-Maas, C., van Thiel de Vries, J. S., and Ranasinghe, R. (2013). A new alternative to saving our beaches from sea-level rise: The sand engine. *Journal of Coastal Research*, 29, 1001–8.

Stoddart, D. R. (1971). Coral reefs and islands and catastrophic storms. In *Applied Coastal Geomorphology*, ed. J. A. Steers. London: Macmillan, pp. 154–97.

Stoffel, M. and Beniston, M. (2006). On the incidence of debris flows from the early Little Ice Age to a future greenhouse climate: a case study from the Swiss Alps. *Geophysical Research Letters*, 33, L16404, doi: 10.1029/2006GL026805.

Stoffel, M. and Huggel, C. (2012). Effects of climate change on mass movements in mountain environments. *Progress in Physical Geography*, 36, 421–39.

Stoffel, M., Tiranti, D., and Huggel, C. (2014). Climate change impacts on mass movements—case studies from the European Alps. *Science of the Total Environment*, 493, 1255–66.

Stokes, A., Atger, C., Bengough, A. G., Fourcaud, T., and Sidle, R. C. (2009). Desirable plant root traits for protecting natural and engineered slopes against landslides. *Plant and Soil*, 324, 1–30.

Stokes, C. R., Popovnin, V., Aleynikov, A., Gurney, S. D., and Shahgedanova, M. (2007). Recent glacier retreat in the Caucasus Mountains, Russia, and associated increase in supraglacial debris cover and supra-/proglacial lake development. *Annals of Glaciology*, 46, 195–203.

Stokes, S., Goudie, A. S, Colls, A., and Al-Farraj, A. (2003). Optical dating as a tool for studying dune reactivation, accretion rates and desertification over decadal, centennial and millennial time-scales. In *Desertification in the Third Millennium*, ed. A. S. Alsharhan, W. W. Wood, A. S. Goudie, A. Fowler, and E. M. Abdellatif. Balkema: Lisse, pp. 53–60.

Stovern, M., Betterton, E. A., Sáez, A. E., Villar, O. I. F., Rine, K. P., Russell, M. R., and King, M. (2014). Modeling the emission, transport and deposition of contaminated dust from a mine tailing site. *Reviews on Environmental Health*, 29, 91–4.

Straneo, F. and Heimbach, P. (2013). North Atlantic warming and the retreat of Greenland's outlet glaciers. *Nature*, 504, 36–43.

Stringer, C. (2003). Out of Africa. *Nature*, 423, 692–9.

Stromberg, J. C., Lite, S. J., Marler, R., Paradzick, C., Shafroth, P. B., Shorrock, D., and White, M. S. (2007). Altered stream-flow regimes and invasive plant species: the Tamarix case. *Global Ecology and Biogeography*, 16, 381–93.

Strong, C. L., Bullard, J. E., Dubois, C., McTainsh, G. H., and Baddock, M. C. (2010). Impact of wildfire on interdune ecology and sediments: an example from the Simpson Desert, Australia. *Journal of Arid Environments*, 74, 1577–81.

Strong, D. R. and Ayres, D. A. (2009). Spartina introduction and consequences in salt marshes. In *Human Impacts on Salt Marshes: a Global Perspective*, ed. B. R. Silliman, E. D. Grosholz, and M. D. Bertness. Vancouver: University of British Columbia Press, pp. 3–22.

Struyf, E., Smis, A., Van Damme, S., Garnier, J., Govers, G., Van Wesemael, B., and Meire, P. (2010). Historical land use change has lowered terrestrial silica mobilization. *Nature Communications*, 1, 129.

Su, Z. A., Zhang, J. H., Qin, F. C., and Nie, X. J. (2012). Landform change due to soil redistribution by intense tillage based on high-resolution DEMs. *Geomorphology*, 175, 190–8.

Suchodoletz, H. von, Oberhänsli, H., Faust, D., Fuchs, M., Blanchet, C., Goldhammer, T., and Zöller, L. (2010). The evolution of Saharan dust input on Lanzarote (Canary Islands) – influenced by human activity during the early Holocene? *The Holocene*, 20, 169–79.

Suckale, J. (2009). Induced seismicity in hydrocarbon fields. *Advances in Geophysics*, 51, 55–106.

Summa-Nelson, M. C. and Rittenour, T. M. (2012). Application of OSL dating to middle to late Holocene arroyo sediments in Kanab Creek, southern Utah, USA. *Quaternary Geochronology*, 10, 167–74.

Sun, G. E., McNulty, S. G., Moore, J., Bunch, C., and Ni, J. (2002). Potential impacts of climate change on rainfall erosivity and water availability in China in the next 100 years. *Proceedings of the 12th International Soil Conservation Conference, Beijing*, 244–50.

Surell, A. (1841). *Étude sur les torrents des Hautes-Alpes par Alexandre Surell*. Paris: Carilian-Goeury et Dalmont.

Surian, N. (1999). Channel changes due to river regulation: the case of the Piave River, Italy. *Earth Surface Processes and Landforms*, 24, 1135–51.

Surian, N. and Rinaldi, M. (2003). Morphological response to river engineering and management in alluvial channels in Italy. *Geomorphology*, 50, 307–26.

Sušnik, J., Vamvakeridou-Lyroudia, L. S., Baumert, N., Kloos, J., Renaud, F. G., La Jeunesse, I., and Zografos, C. (2015). Interdisciplinary assessment of sea-level rise and climate change impacts on the lower Nile Delta, Egypt. *Science of the Total Environment*, 503, 279–88.

Sweatman, H., Delean, S., and Syms, C. (2011). Assessing loss of coral cover on Australia's Great Barrier Reef over two decades, with implications for longer term trends. *Coral Reefs*, 30, 521–31.

Sylla, M. B., Gaye, A. T., Jenkins, G. S., Pal, J. S., and Giorgi, F. (2010). Consistency of projected drought over the Sahel with changes in the monsoon circulation and extremes in a regional climate model projections. *Journal of Geophysical Research*, 115, D16108, doi: 10.1029/2009JD012983.

Syvitski, J. P. M. and Kettner, A. (2011). Sediment flux and the Anthropocene. *Philosophical Transactions of the Royal Society, A*, 369, 957–75.

Syvitski, J. P. M. and Milliman, J. D. (2007). Geology, geograpohy and humans battle for dominance over the delivery of fluvial sediment to the coastal ocean. *Journal of Geology*, 115, 1–19.

Syvitski, J. P. M., and Saito, Y. (2007). Morphodynamics of deltas under the influence of humans. *Global and Planetary Change*, 57, 261–82.

Syvitski, J. P. M., Vörösmarty, C. J., Kettner, A. J., and Green, P. (2005). Impact of humans on the flux of terrestrial sediment to the global coastal ocean. *Science*, 308, 376–80.

Syvitski, J. P. M. and 10 others. (2009). Sinking deltas due to human activities. *Nature Geosciences*, 2, 681–6.

Syvitski, J. P., Kettner, A. J., Overeem, I., Giosan, L., Brakenridge, G. R., Hannon, M., and Bilham, R. (2013). Anthropocene metamorphosis of the Indus Delta and lower floodplain. *Anthropocene*, 3, 24–35.

Szabó, J., Dávid, L., and Lóczy, D. (eds.). (2010) *Anthropogenic Geomorphology: a Guide to Man-made Landforms*. Heidelberg: Springer.

Ta, W., Dong, Z., and Sanzhi, C. (2006). Effect of the 1950s large-scale migration for land reclamation on spring dust storms in Northwest China. *Atmospheric Environment*, 40, 5815–23.

Taborda, R. and Ribeiro, M. A. (2015). A simple model to estimate the impact of sea-level rise on platform beaches. *Geomorphology*, 234, 204–10.

Taheri, K., Gutiérrez, F., Mohseni, H., Raeisi, E., and Taheri, M. (2015). Sinkhole susceptibility mapping using the analytical hierarchy process (AHP) and magnitude–frequency relationships: a case study in Hamadan province, Iran. *Geomorphology*, 234, 64–79.

Talbot, T. and Lapointe, M. (2002). Modes of response of a gravel bed river to meander straightening: the case of the Sainte-Marguerite River, Saguenay Region, Quebec, Canada.*Water Resources Research*, 38, 9–1.

Tan, M. and Li, X. (2015). Does the Green Great Wall effectively decrease dust storm intensity in

China? A study based on NOAA NDVI and weather station data. *Land Use Policy*, 43, 42–7.

Tang, Q. and Lettenmaier, D. P. (2012). 21st century runoff sensitivities of major global river basins. *Geophysical Research Letters*, 39, L06403, doi: 10.1029/2011GL050834.

Tang, Y., Zhong, S., Luo, L., Bian, X., Heilman, W. E., and Winkler, J. (2015). The potential impact of regional climate change on fire weather in the United States. *Annals of the Association of American Geographers*, 105, 1–21.

Tang, Z., Gu, Y., Drahota, J., LaGrange, T., Bishop, A., and Kuzila, M. S. (2015). Using fly ash as a marker to quantify culturally-accelerated sediment accumulation in playa wetlands. *JAWRA Journal of the American Water Resources Association*, 51, 1643–55, doi: 10.1111/1752-1688.12347.

Taniguchi, K. T. and Biggs, T. W. (2015). Regional impacts of urbanization on stream channel geometry: a case study in semiarid southern California. *Geomorphology*, 248, 228–36.

Tarolli, P., Preti, F., and Romano, N. (2014). Terraced landscapes: from an old best practice to a potential hazard for soil degradation due to land abandonment. *Anthropocene*, 6, 10–25.

Tarolli, P., Sofia, G., Calligaro, S., Prosdocimi, M., Preti, F., and Dalla Fontana, G. (2015). Vineyards in terraced landscapes: new opportunities from LiDAR data. *Land Degradation & Development*, 26, 92–102.

Tarriño, A., Elorrieta, I., García-Rojas, M., Orue, I., and Sánchez, A. (2014). Neolithic Flint Mines of Treviño (Basque-Cantabrian Basin, Western Pyrenees, Spain). *Journal of Lithic Studies*, 1, 129–47.

Taylor, M. P. and Little, J. A. (2013). Environmental impact of a major copper mine spill on a river and floodplain system. *Anthropocene*, 3, 36–50.

Teatini, P., Ferronato, M., Gambolati, G., and Gonella, M. (2006). Groundwater pumping and land subsidence in the Emilia-Romagna coastland, Italy: modeling the past occurrence and the future trend. *Water Resources Research*, 42(1), W01406, doi: 10.1029/2005WR004242.

Tegen, I. and Fung, I., (1995) Contribution to the atmospheric mineral aerosol load from land surface modification. *Journal of Geophysical Research*, 100(D9), 18707–26.

Tegen, I., Werner, M., Harrison, S. P., and Kohfeld, K. E. (2004) Relative importance of climate and land use in determining present and future global soil dust emissions. *Geophysical Science Reviews*, article L05105.

Temmerman, S., Meire, P., Bouma, T. J., Herman, P. M., Ysebaert, T., and De Vriend, H. J. (2013). Ecosystem-based coastal defence in the face of global change. *Nature*, 504, 79–83.

Tempany, H. A., Roddan, G. M., and Lord, L. (1944). Soil erosion and soil conservation in the colonial empire. *Empire Forestry Journal*, 23, 142–59.

Teneva, L., Karnauskas, M., Logan, C. A., Bianucci, L., Currie, J. C., and Kleypas, J. A. (2012). Predicting coral bleaching hotspots: the role of regional variability in thermal stress and potential adaptation rates. *Coral Reefs*, 31, 1–12.

Tennant, C. and Menounos, B. (2013). Glacier change of the Columbia Icefield, Canadian Rocky Mountains, 1919–2009. *Journal of Glaciology*, 59, 671–86.

Teo, E. A. and Marren, P. M. (2015). Interaction of ENSO-driven flood variability and anthropogenic changes in driving channel evolution: Corryong/Nariel Creek, Australia. *Australian Geographer*, 46, 339–62.

Ter-Stepanian, G. (1988). Beginning of the Technogene. *Bulletin of the International Association of Engineering Geology*, 38, 133–42.

Ternan, J. L., Williams, A. G., Elmes, A., and Fitzjohn, C. (1996). The effectiveness of bench-terracing and afforestation for erosion control on Rana sediments in central Spain. *Land Degradation and Development*, 7, 337–51.

Thomas, D. S. G. and Twyman, C. (2004). Good or bad rangeland? Hybrid knowledge, science, and local understandings of vegetation dynamics in the Kalahari. *Land Degradation & Development*, 15, 215–31.

Thomas, D. S. G., Knight, M., and Wiggs, G. F. S. (2005). Remobilization of southern African desert dune systems by twenty-first century global warming. *Nature*, 435, 1218–21.

Thomas, J. (2006). On the origins and development of cursus monuments in Britain. *Proceedings of the Prehistoric Society*, 72, 229–41.

Thomas, W. F. (ed.) (1956). *Man's Role in Changing the Face of the Earth*. Chicago: University of Chicago Press.

Thompson, J. R. (1970). Soil erosion in the Detroit metropolitan area. *Journal of Soil and Water Conservation*, 25, 8–10.

Thompson L. M. C. and Shlacher, T. A. (2008). Physical damage to coastal dunes and ecological impacts caused by vehicle tracks associated with beach camping on sandy shores: a case study from Fraser Island, Australia. *Journal of Coastal Conservation*, 12, 67–82.

Thompson, L. G. (2000). Ice core evidence for climate change in the Tropics: implications for our future. *Quaternary Science Reviews*, 19, 19–35.

Thorndycraft, V. R., Pirrie, D., and Brown, A. G. (2004). Alluvial records of medieval and prehistoric tin mining on Dartmoor, southwest England. *Geoarchaeology*, 19, 219–36.

Thornes, J. B. (2007). Modelling soil erosion by grazing: recent developments and new approaches. *Geographical Research*, 45, 13–26.

Tickner, D. P., Angold, P. G., Gurnell, A. M., and Mountford, J. O. (2001). Riparian plant invasions: hydrogeomorphological control and ecological impacts. *Progress in Physical Geography*, 25, 22–52.

Tielidze, L. G., Lomidze, N., and Asanidze, L. (2015). Glaciers retreat and climate change effect during the last one century in the Mestiachala River Basin, Caucasus Mountains, Georgia. *Earth Sciences*, 4, 72–79.

Tiffen, M., Mortimore, M., and Gichuki, F. (1994). *More People, Less Erosion. Environmental Recovery in Kenya*. Chichester: Wiley.

Tihansky, A. B. (1999). Sinkholes, west-central Florida. *Land Subsidence in the United States. United States Geological Survey Circular*, 1,182, 121–40.

Tolksdorf, J. F. and Kaiser, K. (2012). Holocene aeolian dynamics in the European sand-belt as indicated by geochronological data. *Boreas*, 41, 408–21.

Tolksdorf, J. F., Klasen, N., and Hilgers, A. (2013). The existence of open areas during the Mesolithic: evidence from aeolian sediments in the Elbe–Jeetzel area, northern Germany. *Journal of Archaeological Science*, 40, 2813–23.

Tomás, R., Herrera, G., Delgado, J., Lopez-Sanchez, J. M., Mallorquí, J. J., and Mulas, J. (2010). A ground subsidence study based on DInSAR data: calibration of soil parameters and subsidence prediction in Murcia City (Spain). *Engineering Geology*, 111, 19–30.

Tomás, R., Márquez, Y., Lopez-Sanchez, J. M., Delgado, J., Blanco, P., Mallorquí, J. J., and Mulas, J. (2005). Mapping ground subsidence induced by aquifer overexploitation using advanced Differential SAR Interferometry: Vega Media of the Segura River (SE Spain) case study. *Remote Sensing of Environment*, 98, 269–83.

Tonjes, D. J. (2013). Impacts from ditching salt marshes in the mid-Atlantic and northeastern United States. *Environmental Reviews*, 21, 116–26.

Torri, D., Santi, E., Marignani, M., Rossi, M., Borselli, L., and Maccherini, S. (2013). The recurring cycles of "biancana" badlands: Erosion, vegetation and human impact. *Catena*, 106, 22–30.

Tóth, C. (2004). Functional changes of the tumuli at the different stages of history. *Anthropogenic Aspects of Landscape Transformations*, 3, 93–102.

Tourian, M. J., Elmi, O., Chen, Q., Devaraju, B., Roohi, S., and Sneeuw, N. (2015). A spaceborne multisensor approach to monitor the desiccation of Lake Urmia in Iran. *Remote Sensing of Environment*, 156, 349–60.

Touysinhthiphonexay, K. C. and Gardner, T. W. (1984). Threshold response of small streams to surface coal mining, bituminous coal fields, Central Pennsylvania. *Earth Surface Processes and Landforms*, 9, 43–58.

Townsend-Small, A., Pataki, D. E., Liu, H., Li, Z., Wu, Q., and Thomas, B. (2013). Increasing summer river discharge in southern California, USA, linked to urbanization. *Geophysical Research Letters*, 40, 4643–7.

Tréguer, P. J. and De La Rocha, C. L. (2013). The world ocean silica cycle. *Annual Review of Marine Science*, 5, 477–501.

Trenhaile, A. S. (2014). Climate change and its impact on rock coasts. *Geological Society, London, Memoirs*, 40, 7–17.

Trimble, S. W. (1974). *Man-induced Soil Erosion on the Southern Piedmont*. Ankeny, IA: Soil Conservation Society of America.

(1988). The impact of organisms on overall erosion rates within catchments in temperate regions. In *Biogeomorphology*, ed. H. A. Viles. Oxford: Basil Blackwell, pp. 83–142.

(1997). Stream channel erosion and change resulting from riparian forests. *Geology*, 25, 467–9.

(2003). Historical hydrographic and hydrologic changes in the San Diego Creek watershed, Newport Bay, California. *Journal of Historical Geography*, 29, 422–44.

(2004). Effects of riparian vegetation on stream channel stability and sediment budgets. *Water Science and Application*, 8, 153–69.

(2008a). *Man-induced soil erosion on the southern Piedmont* (2nd edn). Ankeny, IA: Soil Conservation Society of America.

(2008b). The use of historical data and artifacts in geomorphology. *Progress in Physical Geography*, 32, 3–29.

(2013). *Historical Agriculture and Soil Erosion in the Upper Mississippi Valley Hill Country*. Boca Raton: CRC Press.

Trimble, S. W. and Cooke, R. U. (1991). Historical sources for geomorphological research in the United States. *Professional Geographer*, 43, 212–28.

Trimble, S. W. and Crosson, S. (2000). US soil erosion rates – myth and reality. *Science*, 289, 248–50.

Trimble, S. W. and Mendel, A. C. (1995). The cow as a geomorphic agent: a critical review. *Geomorphology*, 13, 233–53.

Triplett, L. D., Engstrom, D. R., Conley, D. J., and Schelhaass, S. M. (2008). Silica fluxes and trapping in two contrasting natural impoundments of the upper Mississippi River. *Biogeochemistry*, 87, 217–30.

Trnka, M., Kersebaum, K. C., Eitzinger, J., Hayes, M., Hlavinka, P., Svoboda, M., and Žalud, Z. (2013). Consequences of climate change for the soil climate in Central Europe and the central plains of the United States. *Climatic Change*, 120, 405–18.

Tropeano, D. (1984). Rate of soil erosion processes on vineyards in central Piedmont (NW Italy). *Earth Surface Processes and Landforms*, 9, 253–66.

Trout, T. and Neibling, W. (1993). Erosion and sedimentation processes on irrigated fields. *Journal of Irrigation and Drainage Engineering*, 119, 947–63.

Trzhtsinsky, Y. B. (2002). Human-induced activation of gypsum karst in the southern Priangaria (East

Siberia, Russia). *Carbonates and Evaporites*, 17, 154–58.

Tsai, J. S., Venne, L. S., McMurry, S. T., and Smith, L. M. (2007). Influences of land use and wetland characteristics on water loss rates and hydroperiods of playas in the Southern High Plains, USA. *Wetlands*, 27, 683–92.

Tsoar, H. and Blumberg, D. G. (2002). Formation of parabolic dunes from barchans and transverse dunes along Israel's Mediterranean coast. *Earth Surface Processes and Landforms*, 27, 1147–61.

Tsuboki, K., Yoshioka, M., Shinoda, T., Kato, M., Kanada, S., and Kitoh, A. (2014). Future increase of super-typhoon intensity associated with climate change. *Geophysical Research Letters*, 42, 646–52.

Turetsky, M. R., Benscoter, B., Page, S., Rein, G., van der Werf, G. R., and Watts, A. (2015). Global vulnerability of peatlands to fire and carbon loss. *Nature Geoscience*, 8, 11–4.

Turnbull, L., Wainwright, J., and Brazier, R. E. (2010). Changes in hydrology and erosion over a transition from grassland to shrubland. *Hydrological Processes*, 24, 393–414.

Turner, B. L., Kasperson, R. E., Meyer, W. B., Dow, K. M., Golding, D., Kasperson, J. X., Mitchell, R. C. and Ratick, S. J., 1990, Two types of global environmental change: definitional and spatial-scale issues in their human dimensions. *Global Environmental Change*, 1, 14–22.

Turvey, S. T. (ed.) (2009). *Holocene Extinctions*. Oxford: Oxford University Press.

Tuyet, D. (2001). Characteristics of karst ecosystems of Vietnam and their vulnerability to human impact. *Acta Geologica Sinica*, 75, 325–9.

Tweel, A. W. and Turner, R. E. (2012). Watershed land use and river eigineering drive wetland formation and loss in the Mississippi birdfoot delta. *Limnology and Oceanography*, 57, 18–28.

Tylmann, W. (2005). Lithological and geochemical record of anthropogenic changes in recent sediments of a small and shallow lake (Lake Pusty Staw, northern Poland). *Journal of Paleolimnology*, 33, 313–25.

Urich, P. B., Day, M. J., and Lynagh, F. (2001). Policy and practice in karst landscape protection: Bohol, the Philippines. *Geographical Journal*, 167, 305–23.

Unger, P. W., Stewart, B. A., Parr, J. F., and Singh, R. P. (1991). Crop residue management and tillage methods for conserving soil and water in semi-arid regions. *Soil and Tillage Research*, 20, 219–40.

Urban, M. A. and Rhoads, B. L. (2003). Catastrophic human-induced change in stream-channel planform and geometry in an agricultural watershed, Illinois, USA. *Annals of the Association of American Geographers*, 93, 783–96.

Vacca, A., Loddo, S., Ollesch, G., Puddu, R., Serra, G., Tomasi, D., and Aru, A. (2000). Measurement of runoff and soil erosion in three areas under different land use in Sardinia (Italy). *Catena*, 40, 69–92.

Valentin, C., Rajot, J.-L., and Mitja, D. (2004). Response of soil crusting, runoff and erosion to fallowing in the sub-humid and semi-arid regions of West Africa. *Agriculture, Ecosystems and Environment*, 104, 287–302.

Valese, E., Conedera, M., Held, A. C., and Ascoli, D. (2014). Fire, humans and landscape in the European Alpine region during the Holocene. *Anthropocene*, 6, 63–74.

van Andel, T. H., Zangger, E., and Demitrack, A. (1990). Land use and soil erosion in prehistoric and historical Greece. *Journal of Field Archaeology*, 17, 379–96.

van Beynen, P. E. (Ed.). (2011). *Karst management*. Dordrecht: Springer Science & Business Media.

Van Dam, P. J. (2001). Sinking peat bogs: environmental change Holland, 1350–1550. *Environmental History*, 6, 32–45.

Van der Broeke, M. and 8 others. (2009). Partitioning recent Greenland mass loss. *Science*, 326, 984–6.

Van der Noort, R. (2013). *Climate Change Archaeology. Building Resilence from Research in the World's Coastal Wetlands*. Oxford: Oxford University Press.

Van der Pluijm, B. (2014). Hello Anthropocene, goodbye Holocene. *Earth's Future*, 2, 566–8.

Van der Post, K. D., Oldfield, F., Haworth, E. Y., Crooks, P. R. J., and Appleby, P. G. (1997). A record of accelerated erosion in the recent sediments of Blelham Tarn in the English Lake District. *Journal of Paleolimnology*, 18, 103–20.

Van der Wal, D., Pye, K., and Neal, A. (2002). Long-term morphological change in the Ribble Estuary, northwest England. *Marine Geology*, 189, 249–66.

Van der Ween, C. J. (2002). Polar ice sheets and global sea-level: how well can we predict the future? *Global and Planetary Change*, 32, 165–94.

Van Donk, S. J., Huang, X., Skidmore, E. L., Anderson, A. B., Gebhart, D. L., Prehoda, V. E., and Kellogg, E. M. (2003). Wind erosion from military training lands in the Mojave Desert, California, USA. *Journal of Arid Environments*, 54, 687–703.

Van Manh, N., Dung, N. V., Hung, N. N., Kummu, M., Merz, B., and Apel, H. (2015). Future sediment dynamics in the Mekong Delta floodplains: impacts of hydropower development, climate change and sea level rise. *Global and Planetary Change*, 127, 22–33.

Van Oost, K., Govers, G., de Alba, S., and Quine, T. A. (2006). Tillage erosion: a review of controlling factors and implications for soil quality. *Progress in Physical Geography*, 30, 443–66.

Van Rompaey, A. J., Govers, G., and Puttemans, C. (2002). Modelling land use changes and their impact on soil erosion and sediment supply to rivers. *Earth Surface Processes and Landforms*, 27, 481–94.

Van Vliet, M. T., Franssen, W. H., Yearsley, J. R., Ludwig, F., Haddeland, I., Lettenmaier, D. P., and Kabat, P. (2013). Global river discharge and water temperature under climate change. *Global Environmental Change*, 23, 450–64.

van Wesenbeeck, B. K., Mulder, J. P., Marchand, M., Reed, D. J., de Vries, M. B., de Vriend, H. J., and Herman, P. M. (2014). Damming deltas: a practice of the past? Towards nature-based flood defenses. *Estuarine, Coastal and Shelf Science*, 140, 1–6.

Vannière, B., Bossuet, G., Walter-Simonnet, A. V., Gauthier, E., Barral, P., Petit, C., and Daubigney, A. (2003). Land use change, soil erosion and alluvial dynamic in the lower Doubs Valley over the 1st millenium AD (Neublans, Jura, France). *Journal of Archaeological Science*, 30, 1283–99.

Vanwalleghem, T., Amate, J. I., de Molina, M. G., Fernández, D. S., and Gómez, J. A. (2011). Quantifying the effect of historical soil management on soil erosion rates in Mediterranean olive orchards. *Agriculture, Ecosystems & Environment*, 142, 341–51.

Vanwalleghem, T., Laguna, A., Giráldez, J. V., and Jiménez-Hornero, F. J. (2010). Applying a simple methodology to assess historical soil erosion in olive orchards. *Geomorphology*, 114, 294–302.

Vařilová, Z., Přikryl, R., and Zvelebil, J. (2015). Factors and processes in deterioration of a sandstone rock form (Pravčická brána Arch, Bohemian Switzerland NP, Czech Republic). *Zeitschrift für Geomorphologie, Supplementary Issues*, 59, 81–101.

Vaudour, J. (1986). Travertins holocènes et pression anthropique. *Mediterranée*, 10, 168–73.

Vaughan, D. G. and Doake, C. S. M. (1996). Recent atmospheric warming and retreat of ice shelves on the Antarctic Peninsula. *Nature*, 379, 328–31.

Vaughan, D. G. and Spouge, J. R. (2002). Risk estimation of collapse of the west Antarctic ice sheet. *Climate Change*, 52, 65–91.

Vautard, R., Cattiaux, J., Yiou, P., Thépaut, J.-N., and Ciais, P. (2010). Northern Hemisphere atmospheric stilling partly attributed to an increase in surface roughness. *Nature Geoscience*, 3, 756–61.

Veni, G. (1999). A geomorphological strategy for conducting environmental impact assessments in karst areas. *Geomorphology*, 31,151–80.

Venteris, E. R. (1999). Rapid tide water glacier retreat: a comparison between Columbia Glacier, Alaska and Patagonian calving glaciers. *Global and Planetary Change*, 22, 131–8.

Verheijen, F. G., Jones, R. J., Rickson, R. J., and Smith, C. J. (2009). Tolerable versus actual soil erosion rates in Europe. *Earth-Science Reviews*, 94, 23–38.

Vermaire, J. C., Pisaric, M. F., Thienpont, J. R., Courtney Mustaphi, C. J., Kokelj, S. V., and Smol, J. P. (2013). Arctic climate warming and sea ice declines lead to increased storm surge activity. *Geophysical Research Letters*, 40, 1386–90.

Vermeer, M. and Rahmstorf, S. (2009). Global sea level linked to global temperature. *Proceedings of the National Academy of Sciences*, 106, 21527–32.

Verstraeten, G., Poesen, J., de Vente, J., and Koninckx, X. (2003). Sediment yield variability in Spain: a quantitative and semiqualitative analysis using reservoir sedimentation rates. *Geomorphology*, 50, 327–48.

Vice, R. B., Guy, H. P., and Ferguson, G. E. (1969). Sediment movement in an area of suburban highway construction, Scott Run Basin, Fairfax, County, Virginia, 1961–64. *United States Geological Survey Water Supply Paper*, 1591-E.

Vieira, D. A. and Dabney, S. M. (2011). Modeling edge effects of tillage erosion. *Soil and Tillage Research*, 111, 197–207.

Viles, H. A. (2002). Implications of future climate change for stone deterioration. In *Natural Stone, Weathering Phenomena, Conservation Strategies and Case Studies*, ed. S. Siegesmund, T. Weiss, and J. A. Vollbrecht. Geological Society of London Special Publication 205, pp. 407–18.

(2003). Conceptual modelling of the impacts of climate change on karst geomorphology in the UK and Ireland. *Journal for Nature Conservation*, 11, 59–66.

Viles, H. A. and Cutler, N. A. (2012). Global environmental change and the biology of heritage structures. *Global Change Biology*, 18, 2406–18.

Viles H. A. and Spencer, T. (1995). *Coastal Problems. Geomorphology, Ecology and Society at the Coast*. London: .Arnold

Vilímek, V., Zapata, M. L., Klimeš, J., Patzelt, Z., and Santillán, N. (2005). Influence of glacial retreat on natural hazards of the Palcacocha Lake area, Peru. *Landslides*, 2, 107–15.

Villmoare, B., Kimbel, W. H., Seyoum, C., Campisano, C. J., DiMaggio, E. N., Rowan, J., and Reed, K. E. (2015). Early Homo at 2.8 Ma from Ledi-Geraru, Afar, Ethiopia. *Science*, 347, 1352–5.

Vincent, K. R., Friedman, J. M., and Griffin, E. R. (2009). Erosional consequence of saltcedar control. *Environmental Management*, 44, 218–27.

Visconti, G. (2014). Anthropocene: another academic invention? *Rendiconti Lincei*, 25, 381–92.

Vita-Finzi, C. (1969). *The Mediterranean Valleys*. Cambridge: Cambridge University Press.

Voarintsoa, N. R. G., Cox, R., Razanatseheno, M. O. M., and Rakotondrazafy, A. F. M. (2012).

Relation between bedrock geology, topography and lavaka distribution in Madagascar. *South African Journal of Geology*, 115, 225–50.

Völkel, J., Leopold, M., Dötterl, S., Schneider, A., Hürkamp, K., and Hilgers, A. (2011). Origin and age of the Lower Bavarian sand dune landscape around Abensberg and Siegenburg. *Zeitschrift für Geomorphologie*, 55, 515–36.

Vörösmarty, C. J., Meybeck, M., Fekete, B., Sharma, K., Green, P., and Syvitski, J. P. (2003). Anthropogenic sediment retention: major global impact from registered river impoundments. *Global and Planetary Change*, 39, 169–90.

Wada, Y., van Beek, L. P. H., van Kempen, C. M., Reckman, J. W. T. M., Vasak, S., and Bierkens, M. F. P. (2010). Global depletion of groundwater resources. *Geophysical Research Letters*, 37, L20402, doi:10.1029/2010GL044571.

Wakindiki, I. I. C. and Ben-Hur, M. (2002). Indigenous soil and water conservation techniques: effects on runoff, erosion, and crop yields under semi-arid conditions. *Australian Journal of Soil Research*, 40, 367–79.

Walker, H. J. (1988). *Artificial Structures and Shorelines*. Dordrecht: Kluwer.

Walker, H. J., Coleman, J. M., Roberts, H. H., and Tye, R. S. (1987). Wetland loss in Louisiana. *Geografiska Annaler*, 69A, 189–200.

Walker, M., Gibbard, P., and Lowe, J. (2015). Comment on "When did the Anthropocene begin? A mid-twentieth century boundary is stratigraphically optimal" by Jan Zalasiewicz et al. (2015), *Quaternary International*, 383, 196–203.

Walling, D. E. (2006). Human impact on land–ocean sediment transfer by the world's rivers. *Geomorphology*, 79, 192–216.

Walling, D. E. and Gregory, K. J. (1970). The measurement of the effects of building construction on drainage basin dynamics. *Journal of Hydrology*, 11, 129–44.

Wallwork, K. L. (1974). *Derelict Land*. Newton Abbott: David and Charles.

Walsh, R. P. and Blake, W. H. (2009). Tropical rainforests. In *Geomorphology and Global Environmental Change*, ed. O. Slaymaker, T. Spencer, and C. Embleton-Hamman. Cambridge: Cambridge University Press, pp. 214–47.

Walter, R. C. and Merritts, D. J. (2008). Natural streams and the legacy of water-powered mills. *Science*, 2319, 299–304.

Wan, S., Toucanne, S., Clift, P. D., Zhao, D., Bayon, G., Yu, Z., and Li, T. (2015). Human impact overwhelms long-term climate control of weathering and erosion in southwest China. *Geology*, 43, 439–42.

Wang, C., Dong, S., Evan, A. T., Foltz, G. R., and Lee, S. K. (2012). Multidecadal covariability of North Atlantic sea surface temperature, African dust, Sahel rainfall, and Atlantic hurricanes. *Journal of Climate*, 25, 5404–15.

Wang, F., Zhang, Y., Huo, Z., Peng, X., Araiba, K., and Wang, G. (2008). Movement of the Shuping landslide in the first four years after the initial impoundment of the Three Gorges Dam Reservoir, China. *Landslides*, 5, 321–9.

Wang, H., Saito, Y., Zhang, Y., Bi, N., Sun, X., and Yang, Z. (2011). Recent changes of sediment flux to the western Pacific Ocean from major rivers in East and Southeast Asia. *Earth-Science Reviews*, 108, 80–100.

Wang, P., Li, Z., Luo, S., Bai, J., Huai, B., Wang, F., and Wang, L. (2015). Five decades of changes in the glaciers on the Friendship Peak in the Altai Mountains, China: changes in area and ice surface elevation. *Cold Regions Science and Technology*, 116, 24–31.

Wang, T., Belle, I., and Hassler, U. (2015). Modelling of Singapore's topographic transformation based on DEMs. *Geomorphology*, 231, 367–75.

Wang, W., Liu, H., Li, Y., and Su, J. (2014). Development and management of land reclamation in China. *Ocean & Coastal Management*, 102, 415–28.

Wang, X., Eerdun, H., Zhou, Z., and Liu, X. (2007). Significance of variations in the wind energy environment over the past 50 years with respect to dune activity and desertification in arid and semiarid northern China. *Geomorphology*, 86, 252–66.

Wang, X., Siegert, F., Zhou, A. G., and Franke, J. (2013). Glacier and glacial lake changes and their relationship in the context of climate change, Central Tibetan Plateau 1972–2010. *Global and Planetary Change*, 111, 246–57.

Wang, X., Thompson, D. K., Marshall, G. A., Tymstra, C., Carr, R., and Flannigan, M. D. (2015). Increasing frequency of extreme fire weather in Canada with climate change. *Climatic Change*, 130, 573–86.

Wang, X. M., Zhang, C. X., Hasi, E., and Dong, Z. B. (2010). Has the Three Norths Forest Shelterbelt Program solved the desertification and dust storm problems in arid and semiarid China? *Journal of Arid Environments*, 74, 13–22.

Warburton, J. and Evans, M. (2011). Geomorphic, sedimentary, and potential palaeoenvironmental significance of peat blocks in alluvial river systems. *Geomorphology*, 130, 101–14.

Warner, R. C. and Budd, W. F. (1990). Modelling the long-term response of the Antarctic Ice Sheet to global warming. *Annals of Glaciology*, 27, 161–68.

Warren, A. (2013). *Dunes*. Chichester: Wiley.

Warrick, J. A., Madej, M. A., Goñi, M. A., and Wheatcroft, R. A. (2013). Trends in the suspended-sediment yields of coastal rivers of northern California, 1955–2010. *Journal of Hydrology*, 489, 108–23.

Washington R. and 9 others. (2009). Dust as a tipping element: the Bodélé Depression, Chad.

Proceedings of the National Academy of Sciences, 106, 20564–71.

Wasson, R. J. (2012). Geomorphic histories for river and catchment management. *Philosophical Transactions of the Royal Society, A*, 370, 2240–63.

Watanabe, T., Lamsal, D., and Ives, J. D. (2009). Evaluating the growth characteristics of a glacial lake and its degree of danger of outburst flooding: Imja Glacier, Khumbu Himal, Nepal. *Norsk Geografisk Tidsskrift*, 63, 255–67.

Waters, M. R. and Haynes, C. V. (2001). Late Quaternary arroyo formation and climate change in the American southwest. *Geology*, 29, 399–402.

Watkins, T. (2010). New light on Neolithic revolution in south-west Asia. *Antiquity*, 84, 621–34.

Watson, A. (1990). The control of blowing sand and mobile desert dunes. In *Techniques for Desert Reclamation*, ed. A. S. Goudie. Chichester: Wiley, pp. 35–85.

Waycott, M., Duarte, C. M., Carruthers, T. J., Orth, R. J., Dennison, W. C., Olyarnik, S., and Williams, S. L. (2009). Accelerating loss of seagrasses across the globe threatens coastal ecosystems. *Proceedings of the National Academy of Sciences*, 106, 12377–81.

Webb, E. L., Friess, D. A., Krauss, K. W., Cahoon, D. R., Guntenspergen, G. R., and Phelps, J. (2013). A global standard for monitoring coastal wetland vulnerability to accelerated sea-level rise. *Nature Climate Change*, 3, 458–65.

Webb, N. P., Chappell, A., Strong, C. L., Marx, S. K., and McTainsh, G. H. (2012). The significance of carbon-enriched dust for global carbon accounting. *Global Change Biology*, 18, 3275–8.

Webb, R. H., Boyer, D. E., and Turner, R. M. (eds.) (2010). *Repeat Photography: Methods and Applications in the Natural Sciences*. Washington, DC: Island Press.

Weber, K. A. and Perry, R. G. (2006). Groundwater abstraction impacts on spring flow and base flow in the Hillsborough River Basin, Florida, USA. *Hydrogeology Journal*, 14, 1252–64.

Webster, P. J., Holland, G. J., Curry, J. A., and Chang, H.-R. (2005). Changes in tropical cyclone number, duration, and intensity in a warming environment. *Science*, 309, 1844–6.

Wehrli, M., Mitchell, E. A., van der Knaap, W. O., Ammann, B., and Tinner, W. (2010). Effects of climatic change and bog development on Holocene tufa formation in the Lorze Valley (central Switzerland). *The Holocene*, 20, 325–36.

Weinhold, B. (2012). Energy development linked with earthquakes. *Environmental Health Perspectives*, 120, a388.

Weisrock, A. (1986). Variations climatiques et periodes de sedimentation carbonatée a l'Holocene - l'age des depôts. *Mediterranée*, 10, 165–7.

Wells, J. T. (1995). Effects of sea level rise on coastal sedimentation and erosion. In *Climate Change Impact on Coastal Habitation*, ed. D. Eisma. Boca Raton: Lewis, pp. 111–36.

Wells, N. A. and Andriamihaja, B. (1993). The Initiation and growth of gullies in Madagascar-are humans to blame. *Geomorphology*, 8, 1–46.

Wemple, B. C. and Jones, J. A. (2003). Runoff production on forest roads in a steep, mountain catchment. *Water Resources Research*, 39, doi: 10.1029/2002WR001744.

Wemple, B. C., Swanson, F. J., and Jones, J. A. (2001). Forest roads and geomorphic process interactions, Cascade Range, Oregon. *Earth Surface Processes and Landforms*, 26, 191–204.

Wen, F. and Chen, X. (2006). Evaluation of the impact of groundwater irrigation on streamflow in Nebraska. *Journal of Hydrology*, 327, 603–17.

Westerling, A. L., Turner, M. G., Smithwick, E. A., Romme, W. H., and Ryan, M. G. (2011). Continued warming could transform Greater Yellowstone fire regimes by mid-21st century. *Proceedings of the National Academy of Sciences*, 108, 13165–70.

Westing, A. and Pfeiffer, E. W. (1972). The cratering of Indochina. *Scientific American*, 226, 5, 21–9.

Wharton, G. and Gilvear, D. J. (2007). River restoration in the UK: meeting the dual needs of the European Union Water Framework Directive and flood defence? *International Journal of River Basin Management*, 5, 143–54.

Wheatcroft, R. A., Goñi, M. A., Richardson, K. N., and Borgeld, J. C. (2013). Natural and human impacts on centennial sediment accumulation patterns on the Umpqua River margin, Oregon. *Marine Geology*, 339, 44–56.

Wheaton, E. E. (1990). Frequency and severity of drought and dust storms. *Canadian Journal of Agricultural Economics*, 38, 695–700.

White A. F. and Blum, A. E. (1995). Climatic effects on chemical weathering in watersheds: application of mass balance approaches. In *Solute Modelling in Catchment Systems*, ed. S. Trudgill. Chichester: John Wiley and Sons, pp. 101–31.

Whitmore T. M. and Turner, B. L. (2001). *Cultivated Landscapes of Middle America on the Eve of Conquest*. Oxford: Oxford University Press.

Whitaker, J. R. (1940). World view of destruction and conservation of natural resources. *Annals of the Association of American Geographers*, 30, 143–62.

Whittington, G. (1962). The distribution of strip lynchets. *Transactions and Papers of the Institute of British Geographers*, 31, 115–30.

Wienhold, M. L. (2013). Prehistoric land use and hydrology: a multi-scalar spatial analysis in central Arizona. *Journal of Archaeological Science*, 40, 850–9.

Wilby, R. L., Dalgleish, H. Y., and Foster, I. D. L. (1997). The impact of weather patterns on historic and contemporary catchment sediment yields. *Earth Surface Processes and Landforms*, 22, 353–63.

Wildman, L. A. S. and Macbroom, J. G. (2005). The evolution of gravel bed channels after dam removal: case study of the Anaconda and Union

City dam removals. *Geomorphology*, 71, 245–62.

Wilkinson, B. H. and McElroy, B. J. (2007). The impact of humans on continental erosion and sedimentation. *Bulletin of the Geological Society of America*, 119, 140–56.

Wilkinson, J. C., (1977) *Water and Tribal Settlement in South-East Arabia: a Study of the Aflāj of Oman*. Oxford: Clarendon Press

Wilkinson, T. J. and Rayne, L. (2010). Hydraulic landscapes and imperial power in the Near East. *Water History*, 2, 115–44.

Wilkinson, T. J., French, C., Ur, J. A., and Semple, M. (2010). The geoarchaeology of route systems in northern Syria. *Geoarchaeology*, 25, 745–71.

Williams, M. (2003). *Deforesting the Earth: From Prehistory to Global Crisis*. Chicago: The University of Chicago Press.

Williams, M., Zalasiewicz, J., Davies, N., Mazzini, I., Goiran, J. P., and Kane, S. (2014). Humans as the third evolutionary stage of biosphere engineering of rivers. *Anthropocene*, 7, 57–63.

Williams, P., Biggs, J., Crowe, A., Murphy, J., Nicolet, P., Waetherby, A., and Dunbar, M. (2010). Ponds report from 2007. *Countryside Survey Technical Report* No. 7/07.

Williams, P. W. (ed.) (1993). Karst terrains: environmental changes and human impact. *Catena supplement*, 25.

(2008). *World heritage caves and karst*. Gland, Switzerland: IUCN.

Williams, S. J. (2013). Sea-level rise implications for coastal regions. *Journal of Coastal Research*, 63(sp1), 184–96.

Williams, T. J., Quinton, W. L., and Baltzer, J. L. (2013). Linear disturbances on discontinuous permafrost: implications for thaw-induced changes to land cover and drainage patterns. *Environmental Research Letters*, 8(2), 025006.

Willis, C. M. and Griggs, G. B. (2003). Reductions in fluvial sediment discharge by coastal dams in California and implications for beach sustainability. *Journal of Geology*, 111, 167–82.

Willis, K. J., Gillson, L., and Brncic, T. M. (2004). How "virgin" is virgin rainforest? *Science*, 304, 402–3.

Wilson, C. J. (1999). Effects of logging and fire on runoff and erosion on highly erodible granitic soils in Tasmania. *Water Resources Research*, 35, 3531–46.

Wilson, L., Wilson, J., Holden, J., Johnstone, I., Armstrong, A., and Morris, M. (2011). Ditch blocking, water chemistry and organic carbon flux: evidence that blanket bog restoration reduces erosion and fluvial carbon loss. *Science of the Total Environment*, 409, 2010–8.

Winkler, E. M. (1970). The importance of air pollution in the corrosion of stone and metals. *Engineering Geology*, 4, 327–34.

Winter, M. G., Dixon, N., Wasowski, J., and Dijkstra, T. A. (2010). Introduction to land-use and climate change impacts on landslides. *Quarterly Journal of Engineering Geology and Hydrogeology*, 43, 367–70.

Winter, R. C. de, Sterl, A., de Vries, J. W., Weber, S. L., and Ruessink, G. (2012). The effect of climate change on extreme waves in front of the Dutch coast. *Ocean Dynamics*, 62, 1139–52.

Wishart, D. and Warburton, J. (2001). An assessment of blanket mire degradation and peatland gully development in the Cheviot Hills, Northumberland. *Scottish Geographical Magazine*, 117, 185–206.

Wisser, D., Frolking, S., Hagen, S., and Bierkens, M. F. (2013). Beyond peak reservoir storage? A global estimate of declining water storage capacity in large reservoirs. *Water Resources Research*, 49, 5732–9.

Woeikof, A. (1901). De l'influence de l'homme sur la terre. *Annales de Géographie*, 10, 97–114, 193–215.

Wohl, E. (2013). Wilderness is dead: whither critical zone studies and geomorphology in the Anthropocene. *Anthropocene*, 2, 4–15.

(2015). Legacy effects on sediments in river corridors. *Earth-Science Reviews*, 147, 30–53.

Wohl, E. and Merritts, D. J. (2007). What is a natural river? *Geography Compass*, 1, 871–900.

Wohl, E., Angermeier, P. L., Bledsoe, B., Kondolf, G. M., MacDonnell, L., Merritt, D. M., and Tarboton, D. (2005). River restoration. *Water Resources Research*, 41, W10301, doi: 10.1029/2005WR003985.

Wolanski, E., Moore, K., Spagnol, S., D'adamo, N., and Pattiaratchi, C. (2001). Rapid, human-induced siltation of the macro-tidal Ord River Estuary, Western Australia. *Estuarine, Coastal and Shelf Science*, 53, 717–32.

Wolfe S. A., Hugenholtz, C. H., Evans, C. P., Huntley, D. J., and Ollerhead, J. (2007). Potential aboriginal-occupation-induced dune activity, Elbow Sand Hills, northern Great Plains, Canada. *Great Plains Research*, 17, 173–92.

Wolff, W. J. (1992). The end of a tradition: 1000 years of embankment and reclamation of wetlands in the Netherlands. *Ambio*, 21, 287–91.

Wolman, M. G. (1967). A cycle of sedimentation and erosion in urban river channels. *Geografiska Annaler*, 49A, 385–95.

Wolman, M. G. and Schick, A. P. (1967). Effects of construction on fluvial sediment, urban and suburban areas of Maryland. *Water Resources Research*, 3, 451–64.

Wondzell, S. M. and King, J. G. (2003). Postfire erosional processes in the Pacific Northwest and Rocky Mountain regions. *Forest Ecology and Management*, 178, 75–87.

Wong, S., Dessler, A. E., Mahowald, N., Colarco, P. R., and da Silva, A. (2008). Long-term variability in Saharan dust transport and its link to North Atlantic sea surface temperature. *Geophysical*

Research Letters, 35, L07812, doi: 10.1029/2007GL032297.

Woo, M.-K., Lewkowicz, A. G., and Rouse, W. R. (1992). Response of the Canadian permafrost environment to climate change. *Physical Geography*, 13, 287–317.

Wood, C. (2009). *World Heritage Volcanoes: Thematic Study, Global Review of Volcanic World Heritage Properties: Present Situation, Future Prospects and Management Requirements*. Gland: IUCN.

Woodroffe, C. D. (1990). The impact of sea-level rise on mangrove shorelines. *Progress in Physical Geography*, 14, 483–520.

(2008). Reef-island topography and the vulnerability of atolls to sea-level rise. *Global and Planetary Change*, 62, 77–96.

Woodruff, J. D., Irish, J. L., and Camargo, S. J. (2013). Coastal flooding by tropical cyclones and sea-level rise. *Nature*, 504, 44–52.

Woodruff, J. D., Martini, A. P., Elzidani, E. Z., Naughton, T. J., Kekacs, D. J., and MacDonald, D. G. (2013). Off-river waterbodies on tidal rivers: Human impact on rates of infilling and the accumulation of pollutants. *Geomorphology*, 184, 38–50.

Woodward, C., Shulmeister, J., Larsen, J., Jacobsen, G. E., and Zawadzki, A. (2014). The hydrological legacy of deforestation on global wetlands. *Science*, 346, 844–7.

Woolsey, S., Capelli, F., Gonser, T. O. M., Hoehn, E., Hostmann, M., Junker, B., and Peter, A. (2007). A strategy to assess river restoration success. *Freshwater Biology*, 52, 752–69.

World Glacier Monitoring Service. (2008). Global glacier changes: facts and figures, www.grid.unep.ch/glaciers/ (Accessed October 29, 2011).

Wosten, J. H. M., Ismail, A. B., andVan Wijk, A. L. M. (1997). Peat subsidence and its practical implications: a case study in Malaysia. *Geoderma*, 78, 25–36.

Woth, K., Weisse, R., and von Storch, H. (2006). Climate change and North Sea storm surge extremes: an ensemble study of storm surge extremes expected in a changed climate projected by four different regional climate models. *Ocean Dynamics*, 56, 3–15.

Wright, S. A. and Schoellhamer, D. H. (2004). Trends in the sediment yield of the Sacramento River, California, 1957–2001. *San Francisco Estuary and Watershed Science* (on line serial), 2(2), Article 2.

Wu, C. S., Yang, S. L., and Lei, Y. P. (2012). Quantifying the anthropogenic and climatic impacts on water discharge and sediment load in the Pearl River (Zhujiang), China (1954–2009). *Journal of Hydrology*, 452, 190–204.

Wu, J., Shi, X., Xue, Y., Zhang, Y., Wei, Z., and Yu, J. (2008). The development and control of the land subsidence in the Yangtze Delta, China. *Environmental Geology*, 55, 1725–35.

Wu, Q., Hou, Y., Yun, H., and Liu, Y. (2015). Changes in active-layer thickness and near-surface permafrost between 2002 and 2012 in alpine ecosystems, Qinghai–Xizang (Tibet) Plateau, China. *Global and Planetary Change*, 124, 149–55.

Wu, Q. Jiewu, P., Shanzhong, Q., Yiping, L., Congcong, H., Tingxiang, L., and Limei, H. (2009). Impacts of coal mining subsidence on the surface landscape in Longkou city, Shandong Province of China. *Environmental Earth Sciences*, 59, 783–91.

Wu, R. S., Sue, W. R., Chien, C. B., Chen, C. H., Chang, J. S., and Lin, K. M. (2001). A simulation model for investigating the effects of rice paddy fields on the runoff system. *Mathematical and Computer Modelling*, 33, 649–58.

Wu, Y. and Zhu, L. (2008). The response of lake-glacier variations to climate change in Nam Co Catchment, central Tibetan Plateau, during 1970–2000. *Journal of Geographical Sciences*, 18, 177–89.

Wüst, R. A. J. and Schlüchter, C. (2000). The origin of soluble salts in rocks of the Thebes Mountains, Egypt: the damage potential to ancient Egyptian wall art. *Journal of Archaeological Science*, 27, 1161–72.

Wyzga, B. (1996). Changes in the magnitude and transformation of flood waves subsequent to the channelization of the Raba River, Polish Carpathians. *Earth Surface Processes and Landforms*, 21, 749–63.

Xu, K. and Milliman, J. D. (2009). Seasonal variations of sediment discharge from the Yangtze River before and after impoundment of the Three Gorges Dam. *Geomorphology*, 104, 276–83.

Xu, K., Milliman, J. D., Yang, Z., and Xu, H. (2007). Climatic and anthropogenic impacts on water and sediment discharges from the Yangtze River (Changjiang), 1950–2005. In *Large Rivers: Geomorphology and Management*, ed. A. Gupta. Chichester: Wiley, pp. 609–26.

Xu, Y. S., Shen, S. L., Cai, Z. Y., and Zhou, G. Y. (2008). The state of land subsidence and prediction approaches due to groundwater withdrawal in China. *Natural Hazards*, 45, 123–35.

Xue, Y. Q., Zhang, Y., Ye, S. J., Wu, J. C., and Li, Q. F. (2005). Land subsidence in China. *Environmental Geology*, 48, 713–20.

Yamano, H., Sugihara, K., and Nomura, K. (2011). Rapid poleward range expansion of tropical reef corals in response to rising sea surface temperatures. *Geophysical Research Letters*, 38, L04601, doi:10.1029/2010GL046474.

Yang, D., Kanae, S., Oki, T., Koike, T., and Musiake, K. (2003). Global potential soil erosion with reference to land use and climate changes. *Hydrological Processes*, 17, 2913–28.

Yang, Q., Wang, K., Zhang, C., Yue, Y., Tian, R., and Fan, F. (2011). Spatio-temporal evolution of rocky desertification and its driving forces in karst areas of Northwestern Guangxi, China. *Environmental Earth Science*, 64, 383–93.

Yang, X. and Lu, X. (2014). Drastic change in China's lakes and reservoirs over the past decades. *Scientific Reports*, 4, article number: 6041, doi: 10.1038/srep06041.

Yang, X., Ding, Z., Fan, X., Zhou, Z., and Ma, N. (2007). Processes and mechanisms of desertification in northern China during the last 30 years, with a special reference to the Hunshandake Sandy Land, eastern Inner Mongolia. *Catena*, 71, 2–12.

Yang, X., Shen, S., Yang, F., He, Q., Ali, M., Huo, W., and Liu, X. (2015). Spatial and temporal variations of blowing dust events in the Taklimakan Desert. *Theoretical and Applied Climatology*, doi: 10.1007/s00704-015-1537-4.

Yang, Z., Wang, T., Leung, R., Hibbard, K., Janetos, T., Kraucunas, I., and Wilbanks, T. (2014). A modeling study of coastal inundation induced by storm surge, sea-level rise, and subsidence in the Gulf of Mexico. *Natural Hazards*, 71, 1771–94.

Yao, T., Pu, J., Lu, A., Wang, Y., and Yu, W. (2007). Recent glacial retreat and its impact on hydrological processes on the Tibetan Plateau, China, and surrounding regions. *Arctic, Antarctic, and Alpine Research*, 39, 642–50.

Yarushina, V. M. and Bercovici, D. (2013). Mineral carbon sequestration and induced seismicity. *Geophysical Research Letters*, 40, 814–8.

Yechieli, Y., Abelson, M., Bein, A., Crouvi, O., and Shtivelman, V. (2006). Sinkhole "swarms" along the Dead Sea coast: reflection of disturbance of lake and adjacent groundwater systems. *Bulletin of the Geological Society of America*, 118, 1075–87.

Yeck, W. L., Block, L. V., Wood, C. K., and King, V. M. (2015). Maximum magnitude estimations of induced earthquakes at Paradox Valley, Colorado, from cumulative injection volume and geometry of seismicity clusters. *Geophysical Journal International*, 200, 322–36.

Yesner, D. R. (2001). Human dispersal into interior Alaska: antecedent conditions, mode of colonization, and adaptations. *Quaternary Science Reviews*, 20, 315–27.

Yin, Y., Zhang, K., and Li, X. (2006). Urbanization and land subsidence in China. *IAEG2006 Paper* 31, Geological Society of London.

Yonggui, Y., Xuefa, S., Houjie, W., Chengkun, Y., Shenliang, C., Yanguang, L., and Shuqing, Q. (2013). Effects of dams on water and sediment delivery to the sea by the Huanghe (Yellow River): the special role of water-sediment modulation. *Anthropocene*, 3, 72–82.

Yorke, T. H. and Herb, W. J. (1978). Effects of urbanization on streamflow and sediment transport in the Rock Creek and Anacostia River Basins, Montgomery County, Maryland, 1962–74. *U.S. Geological Survey Professional Paper*, 1003.

Yoshikawa, K. and Hinzman, L. D. (2003). Shrinking thermokarst ponds and groundwater dynamics in discontinuous permafrost near Council, Alaska. *Permafrost and Periglacial Processes*, 14, 151–60.

Young, J. E. (1992). Mining the Earth. *Worldwatch Institute Paper*, 109, 1–53.

Yu, K., D'Odorico, P., A. Bhattachan, Okin, G. S., and Evan, A. T., (2015), Dust-rainfall feedback in West African Sahel, *Geophysical Research Letters*, 42, 7563–71.

Yuill, B., Lavoie, D., and Reed, D. (2009). Understanding subsidence processes in coastal Louisiana. *Journal of Coastal Research, SI*, 54, 23–36.

Zaimes, G. N. and Schultz, R. C. (2015). Riparian land-use impacts on bank erosion and deposition of an incised stream in north-central Iowa, USA. *Catena*, 125, 61–73.

Zalasiewicz, J., Kryza, R., and Williams, M. (2014). The mineral signature of the Anthropocene in its deep-time context. *Geological Society, London, Special Publications*, 395, 109–17.

Zalasiewicz, J., Waters, C. N., and Williams, M. (2014). Human bioturbation, and the subterranean landscape of the Anthropocene. *Anthropocene*, 6, 3–9.

Zalasiewicz, J., Williams, M., Haywood. A., and Ellis, M. (2011a). The Anthropocene: a new epoch of geological time? *Philosophical Transactions of the Royal Society*, 369A: 835–41.

Zalasiewicz J., Williams, M., Fortey, R., et al. (2011b). Stratigraphy of the Anthropocene. *Philosophical Transactions of the Royal Society*, 369A, 1036–55.

Zalasiewicz, J., Waters, C. N., Williams, M., Barnosky, A. D., Cearreta, A., Crutzen, P., and Oreskes, N. (2015). When did the Anthropocene begin? A mid-twentieth century boundary level is stratigraphically optimal. *Quaternary International*, 383, 196–203.

Zanello, F., Teatini, P., Putti, M., and Gambolati, G. (2011). Long term peatland subsidence: Experimental study and modeling scenarios in the Venice coastland. *Journal of Geophysical Research: Earth Surface*, 116(F4), F04002, doi: 10.1029/2011JF002010.

Zang, A., Oye, V., Jousset, P., Deichmann, N., Gritto, R., McGarr, A., and Bruhn, D. (2014). Analysis of induced seismicity in geothermal reservoirs – An overview. *Geothermics*, 52, 6–21.

Zarfl, C., Lumsdon, A. E., Berlekamp, J., Tydecks, L., and Tockner, K. (2015). A global boom in hydropower dam construction. *Aquatic Sciences*, 77, 161–70.

Zarnetske, P. L., Hacker, S. D., Seabloom, E. W., Ruggiero, P., Killian, J. R., Maddux, T. B., and Cox, D. (2012). Biophysical feedback mediates effects of invasive grasses on coastal dune shape. *Ecology*, 93, 1439–50.

Zavada, M. S., Wang, Y., Rambolamanana, G., Raveloson, A., and Razanatsoa, H. (2009). The significance of human induced and natural erosion features (lavakas) on the central highlands of Madagascar. *Madagascar Conservation & Development*, 4, 120–7.

Zawiejska, J. and Wyżga, B. (2010). Twentieth-century channel change on the Dunajec River, southern Poland: patterns, causes and controls. *Geomorphology*, 117, 234–46.

Zeeberg, J. and Forman, S. L. (2001). Changes in glacier extent of north Novaya Zemlya in the twentieth century. *The Holocene*, 11, 161–75.

Zektser, S., Loáiciga, H. A., and Wolf, J. T. (2005). Environmental impacts of groundwater overdraft: selected case studies in the southwestern United States. *Environmental Geology*, 47, 396–404.

Zemp, M., Frey, H., Gärtner-Roer, I., Nussbaumer, S., Hoelzle, M., Paul, F., and Vincent, C. (2015). Historically unprecedented global glacier decline in the early 21st century. *Journal of Glaciology*, 61, 745–62, doi: 10.3189/2015JoG15J017.

Zeng, N. and Yoon, J. (2009). Expansion of the world's deserts due to vegetation-albedo feedback under global warming. *Geophysical Research Letters*, 36, L17401, doi: 10.1029/2009GL039699.

Zennaro, P., Kehrwald, N., Marlon, J., Ruddiman, W., Brücher, T., Agostinelli, C., and Barbante, C. (2015). Europe on fire three thousand years ago: Arson or climate?. *Geophysical Research Letters*. DOI: 10.1002/2015GL064259.

Zhang, L., Wu, B., Yin, K., Li, X., Kia, K., and Zhu, L. (2014). Impacts of human activities on the evolution of estuarine wetland in the Yangtze Delta from 2000 to 2010. *Environmental Earth Sciences*, 73, 435–47, doi: 10.1007/s12665-014-3565-2.

Zhang, T.-H., Zhao, H.-L., Li, S.-G., Li, F.-R., Shirato, Y., Ohkuro, T., and Taniyama, I. (2004). A comparison of different measures for stabilizing moving sand dunes in the Horqin Sandy Land of Inner Mongolia, China. *Journal of Arid Environments*, 58, 203–14.

Zhang, Y., Person, M., Rupp, J., Ellett, K., Celia, M. A., Gable, C. W., and Elliot, T. (2013). Hydrogeologic controls on induced seismicity in crystalline basement rocks due to fluid injection into basal reservoirs. *Groundwater*, 51, 525–38.

Zhang, Y., Xue, Y. Q., Wu, J. C., Shi, X. Q., and Yu, J. (2010). Excessive groundwater withdrawal and resultant land subsidence in the Su-Xi-Chang area, China. *Environmental Earth Sciences*, 61, 1135–43.

Zhang, Y., Xue, Y. Q., Wu, J. C., Yu, J., Wei, Z. X., and Li, Q. F. (2008). Land subsidence and earth fissures due to groundwater withdrawal in the Southern Yangtse Delta, China. *Environmental Geology*, 55, 751–62.

Zhang, Z. and Wu, Q. (2012). Thermal hazards zonation and permafrost change over the Qinghai–Tibet Plateau. *Natural Hazards*, 61, 403–23.

Zhao, S., Fang, J., Miao, S., Gu, B., Tao, S., Peng, C., and Tang, Z. (2005). The 7-decade degradation of a large freshwater lake in Central Yangtze River, China. *Environmental Science & Technology*, 39, 431–36.

Zhou, Z. C., Gan, Z. T., Shangguan, Z. P., and Dong, Z. B. (2010). Effects of grazing on soil physical properties and soil erodibility in semiarid grassland of the Northern Loess Plateau (China). *Catena*, 82, 87–91.

Zhu, C., Wang, B., and Qian, W. (2008). Why do dust storms decrease in northern China concurrently with the recent global warming? *Geophysical Research Letters*, 35, L18702, doi: 10.1029/2008GL034886.

Zhu, D., Tian, L., Wang, J., Wang, Y., and Cui, J. (2014). Rapid glacier retreat in the Naimona'Nyi region, western Himalayas, between 2003 and 2013. *Journal of Applied Remote Sensing*, 8, 083508–083508.

Zhu, J. and Olsen, C. R. (2014). Sedimentation and Organic Carbon Burial in the Yangtze River and Hudson River Estuaries: implications for the Global Carbon Budget. *Aquatic Geochemistry*, 20, 325–42.

Zhu, R. X., Potts, R., Pan, Y. X., Yao, H. T., Lü, L. Q., Zhao, X., Gao, X., Chen, L. W., Gao, F., and Deng, C. L. (2008). Early evidence of the genus Homo in East Asia. *Journal of Human Evolution*, 55, 1075–85.

Zhuang, Y. and Kidder, T. R. (2014). Archaeology of the Anthropocene in the Yellow River region, China, 8000–2000 cal. BP. *The Holocene*, 24, 1602–23.

Ziegler, A. D. and Giambelluca, T. W. (1997). Importance of rural roads as source areas for runoff in mountainous areas of northern Thailand. *Journal of Hydrology*, 196, 204–29.

Ziegler, A. D., Bruun, T. B., Guardiola-Claramonte, M., Giambelluca, T. W., Lawrence, D., and Lam, N. T. (2009). Environmental consequences of the demise in swidden cultivation in montane mainland Southeast Asia: hydrology and geomorphology. *Human Ecology*, 37, 361–73.

Zimmermann, A., Francke, T., and Elsenbeer, H. (2012). Forests and erosion: insights from a study of suspended-sediment dynamics in an overland flow-prone rainforest catchment. *Journal of Hydrology*, 428, 170–81.

Zizumbo-Villarreal, D. and Colunga-GarcíaMarín, P. (2010). Origin of agriculture and plant domestication in West Mesoamerica. *Genetic Resources and Crop Evolution*, 57, 813–25.

Zobeck, T. M., Baddock, M. C., and Van Pelt, R. S. (2013). Anthropogenic environments. In: *Treatise on Geomorphology*, ed. J. Shroder, (Editor in Chief), N. Lancaster, D. J. Sherman, A. C. W. Baas. San Diego, CA: Academic Press, vol. 11, *Aeolian Geomorphology*, pp. 395–413.

Zöbisch, M. A. (1993). Erosion susceptibility and soil loss on grazing lands in some semiarid and subhumid locations of eastern Kenya. *Journal of Soil and Water Conservation*, 48, 445–8.

Zolitschka, B., Behre, K. E., and Schneider, J. (2003). Human and climatic impact on the environment as derived from colluvial, fluvial and lacustrine archives—examples from the Bronze Age to the Migration period, Germany. *Quaternary Science Reviews*, 22, 81–100.

Zuazo, V. H. D. and Pleguezuelo, C. R. R. (2008). Soil-erosion and runoff prevention by plant covers: a review. *Agronomy for Sustainable Development*, 28, 65–86.

Zu Ermgassen, P. S., Spalding, M. D., Blake, B., Coen, L. D., Dumbauld, B., Geiger, S., and Brumbaugh, R. (2012). Historical ecology with real numbers: past and present extent and biomass of an imperilled estuarine habitat. *Proceedings of the Royal Society B: Biological Sciences*, 279, 3393–400.

Zwally, H. J., Abdalafi, W., Herring, T., Larson, K., Saba, J., and Steffen, K. (2002). Surface melt-induced acceleration of Greenland ice-sheet flow. *Science*, 297, 218–21.

Index